Proceedings of the 10th International Conference on
THE PROPERTIES OF STEAM

**THE CONFERENCE
WAS ORGANIZED AND SPONSORED BY**

**The International Association for the Properties of Steam
The USSR State Committee on Science and Technology
The Ministry of Power and Electrification of the USSR
The USSR Ministry of Higher and Secondary Specialized Education**

*The authors' typescripts have been
reproduced in their original form*

*Printed in the Union of Soviet Socialist Republics
at the printing press of the Moscow Power Engineering Institute,
Moscow*

Proceedings of the 10th International Conference on
THE PROPERTIES OF STEAM

Moscow, USSR
3-7 September 1984

EDITED BY
V. V. SYTCHEV and A. A. ALEKSANDROV

Volume 2

Published for the International Association
for the Properties of Steam

Mir Publishers
Moscow

Plenum Press
New York and London

CHEMISTRY

ISBN 0-306-42159-3

© 1986 Organizing Committee of the 10th Conference of the International Association for the Properties of Steam

Sole distributor

Plenum Press, New York
A Division of Plenum Publishing Corporation
233 Spring Street, New York, N.Y. 10013

Contents

VOLUME 1

6

7

8

Aqueous Solutions

12

CONFERENCE SECRETARIAT

KRZHIZHANOVSKY STATE RESEARCH INSTITUTE OF ENERGETICS (ENIN) MOSCOW POWER ENGINEERING INSTITUTE

PROGRAMME COMMITTEE

Cochairmen:	Professor U. Grigull	FRG
	Professor V. V. Sytchev	USSR
Members:	Professor A. A. Aleksandrov	USSR
	Professor J. Kestin	USA
	Mr. O. Jonas	USA
	Professor O. I. Martynova	USSR
	Professor K. Watanabe	Japan
Ex officio:	Dr. H. J. White, Jr.	USA
	Secretary of IAPS	

Executive secretary

of the International Association
for the Properties of Steam (IAPS)

Dr. H. J. White, Jr.

Office of Standard Reference Data
National Bureau of Standards
Washington, D. C., 20234, USA

SURFACE, ELECTRICAL
AND OTHER PROPERTIES

Interrelations Among Various Thermophysical Properties of Fluids

T. MAKITA, H. KASHIWAGI, and S. MATSUO

Department of Chemical Engineering, Kobe University,
Rokkodai, Nada-ku, Kobe 657, Japan

SYNOPSIS

Interrelations among density, viscosity, thermal conductivity, dielectric constant and refractive index of liquid H_2O and D_2O are examined phenomenologically, using various methods of correlation obtained for non-polar or weak-polar fluids. It is found that the refractive index of H_2O is represented by a simple relation of density covering all the liquid phase. The Tait-type equation is found to be valid for the expression of effects of temperature and pressure on density, viscosity and thermal conductivity of liquid H_2O and D_2O and on dielectric constant of H_2O. The isotope effects on molar volume and viscosity between H_2O and D_2O are also discussed.

1. INTRODUCTION

From the viewpoint of thermophysical properties, liquid water is one of the most remarkable and interesting fluids. Taking into account its molecular weight, the normal melting and boiling points, as well as the triple and critical points, are anomalously high. The values of the heat capacity, heat of vaporization, surface tension and dielectric constant are also unusually large. At atmospheric pressure, the density exhibits a maximum at 3.98°C for H_2O and 11.19°C for D_2O, there exists a mini-

17

mum near 50°C in the heat capacity and isothermal compressibility, the velocity of sound in water has a maximum near 70°C, the thermal conductivity increases with increasing temperature and shows a maximum near 130°C. Furthermore, at temperatures between 0° and 33°C, the viscosity decreases slightly with increasing pressure and each isotherm has a minimum at a pressure from 100 to 120 MPa. At higher temperatures, the viscosity increases with increasing pressure, although the effect of pressure on viscosity is relatively small.

These anomalous behaviors of liquid water should arise from the facts that water molecule is highly polar (its dipole moment is 1.85 Debye) and that, owing to the existence of hydrogen-bonding between molecules, liquid water is a mixture of "ice-like" clusters and "free" monomolecules, that is, water has partly "structured" region of hydrogen-bonded molecules, surrounded by free water. Although a number of theoretical models have been discussed, the simple "two-fluid model" of Némethy and Scheraga [1] provides quantitative intepretation of the behaviors of water to some extent, from the view that the ice-like clusters are gradually broken up with increasing temperature and pressure.

The International Association for the Properties of Steam (IAPS) has continued the efforts to prepare and release the recommended values on various fundamental properties of ordinary and heavy water. These values are considered to be the most reliable "standard" values confirmed internationally. In this paper, interrelations among such thermophysical properties of liquid H_2O and D_2O in the IAPS releases are examined phenomenologically, employing various methods of correlations found for several non-polar or weak-polar fluids by authors [2-6]. The properties used in this work are density, viscosity, thermal conductivity, dielectric constant and refractive index of liquid phase ranging over temperatures from 0° to 350°C and pressures between 0.1 to 100 MPa.

2. GENERAL BEHAVIOR WITH TEMPERATURE AND PRESSURE

The general behaviors of thermophysical properties of water are compared with those of non-polar or weak-polar organic liquids [2-6]. The similarities and disparities in the behaviors with temperature and pressure could be summarized as follows:

(a) **Density:** Generally, the density ρ increases with decreasing temperature T and increasing pressure P, that is, $(\partial\rho/\partial T)_P < 0$ and $(\partial\rho/\partial P)_T > 0$. Although water is somewhat less compressible than organic liquids, the behavior is almost similar except the existence of a maximum in density, as described above.

(b) **Viscosity:** The effect of pressure on the viscosity η of water is remarkably smaller and $(\partial\eta/\partial P)_T < 0$ at 0-33°C. As $(\partial\eta/\partial T)_P$ is always negative, the isobars intersect one another near 30°C. Such an anomalous behavior has not been found on the viscosity of organic liquids.

(c) **Thermal conductivity:** The thermal conductivity λ of water increases with increasing temperature up to nearly 130°C, $(\partial\lambda/\partial T)_P > 0$,

while most of organic liquids exhibit the negative temperature coefficient, $(\partial\lambda/\partial T)_P<0$. Although the pressure effect for water is somewhat smaller, the pressure coefficient is always positive, $(\partial\lambda/\partial P)_T>0$.

(d) **Dielectric constant:** Excluding the fact that the absolute values of water are very large and the pressure effect is relatively small, general behavior of dielectric constant ε of water is similar to that of organic liquids, that is, $(\partial\varepsilon/\partial T)_P<0$ and $(\partial\varepsilon/\partial P)_T>0$.

(e) **Refractive index:** The refractive index η_D of water behaves regularly in the same manner as organic liquids, that is, $(\partial\eta_D/\partial T)_P<0$ and $(\partial\eta_D/\partial P)_T>0$.

3. DENSITY DEPENDENCE

The effects of temperature and pressure on various thermophysical properties of fluids are represented simply in terms of density. For instance, isotherms of the viscosity of toluene between 298 and 348 K [3] and those of the thermal conductivity of n-hexane between 298 and 373 K [5] gather together closely and are almost parallel to each other. Therefore, each isotherm can be expressed by a simple empirical equation of density. Furthermore, the dielectric constant of nonpolar liquids, such as benzene at temperatures from 298 to 348 K up to the freezing pressures [4], can be represented by a single linear equation of density. These density dependences for nonpolar or weak-polar liquids could be summarized as follows:

$$(\partial\eta/\partial\rho)_T>0, \quad (\partial\lambda/\partial\rho)_T>0, \quad (\partial\varepsilon/\partial\rho)_T>0$$
$$(\partial\eta/\partial T)_\rho<0, \quad (\partial\lambda/\partial T)_\rho>0, \quad (\partial\varepsilon/\partial T)_\rho\leq 0$$

On the other hand, isotherms of these properties of water are different considerably in shape, as seen in Figs. 1 and 2. However, these properties always increase with increasing density and the temperature coefficients of η and λ at constant density show the same tendency as organic liquids and $(\partial\varepsilon/\partial T)_\rho<0$.

An exceptional property of water is the refractive index. According to the electromagnetic theory of light, the molar refraction of the light, whose wave length is sufficiently great, is not affected by the orientation polarization due to permanent dipoles of the molecules. Even using a visible light, such as the Sodium D-line, the molar refraction is almost equal to the electron polarization and does not include the effect of dipoles. In Fig. 3, the refractive index of water for the D-line is plotted against density, where all the points of liquid water are on a single and almost linear curve. This is the same tendency as the case of dielectric constant of nonpolar fluids [4].

The refractive index of liquid water can be expressed by a linear equation:

$$\eta_D=0.96496+3.6828\times10^{-4}\rho \tag{1}$$

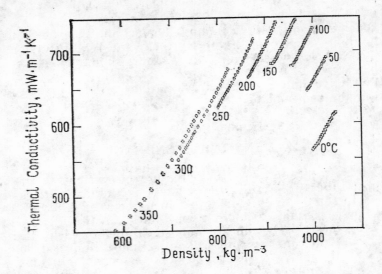

Fig. 1. Plots of thermal conductivity versus density of liquid water

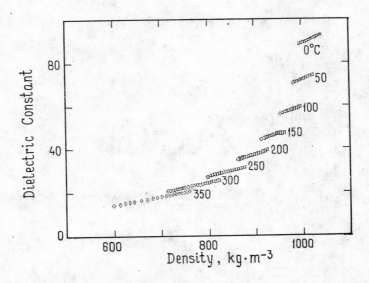

Fig. 2. Plots of dielectric constant versus density of water

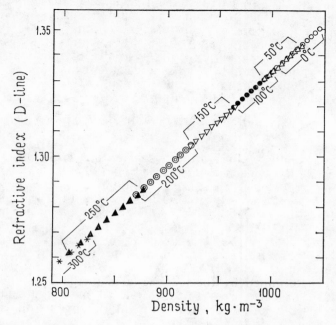

Fig. 3. Refractive index of water as a function of density

with a mean deviation of 0.027% and a maximum one of 0.139%, where ρ is in kg/m^3. If we use a quadratic equation:

$$\eta_D = 1.01350 + 2.5872 \times 10^{-4}\rho + 6.12077 + 10^{-8}\rho^2 \tag{2}$$

it is found that the tabulated values in the release are reproduced with a mean deviation of 0.015% and a maximum of 0.075%.

4. THE TAIT-TYPE EQUATIONS

Among several empirical equations used to represent isothermally the effect of pressure on volume or density of a liquid, the Tait equation is often used, because of its good fit and the simplicity having only two constants. Using density ρ, the equation is written as

$$(\rho - \rho_0)/\rho = C \ln[(B+P)/(B+P_0)] \tag{3}$$

where the subscript "0" denotes an arbitrarily selected reference state, and B and C are constants depending on the fluid and temperature. Although this equation was first used for the compression of water more than 60 years ago, it was found [7, 8] that Eq. (3) represented satisfactorily isothermal changes in volume of various organic liquids with

21

pressure, and that C appeared to be independent of temperature. Furthermore, the same type equation was found to express well the effect of pressure on the dielectric constant of various liquids [9, 10]. The present authors [2, 3] employed the Tait-type equation to represent the isotherms of thermal conductivity and viscosity of nonpolar and weak-polar organic liquids, and also discussed the constants B and C for density and dielectric constant [4].

In this work, the Tait-type equation is written as

$$(X-X_0)/X = C \ln[(B+P)/(B+P_0)] \tag{4}$$

where X denotes ρ, η, λ or ε and the reference states subscripted by "0" are taken as the values at atmospheric pressure (0.1 MPa) or of the saturated liquid at each temperature above the normal boiling point. At first, $(X-X_0)/X$ for η, λ and ε is plotted against $(\rho-\rho_0)/\rho$. For instance, the case of viscosity of H_2O is shown in Fig. 4, where the data at temperatures below 75°C were excluded because of its negative or slight pressure effect. Each isotherm appears to be almost linear, and the same tendency is also found in cases of λ and ε. Therefore, it is supposed that these isotherms could be represented by Eq. (4). Then, the best values of B and C for each isotherm of each property of H_2O and D_2O are determined by the least-squares method. As the next step, the constant C of each property is correlated as a function of temperature. As the behavior of C with temperature appears to change near 100°C, the following formula is adopted:

$$C = c_1 + c_2(T-373.15) \tag{5}$$

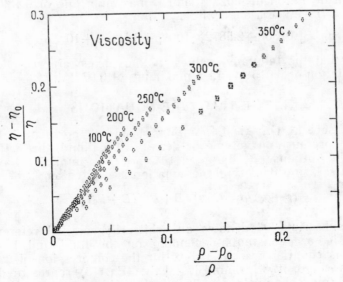

Fig. 4. Plot of $(\eta-\eta_0)/\eta$ versus $(\rho-\rho_0)/\rho$ of H_2O

Table 1. Coefficients of Eqs. (5) and (6) and the reproducibility of Eq. (4) for H_2O

Property	Density		Viscosity
T. range *	I	II	II
b_1	2.3658 $E+2$	2.3658 $E+2$	3.9809 $E+2$
b_2	—1.5009	—1.1505	—4.1083
b_3	3.7682 $E-3$	—7.8815 $E-4$	2.0207 $E-2$
b_4	1.2698 $E-4$	5.8520 $E-6$	—4.4257 $E-5$
c_1	0.1190	0.1190	0.3810
c_2	—3.25 $E-4$	—1.10 $E-4$	—9.96 $E-4$
Av. Dev.	0.004%		0.130%

Property	Thermal conductivity		Dielectric constant	
T. range *	I	II	I	II
b_1	3.9561 $E+2$	3.9561 $E+2$	2.4686 $E+2$	2.4686 $E+2$
b_2	—1.7428 $E-1$	—2.5342	—1.1755	—1.4614
b_3	2.4440 $E-3$	3.1622 $E-3$	—1.2362 $E-4$	1.1473 $E-3$
b_4	1.9022 $E-4$	1.5090 $E-6$	8.1634 $E-5$	2.3328 $E-6$
c_1	0.3200	0.3200	0.1570	0.1570
c_2	7.17 $E-4$	—8.50 $E-4$	6.00 $E-5$	—1.12 $E-4$
Av. Dev.	0.024%		0.043%	

* *Temperature range:* (I) 273.15-373.15 K, (II) 373.15-573.15 K.

Table 2. Coefficients of Eqs. (5) and (6) and the reproducibility of Eq. (4) for D_2O

Prop.	Density		Viscosity	Thermal conductivity	
T*	I	II	II	I	II
b_1	2.5447 $E+2$	2.5447 $E+2$	2.1200 $E+2$	2.0867 $E+2$	2.0867 $E+2$
b_2	—1.3061	—1.2687	—1.9733	—1.0575	—5.7763 $E-1$
b_3	2.8201 $E-4$	—1.1174 $E-3$	8.8814 $E-3$	2.1416 $E-3$	—3.8283 $E-3$
b_4	1.0718 $E-4$	8.1376 $E-6$	—1.9636 $E-5$	1.1015 $E-4$	1.1453 $E-5$
c_1	0.1253	0.1253	0.2385	0.1755	0.1755
c_2	—2.53 $E-4$	—1.32 $E-4$	—5.08 $E-4$	—6.94 $E-4$	—1.44 $E-4$
Dev.	0.008%		0.073%	0.045%	

* *Temperature range:* (I) 276.94-373.15 K, (II) 373.15-573.15 K

where T is temperature in K. The empirical constants c_1 and c_2 are determined in two temperature ranges: 273.15(276.95)-373.15 K and 373.15-573.15 K for each property. Then, using the values of C calculated by Eq. (5), those of B are re-determined for each isotherm of each property. Finally, B-values for each property are fitted to an empirical equation which can reproduce the tabulated values with average deviations comparable to those in the calculation using the best values of B and C in the first step. The correlation formula of B in MPa is

$$B = b_1 + b_2(T-373.15) + b_3(T-373.15)^2 + b_4(T-373.15)^3 \qquad (6)$$

where constants b_1-b_4 are determined also in two temperature ranges as the case of c_1 and c_2.

In Tables 1 and 2, the coefficients in Eqs. (5) and (6) are listed for each property, as well as the average deviation to show the reproducibility of Eq. (4). The percentage deviation is calculated by

$$\text{dev.}[\%] = 100(X_{calc} - X_{tab})/X_{tab} \qquad (7)$$

where X_{calc} is the value calculated by Eqs. (4), (5), (6) and X_{tab} denotes the tabulated values in the IAPS releases. For the typical examples, the deviation plots of the thermal conductivity of H_2O and the density of D_2O, which is calculated from the equation of state [11], are given in Figs. 5 and 6, respectively. From these tables and figures, the Tait-type equation in found to fit satisfactorily the data of ρ, η, λ, ε of liquid water.

5. ISOTOPE EFFECT

As IAPS recently issued a release on the viscosity and thermal conductivity of heavy water, and recommended an equation of state by Hill et al. [11], it becomes possible to consider the isotope effect on these properties between H_2O and D_2O. Therefore, it is interesting to consider the anomalous behavior of water from the viewpoint of the isotope effect.

Generally, the isotope effect [6] is defined by

$$\text{I. E.}[\%] = 100[X(H_2O) - X(D_2O)]/X(H_2O) \qquad (8)$$

where X denotes one of thermophysical properties. The calculated isotope effects on molar volume, viscosity and thermal conductivity at 50°C and 0.1 MPa are -0.25, -19.0 and $+4.0\%$, respectively. In the present pressure range up to 100 MPa, the isotope effects of liquid phase appear to be little affected by pressure and to change obviously with temperature. Therefore, the temperature dependence on the isotope effects is discussed in this paper.

(a) *Molar volume*: In case of the nonpolar molecules, it is well-known [6] that the light molecules have somewhat larger volume than the heavy molecules, that is, I. E. is positive. On the contrary, water

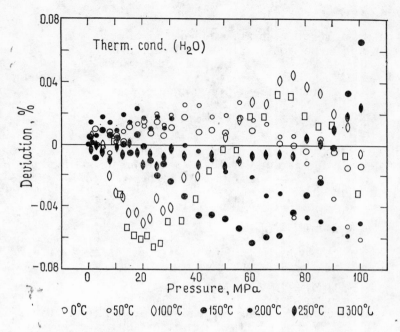

Fig. 5. Deviation plot for thermal conductivity of H_2O

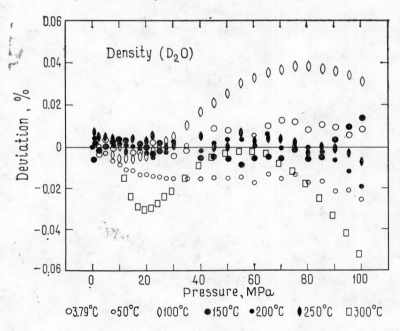

Fig. 6. Deviation plot for density of D_2O

gives the negative effect as mentioned above. It is supposed that this anomaly would come from the highly structured molecules of D_2O. Then, we would like to go back to the "two-fluid model" theory [1], that is, liquid water is a mixture of ice-like clusters and free monomolecules. According to this theory, the molar volume of water V can be written as

$$V = \chi V_{cl} + (1-\chi) V_{mon} \qquad (9)$$

where V_{cl} and V_{mon} are the hypothetic molar volume of clusters and free molecules, respectively, and χ denotes the mole fraction of clusters in liquid water. On the assumption that V_{cl} is equal to the molar volume of Ice-I and that its thermal expansion can be extrapolated up to 100°C, the temperature dependence of V_{cl} is determined by the following equation:

$$V_{cl}[\mathrm{cm^3/mol}] = 19.441 - 1.97 \times 10^{-3}T + 9.84 \times 10^{-6}T^2 \qquad (10)$$

where T is in K. Change in χ with temperature is also expressed by an empirical formula based on the theoretical values [1] as follows:

$$\chi(H_2O) = 2.185 - 8.74 \times 10^{-3}T + 9.94 \times 10^{-6}T^2 \qquad (11)$$

Using the values of V_{cl} and χ calculated from Eqs. (10) and (11), respectively, and $V(H_2O)$ derived from the skeleton table, the values of V_{mon} are obtained at each temperature and expressed by the following equation:

$$V_{mon} = 11.354 + 1.471 \times 10^{-2}T + 1.006 \times 10^{-5}T^2 \qquad (12)$$

As it is supposed that D_2O is more structured than H_2O, assuming that D_2O has the same V_{cl} and V_{mon} as those of H_2O, the mole fraction of clusters in D_2O is calculated using the $V(D_2O)$ obtained from the equation of state [11]. $\chi(D_2O)$ obtained are fitted to the following formula:

$$\chi(D_2O) = 2.591 - 1.103 \times 10^{-2}T + 1.329 \times 10^{-5}T^2 \qquad (13)$$

The calculated results on the volumetric behaviors of H_2O and D_2O at 0.1 MPa are given in Table 3.

Table 3. Volumetric values of H_2O and D_2O

$T/°C$	$V(H_2O)$	$V(D_2O)$	V_{cl}	V_{mon}	$\chi(H_2O)$	$\chi(D_2O)$
0	18.014	—	19.636	16.125	0.538	—
3.79	18.011	18.112	19.649	16.193	0.526	0.555
25	18.064	18.129	19.727	16.636	0.462	0.483
50	18.228	18.274	19.831	17.173	0.397	0.414
75	18.475	18.514	19.946	17.697	0.346	0.363
100	18.797	18.829	20.074	18.230	0.306	0.323

* Unit of V's is in cm³/mol

Because there is a large change in volume between V_{cl} and V_{mon}, as seen from Table 3, the anomalous behaviors in volumetric properties of H_2O and D_2O, such as the shrinkage on melting, the density maximum. the compressibility minimum, as well as the negative isotope effects on molar volume and isothermal compressibility, can be interpreted well, even using simple assumptions mentioned above.

(b) *Viscosity*: If we assume that the viscous flow in water occurs only when the adjacent molecules are free monomolecules, $(1-\chi)^{-1}$ term should be considered in the famous Eyring's theory of viscosity [12], that is,

$$\frac{\eta(D_2O)}{\eta(H_2O)} = \frac{1-\chi(H_2O)}{1-\chi(D_2O)} \cdot \frac{A(D_2O)}{A(H_2O)} \cdot \exp\frac{\Delta E(D_2O) - \Delta E(H_2O)}{RT} \qquad (14)$$

and

$$A = C(M^{1/2} \cdot T^{3/2}/V^{3/2}\Delta H_v) \qquad (15)$$

In these equations, ΔE is an activation energy for viscous flow, R is the gas constant, C an empirical constant, M molecular weight, ΔH_v heat of vaporization. The exponential term in Eq. (14) is estimated empirically using the heats of vaporization. The viscosity ratio is computed by Eqs. (14) and (15) using the values of χ in Table 3. The results obtained are shown in Table 4, where it is found that the isotope effects calculated from Eqs. (14), (15) agree well with the values obtained from the tables in the IAPS releases, taking into account their uncertainty.

Table 4. The isotope effect on viscosity of water

Temp. °C	Terms in Equations (14), (15)				Isotope Effect, %	
	$\dfrac{A(D_2O)}{A(H_2O)}$	$\dfrac{1-\chi(H_2O)}{1-\chi(D_2O)}$	Exp. term	$\dfrac{\eta(D_2O)}{\eta(H_2O)}$	IE(calc)	IE(tab)
25	1.029	1.041	1.153	1.235	—23.5	—22.9
50	1.030	1.041	1.122	1.203	—20.3	—19.0
75	1.030	1.041	1.101	1.181	—18.1	—17.4
100	1.030	1.041	1.085	1.164	—16.4	—16.4
150*	1.030	1.041	1.064	1.141	—14.1	—15.4

* Extrapolated by means of empirical equations.

(c) *Thermal conductivity*: Due to the lack of reliable theories on thermal conductivity of liquids, we have not obtained satisfactory results, although we tried to calculate using some theories.

The details of the consideration on the isotope effects will be published elsewhere.

6. CONCLUSION

Using the recommended values in the IAPS releases on density, viscosity, thermal conductivity, dielectric constant and refractive index of liquid H_2O and D_2O, various correlations have been examined by means of methods for non-polar or weak-polar fluids in authors' experience. However, most of the methods are found to be unsuccessful for the thermophysical properties of water, due to its high polarity and partial structure in liquid. In this paper, only a few satisfactory results are described. As the anomalous behaviors in various properties and complex structural effects are more attractive, the authors hope for the development of investigations in this field.

REFERENCES

1. G. Némethy and H. A. Scheraga, *J. Chem. Phys.*, **36**, *3382* (1962).
2. H. Kashiwagi, T. Hashimoto, Y. Tanaka, H. Kubota, and T. Makita, *Int. J. Thermophys.*, **3**, 201 (1982).
3. H. Kashiwagi and T. Makita, *Int. J. Thermophys.*, **3**, 289 (1982).
4. H. Kashiwagi, T. Fukunaga, Y. Tanaka, H. Kubota, and T. Makita, *J. Chem. Thermodynamics*, **15**, 567 (1983).
5. T. Makita, *Int. J. Thermophys.*, **5**, 23 (1984).
6. S. Matsuo and A. Van Hook, *J. Phys. Chem.*, **88**, 1032 (1984).
7. R. E. Gibson and O. H. Loeffler, *J. Am. Chem. Soc.*, **61**, 2515 (1939); **63**, 898 (1941).
8. K. E. Weale, "Chemical Reactions at High Pressures", E. & F. N. Spon Ltd., London (1967).
9. B. B. Owen and S. R. Brinkley, *Phys. Rev.*, **42**, 32 (1943).
10. S. D. Hamann, "Physico-Chemical Effects of Pressure", Butterworths, London (1957).
11. P. G. Hill, R. D. C. MacMillan, and V. Lee, *J. Phys. Chem. Ref. Data*, **11**, 1 (1982); **12**, 1065 (1983).
12. S. Glasstone, K. J. Laidler, and H. Eyring, "The Theory of Rate Processes", McGraw-Hill, New York (1941).

Refractive Index and Lorentz-Lorenz Function of Water Vapour up to 25 Bar from 373 to 500 K

H.-J. ACHTERMANN and H. RÖGENER

Institut für Thermodynamik, Universität Hannover,
3000 Hannover, FRG

ABSTRACT

An experimental apparatus to measure the refractive index of water vapor is developed. The apparatus consists of two grating interferometers to measure the refractive index and the pressure.

Experiments are performed at constant temperatures of 100, 125, 150, 175, 200 and 225°C with pressure close to the vapor pressure. Three measurements with different fluid samples were performed and the repeatability was excellent. The resulting values for the refractive-index are presented as a function of pressure. The Lorentz-Lorenz function is calculated for one isotherm.

The uncertainty of the refractive index measurement is less than 2×10^{-7} and the uncertainty of the pressure measurement is 10^{-4}.

KEYWORDS

Refractive index; steam; water; laser interferometer

INTRODUCTION

In 1971 the International Association for the Properties of Steam (IAPS) proposed that the optical properties of water substances should be studied. At the Ottawa meeting of the IAPS-Working Group III in

1975 it was formally decided that the refractive index should be investigated among other properties of water. The objective was to establish an internationally recognized standard of the optical properties of water, and K. Scheffler was asked to compile the available information. This work yielded a new equation for the refractive index of water [11], which was restricted to the liquid phase due to the lack of data in the vapor phase.

Refractive index measurements of water vapor were first performed at the Institut für Thermodynamik, Universität Hannover in 1981. The results of these measurements, the new measurements by Scheffler in the liquid phase [10] and the other available data are used to formulate a representative equation for the refractive index of water and steam [12].

Below we have discussed the method described. We also give the results for the refractive index of water vapor at 100, 125, 150, 175, 200, 225°C with pressures up to 25 bars.

EXPERIMENTAL APPARATUS

The accurate determination of the refractive index as a function of pressure on isotherms is performed by using the interferometric technique for both properties. The absolute refractive index n of water vapor is obtained by counting the changes in the interference fringes while venting the fluid to vacuum. For a known wavelength λ and a precisely measured length l_0 of the test section at 0°C and a known linear coefficient of expansion α the refractive index on isotherms t in °C is given as

$$n = \frac{\lambda K_n}{l_0(1+\alpha t)} + 1$$

where K_n is the number of changes in the fringe count referred to vacuum, i. e. $K_n = 1$ is a complete change (dark-light-dark) in the interference pattern.

The absolute pressure is obtained by fringe counting in a second interferometer specially developed for pressure measurements.

The pressure interferometer was calibrated with a precision standard piston gauge with nitrogen as the pressure measurement fluid at 0°C [1]. The calibration gives a measured relation between the pressure interferometer fringes K_p and the absolute pressure P which is approximated by a fifth-order least-squares polynomial.

With this pressure interferometer the high speed of data taking of the refractive index measurement and the accuracy of the standard piston gauge is maintained.

OPTICAL SET-UP

A grating interferometer is used to measure the refractive index of water vapor as well as the pressure. The experimental apparatus and the procedure are essentially the same as reported before [2, 8]. Therefore

3)

we will only give a brief description of the optical set-up and the measuring cell to point out the details that are specific to this work.

A schematic diagram of the optical set-up is shown in Figure 1. A linearly polarized He-Ne laser light is used to illuminate the grating interferometer. The grating (G_1) produces numerous diffraction orders where the two symmetrical beams of the first diffraction order are selected as a reference beam (R) in one case and as a measuring beam (M) in the other case. The planes of polarization of the measuring beam and the reference beam are made perpendicular by including the $\lambda/2$-plate (P_1) into one of the beams. The two beams are refracted parallel by the first objective (O_1) and then enter the cell, where one goes through the measuring boring and the other through the reference boring. This design ensures that the variation in length of the cell due to change in pressure is compensated to a great extent. The two borings in the cell are arranged symmetrically around the axis of the cylindrical cell and are immersed into a temperature-controlled bath. A second objective (O_2) is used to focus the two beams on the second grating (G_2). The reference beam (R) and the measuring beam (M) produce two diffraction patterns superimposed on one another. Among the resultant diffractions only three superposed orders $(0, \pm 2)$ have enough contrast for them to be converted to electrical signals at a later stage.

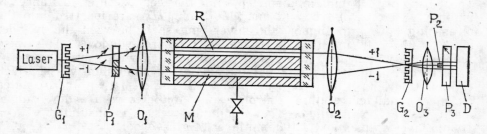

Fig. 1. Grating interferometer

The three beams rendered parallel by the lens (O_3) impinge on a double refracting quartz plate (P_3) in order that three pairs of push-pull modulated signals be produced. The six signals are in phase and antiphase. To produce quadrature signals $(\pm \sin, \pm \cos)$ for the rotating electric field, the $\lambda/4$ wave plate (P_2) is incorporated in the path of the zero order of diffraction.

An array of six photodiodes with similar dependence on dark currents and temperature coefficients is arranged in a matrix (D) to transform the optical signals into electrical signals. This arrangement ensures simple mechanical adjustments and short amplifier connections.

The central pair of the push-pull modulated electrical signals which is in quadrature with the other two pairs is introduced into two inputs of a differential amplifier. We introduce the combination of the remaining pairs of push-pull modulated electrical signals into the inputs of another

differential amplifier. The amplitude of the quadrature signal is approximately equal to the sum of the amplitudes of the other two signals. The signals introduced in the differential amplifiers are combined in such a way that the d.c. parts of the signals are compensated. The two outputs of the differential amplifiers are components of a d.c. compensated rotating electrical field which operates a reversible counter. Each change of one wavelength is digitized into 40 parts for counting.

The grating interferometer has substantial advantages compared to the other double beam interferometers, such as:

(1) favourable production of push-pull signals to generate a rotating electrical field;

(2) simple adjustment with 6 photo diodes arranged in a matrix;

(3) symmetrical and compact arrangement;

(4) error compensation between the beam of the test section and the reference beam.

MECHANICAL SET-UP

A technical drawing of the cell for the refractive index measurement with the vessel of the temperature-controlled bath is given in Figure 2. The cylinder block of the cell is made of Inconel 600. It is about 250 mm in length between the end planes with an outside diameter of 40 mm. The borings are 8 mm in diameter and the center-center distance is 9.5 mm. 25 pieces of tubes (8 mm OD, 3 mm ID) are inserted into the measuring boring to support the silver ring seals between the cylinder block of the cell and the plan parallel sapphire windows. This also reduces the volume measuring section to be filled with sample gas.

The cell is incorporated about the axis of an aluminium vessel which contains the temperature-controlled oil bath. The oil-bath is thermostated by an external temperature controller and circulates with a flow rate of about 8 liters per minute. The aluminium vessel is insulated with a 10-mm layer of mica and a 100-mm layer of glass wool.

The cell for the pressure measurement is essentially the same in geometry. Because of the inert gas nitrogen less expense is required for materials of this cell as compared to the materials the refractive index cell. The cylinder block is made out of brass, and Teflon O-rings are used to seal the windows. The cell is placed in an open ice-bath.

EXPERIMENTAL PROCEDURE

Figure 3 gives a schematic diagram with the various containers, valves and connecting tubes of the cells for the coupled refractive index and pressure measurements.

At the beginning of a measurement the two test sections are evacuated (V_3, V_7 closed and V_1, V_2 opened). Then V_2 is closed and the test sections are charged. The measuring boring of the refractive index interferometer (RI) and the buffer containment C_2 are filled from the supply vessel C_1 by opening V_3. The water in C_1 is purified by the

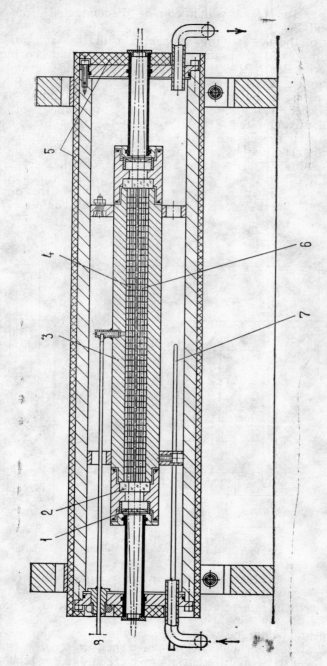

Fig. 2. Cell and vessel of the temperature-controlled bath for refractive index measurements on water vapor

degasifying apparatus (DA) which can be disconnected from C_1 using the valves V_4 and V_5.

The supply of the degasifying apparatus with research-quality water is given through V_6. At the same time the measuring boring of the pressure interferometer (PI), the buffer containment C_3, and pressure equilibrium chamber (PEC) are charged with commercially available research-grade nitrogen by opening valve V_7. The maximum pressure in the water test section is the vapor pressure for the isotherm. The pressure in the nitrogen test section is adjusted to a pressure slightly higher than the vapor pressure in the water test section using gauge G. The experiment starts just by opening valve V_1. We then very carefully open and close valve V_2 to check on an oscilloscope whether the direction of the rotating electrical field produced at the RI is the same as that produced at the PI. When the two rotating fields on the oscilloscopes move in the same direction valve V_2 can be opened completely.

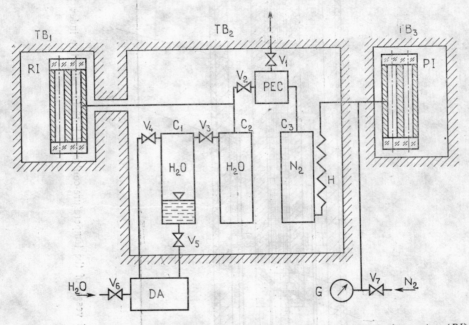

Fig. 3. Coupled refractive index interferometer (RI) and pressure interferometer (PI)

Valve V_1 is then closed and the fringe count at the RI and at the PI starts when equilibrium is reached in both cells. Further pressure changes and settling to thermal equilibrium for data collecting are carried out solely by manipulating valve V_1.

This coupling technique with the PEC becomes adaptable in combination with the interferometric pressure measurement. Its advantage compared to the use of a differential pressure indicator (DPI) is that it

enables us to carry out high-speed measurements of the refractive index as a function of pressure.

The possible mixing of gases by diffusion between two cells (valve V_1 is closed) is avoided by using long and thin tubes. Since the duration of the time when V_1 is closed is only a small fraction of the total time of the experiment, the problem of diffusion is certainly negligible.

The nitrogen that is vented from the test section of the PI passes through a heat exchanger H to adapt the temperature of the isotherm. This prevents condensation of the water vapor in the PEC. The temperature baths TB_1 and TB_2 are held at the temperature of the investigated isotherm and TB_3 is the ice bath of the PI.

RESULTS AND DISCUSSION

Measurements were performed on six isotherms, at 100, 125, 150, 175, 200 and 225°C with pressures up to the vapor pressure P_s. Figure 4 shows a P, T-diagram with the H_2O-refractive index data of our work and of other authors [10, 13]. This work presents the only measurements in the vapor region. Three different samples of water vapor were investigated for each temperature to check the sample-to-sample repeatability. In Table 1 the refractive index n of water vapor is given as a function of pressure P. The curves $n=n(P)$ and standard deviations of the three runs are consistent for each isotherm, therefore only one set of data is presented per temperature. The maximum uncertainty of the refractive index is $\pm 2 \times 10^{-7}$ for three standard deviations. All measured $n(P)$-values are within the bounds of estimated uncertainty. The highest pressure is about 0.3 bar below P_s due to pressure losses during the coupling procedure.

Precise refractive index measurements may be important not only for the completeness of the physical properties. Refractive index measurements are also used for the determination of density ρ [3, 4, 5, 9]. The functional relationship between ρ and n is given by the Lorentz-Lorenz (LL) equation

$$\frac{n^2-1}{n^2+2} \frac{1}{P} = A_n + B_n \rho + C_n \rho^2 + \dots$$

where A_n, B_n, C_n are respectively the first, the second and the third refractivity virial coefficients. The terms $B_n \rho$, $C_n \rho^2$, ... only account for very small (less than 1%) contributions so that the ideal gas Lorentz-Lorenz function $LL=A_n$ gives a good approximation for ρ over a wide range of pressure and temperature. A_n can be determined from $n=n(p)$ measurements as the intercept of a least-squares approximation for pressure dependence of the ideal gas LL-function.

$$A_n = R_m T \lim_{P \to 0} \frac{n^2-1}{n^2+2} \frac{1}{P}$$

where R_m is the molar gas constant and T the absolute temperature.

Very precise values of the density can be determined if, in addition to the absolute measurements of the refractive index as a function of pressure, the second and third refractivity virial coefficients are measured. Such measurements are performed with an expansion technique [4, 5]. This method consists in measuring the sum of optical path lengths of two similar cells, where one of them is filled with gaseous fluid at density ρ and the other is evacuated. After the expansion, the density is nearly halved, and the sum of the optical path lengths is measured again.

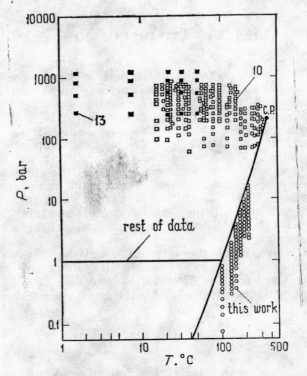

Fig. 4. Location of refractive index data of H_2O in a P, T-diagram

Because the linear term in density remains the same before and after the expansion and only the quadratic and higher-order terms change, we can determine B_n and C_n from the change of the optical lengths. We have used this technique to measure B_n and C_n for methane and ethylene and we intend to use it to measure with water vapor in the near future. If the density is known from other measurements or an equation of state, LL can be determined.

In preparation for the planned water vapor expansion measurements we have calculated *LL* using the absolute method, where ρ is taken from other measurements or equation of state. In Figure 5 various values of the Lorentz-Lorenz function are given as a function of pressure for one isotherm. In LL_{IFC67}, LL_{IFC68}, LL_{POL}, and LL_{HGK} our measured refractive index values are combined with density values from equations of state by the 1967 IFC Formulation for Industrial Use, the 1968 IFC Formulation for Scientific and General Use, Pollak [7], and Haar et al. [6], respectively.

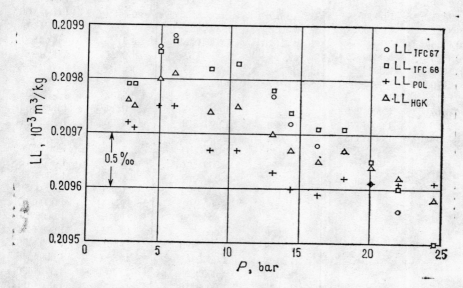

Fig. 5. Lorentz-Lorenz function *vs* pressure for water vapor at 225° C

The comparison of the *LL*-values in Figure 5 shows similar trends for the four cases with variations that are within 0.1%. Up to a pressure of 10 bars *LL* is nearly constant. Beyond 10 bars a slight decrease can be observed. For the other isotherms the pressure is too small to cause significant changes in *LL*.

Table 1. Refractive index *n* of

P bar $T=100°C$	n	P bar $T=125°C$	n	P bar $T=150°C$	n
0.0737	1.0000133	0.2869	1.0000492	0.3550	1.0000577
0.1328	1.0000238	0.3464	1.0000593	0.4646	1.0000756
0.1782	1.0000322	0.3957	1.0000678	0.5725	1.0000931
0.2361	1.0000427	0.4545	1.0000781	0.7320	1.0001191
0.2895	1.0000524	0.5079	1.0000871	0.8409	1.0001374
0.3427	1.0000623	0.5628	1.0000965	0.8958	1.0001461
0.8935	1.0000716	0.6161	1.0001059	0.9496	1.0001550
0.4523	1.0000825	0.6704	1.0001152	1.0021	1.0001636
0.5055	1.0000923	0.7240	1.0001243	1.0569	1.0001727
0.5470	1.0001000	0.7787	1.0001340	1.1615	1.0001897
0.6116	1.0001119	0.8307	1.0001430	1.2732	1.0002084
0.6667	1.0001220	0.8865	1.0001528	1.3811	1.0002262
0.7223	1.0001326	0.9383	1.0001618	1.4790	1.0002421
0.7774	1.0001438	0.9947	1.0001718	1.5960	1.0002616
		1.0479	1.0001809	1.7031	1.0002794
		1.1025	1.0001907	1.8130	1.0002978
		1.1548	1.0002000	1.9207	1.0003158
		1.2103	1.0002099	2.0288	1.0003338
		1.2637	1.0002191	2.1365	1.0003519
		1.3278	1.0002303	2.2446	1.0003701
		1.3706	1.0002387	2.3523	1.0003884
		1.4258	1.0002490	2.4601	1.0004066
		1.4798	1.0002585	2.5680	1.0004246
		1.5442	1.0002698	2.6767	1.0004430
		1.5876	1.0002776	2.7822	1.0004611
		1.6420	1.0002870	2.8648	1.0004753
		1.6962	1.0002969	2.9935	1.0004973
		1.7504	1.0003067	3.1063	1.0005166
		1.8033	1.0003171	3.2145	1.0005357
		1.8585	1.0003264	3.3175	1.0005536
				3.4305	1.0005734
				3.5249	1.0005900
				3.6477	1.0006111
				3.7438	1.0006288
				3.8604	1.0006486
				3.9709	1.0006678
				4.0766	1.0006871
				4.1850	1.0007064
				4.2959	1.0007256
				4.3999	1.0007457

steam as a function of pressure P ($\lambda = 632.99$ nm)

P bar $T = 175°C$	n	P bar $T = 200°C$	n	P bar $T = 225°C$	n
1.2107	1.0001851	2.2933	1.0003382	2.3588	1.0003263
1.3224	1.0002024	2.6090	1.0003851	2.7731	1.0003841
1.4291	1.0002192	2.9422	1.0004348	3.2130	1.0004459
1.6191	1.0002484	3.2604	1.0004821	3.6367	1.0005056
1.8084	1.0002778	3.5801	1.0005301	4.0635	1.0005661
1.9880	1.0003058	3.8817	1.0005756	4.5106	1.0006297
2.1641	1.0003337	4.2352	1.0006289	4.9411	1.0006913
2.3244	1.0003586	4.5580	1.0006777	5.3632	1.0007518
2.4875	1.0003841	4.8800	1.0007267	6.0031	1.0008441
2.7037	1.0004182	5.3080	1.0007924	6.6571	1.0009393
2.9178	1.0004522	5.7171	1.0008554	7.3039	1.0010330
3.1345	1.0004884	6.1753	1.0009261	7.9281	1.0011250
3.3496	1.0005221	6.6134	1.0009941	8.5917	1.0012229
3.5656	1.0005556	7.0357	1.0010601	9.2357	1.0013188
3.7748	1.0005892	7.4606	1.0011265	9.8752	1.0014146
3.9969	1.0006252	7.8897	1.0011948	10.5228	1.0015125
4.2117	1.0006596	8.3275	1.0012643	11.1597	1.0016090
4.4275	1.0006953	8.7477	1.0013322	11.8059	1.0017078
4.6423	1.0007303	9.2375	1.0014109	12.4458	1.0018062
4.8576	1.0007654	9.6094	1.0014715	13.0889	1.0019066
5.0739	1.0008011	10.0429	1.0015418	13.7051	1.0020030
5.2873	1.0008361	10.4828	1.0016143	14.3619	1.0021064
5.5033	1.0008715	10.9547	1.0016924	15.0111	1.0022097
5.7190	1.0009072	11.3177	1.0017524	15.6470	1.0023118
5.9345	1.0009432	11.7808	1.0018304	16.2860	1.0024147
6.1465	1.0009794	12.1824	1.0018986	16.8953	1.0025146
6.3613	1.0010151	12.5960	1.0019682	17.5348	1.0026191
6.5783	1.0010511	13.0202	1.0020406	18.1846	1.0027271
6.7932	1.0010879	13.5157	1.0021260	18.8371	1.0028360
7.0087	1.0011241	13.8387	1.0021826	19.4374	1.0029373
7.2228	1.0011612	14.2893	1.0022617	20.0939	1.0030489
7.4376	1.0011983	14.7497	1.0023434	20.7451	1.0031610
7.6484	1.0012333	15.1664	1.0024169	21.3854	1.0032720
7.8671	1.0012723			22.0323	1.0033852
8.0818	1.0013094			22.6527	1.0034952
8.2886	1.0013459			23.2721	1.0036058
8.5083	1.0013842			24.5726	1.0038416
8.7254	1.0014201			25.2101	1.0039638

ACKNOWLEDGEMENT

We would like to thank Prof. T. K. Bose and Prof. J. M. St.-Arnaud, Université du Québec à Trois — Rivières, Canada and R. Scharf for fruitful discussions and G. Magnus for technical help.

The authors would like to express their gratitude to the Bundesministerium für Forschung und Technologie for its financial support which made this work possible.

REFERENCES

1. Achtermann, H.-J. and H. Rögener; Ein neues Gerät zur kontinuierlichen Druckmessung — Das Druckmeßinterferometer, *Technisches Messen* **3** (1982) S. 87-92.
2. Achtermann, H.-J.; Entwicklung und Erprobung einer optischen Versuchsanlage zur Bestimmung der Dichte aus Brechungsindex-Messungen, Dissertation, Technische Universität Hannover, 1978.
3. Achtermann, H.-J., F. Klobasa und H. Rögener; Realgasfaktoren von Erdgasen, Teil I: Bestimmung von Realgasfaktoren aus Brechungsindex-Messungen, *Brennstoff-Wärme-Kraft* **34** (1982) Nr. 5, S. 266/271.
4. Achtermann, H.-J., R. Scharf, G. Magnus; Ein Verfahren zur genauen Bestimmung der Dichte von Gasen aus optischen Messungen, *VDI-Forsch.-Heft* **619**, S. 11/16.
5. Achtermann, H.-J., T. K. Bose, R. Scharf, J. M. St.-Arnaud; An improved method for the precise determination of the compressibility factor from refractive index measurements. To be published.
6. Haar, L., J. S. Gallagher and G. S. Kell; National Research Council Canada, NRCC 19178 (1981).
7. Pollak, R.; Eine neue Fundamentalgleichung zur konsistenten Darstellung der thermodynamischen Eigenschaften von Wasser. BWK 27, 210 (1975).
8. Rögener, H. and H.-J. Achtermann; The Experimental Determination of the Refractive Index of Water Vapor with Grating Interferometer, *Proc. 9th Int. Conf. Prop. Water & Steam* 489-493, München 1979.
9. Rögener, H. und H.-J. Achtermann; Realgasfaktoren von Argon-Trifluorbrommethan Gasgemischen aus Brechungsindex-Messungen, *Wärme- und Stoffübertragung* **16** (1982) S. 57-62.
10. Scheffler, K.; Experimentelle Bestimmung der Koexistenzkurve von Wasser im kritischen Gebiet. Dissertation, Techn. Universität München (1981).
11. Thormählen, I., K. Scheffler and J. Straub; Equation for the Refractive Index of Water, *Proc. 9th Int. Conf. Prop. Water & Steam* 477-488, München 1979.
12. Thormählen, I., J. Straub and U. Grigull; Refractive Index of Water and its Dependence on Wavelength, Temperature, and Density. To be published.
13. Waxler, R. M., C. E. Weir, H. R. Schampl; Effect of Pressure and Temperature Upon the Optical Dispersion of Benzene, Carbon Tetrachloride and Water. *J. of Research of NBS* **68**, 489 (1964).

An X-Ray Diffraction Study of Structure
of Liquid and Supercritical Water

Yu. E. GORBATY and Yu. N. DEMIANETS

Institute of Experimental Mineralogy, USSR Academy of Sciences,
USSR

Numerous anomalous properties of water are known to be a result of two features of water structure which make water be different from normal liquids. We mean the essential contribution of hydrogen bonding into intermolecular interactions and unusual tetrahedral short-range order being characteristic for water in the liquid state. Unfortunately, this rather vague statement is the only one which could be voiced with a certain confidence. Any attempt to specify this notion in more quantitative way can hardly be indisputable. The reason for such a situation is a shortage of accurate experimental data especially at high pressures and temperatures.

To fill this gap we have undertaken an X-ray diffraction study of the temperature effect on the water structure at a constant pressure of 100 MPa in the temperature range 398-773 K [1]. The other experiment was aimed at the pressure effect at constant temperature of 393 K and at pressures of up 200 MPa [2]. To achieve such high values, an energy dispersive diffraction method (EDD) was applied to a design of a special X-ray diffractometer. The most useful feature of EDD is the ability to obtain quite large fragments of a diffraction pattern at a constant angle of scattering. This feature allowed us to overcome serious difficulties arising usually at designing high-pressure high-temperature X-ray cells.

The molecular pair correlation functions of water $g(r)$ at increasing temperature are presented in Fig. 1 in the form $r^2[g(r)-1]$. We can see the following changes occurring at isobaric heating:

1. A weakening of the correlation with increasing temperature is observed. However, at temperatures above 623 K the correlation seems to grow again. But even at 773 K the function $r^2[g(r)-1]$ does not correspond to a simple close packing which could be intuitively expected at such a high temperature.

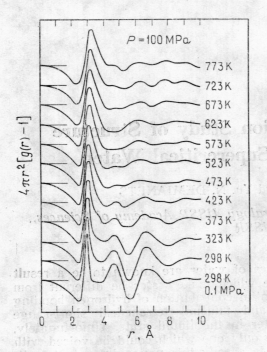

Fig. 1. The functions $r^2[g(r)-1]$ of liquid and supercritical water

2. Hydrogen bonding decreases with an increased temperature. This is evidenced by a lessening of the first peak amplitude and a nearly linear shift of the peak from 2.84 to 3.02 Å.

3. The second peak of $g(r)$ at ~ 4.5 Å, which provides the most strong evidence for tetrahedral ordering in liquid water, diminishes gradually and seems to have completely disappeared at 623 K. So, the ice-like tetrahedral short-range order may be thought to vanish above this temperature. But at the further increase of temperature the peak reappears and shows a tendency to grow. This surprising result reflects most likely the increasing growth of entropy and density fluctuations as temperature increases and comes nearer to the maxima of compressibility and thermocapacity.

4. One can notice some excess of the correlation at ~ 3.3 Å at the ambient temperature. With increasing temperature this feature becomes more prominent and turns into a shoulder on the right-hand side of the first peak. At high temperatures the shoulder becomes less distinct because of strong broadening of the first peak. This interesting feature allows for the possibility of a structural inhomogeneity within the first coordination sphere.

The latter point appears to be very important for the understanding of the water structure. Numerous diffraction experiments have shown that a value of the first coordination number is close to four. On the other hand, it is known that the molecule of water can form four hydrogen bonds. Such a numerical coincidence gives a strong impression that almost all the molecules in liquid water are tetrahedrally coordinated

and included into full number of hydrogen bonds. It seems that many models of water structure have arisen under influence of this coincidence.

However, it should be noted that results of determination of a coordination number may in fact be very different depending on the method being used. Unfortunately, it is quite impossible to present serious physical grounds for a choice of the method or, in other words, a criterion for a boundary of the first coordination sphere in liquid. In the case of water the task becomes more complicated for the chosen method has to take into consideration the shoulder of the main peak of $g(r)$ mentioned above.

To understand the influence of temperature on the nearest ordering, we have decomposed the radial distribution functions $D(r)/r$ (where $D(r) = 4\pi r^2 \rho_0 g(r)$ is the radial distribution function of molecular density) into three gaussians with usual least-square fitting procedure. Two of them approximate the first peak and the shoulder and the third represents some background contributed by the second and following coordination spheres. Fig. 2 shows some of the results of the fitting procedure.

Thus, the first coordination sphere may be represented by the sum of two components corresponding to two types of the molecules in the nearest neighbourhood. It is quite clear that the molecules with an average separation between them of ~2.84 Å can form fairly strong hydrogen bonds and only these molecules can be located in the vertices of a tetrahedron with distances of ~4.5 Å between them. Of course one may suppose that the

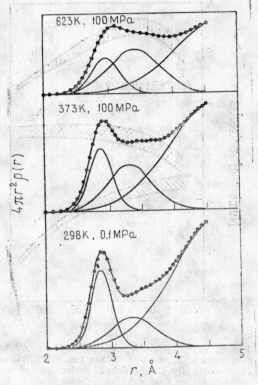

Fig. 2. Dots: experimental $D(r)$, solid lines: the three gaussians and their sum

molecules of the second type with a distance of ~3.3 Å can also be included into hydrogen bonds. But they certainly cannot form a tetrahedron corresponding to the peak at 4.5 Å in $g(r)$.

The sum of the areas under the first and the second gaussians gives the coordination number N. As Fig. 3 shows, N is equal to 4.4-4.7 at ambient conditions in a good agreement with results of the preceding

works. What is more interesting, the contribution of the first nearest neighbours, A_1, (those which may belong to tetrahedral configurations) turns out to be only 2.4-2.5 molecules. With an increasing temperature at a constant pressure of 0.1 MPa, the number of these molecules decreases monotonically and reaches a value of 0.8 at 773 K. The number of the second nearest neighbours is ~ 2 at 0.1 MPa and ~ 2.3 at 100 MPa. So, we may conclude that the tetrahedral configurations have roughly an equal probability with the other possible types of nearest ordering.

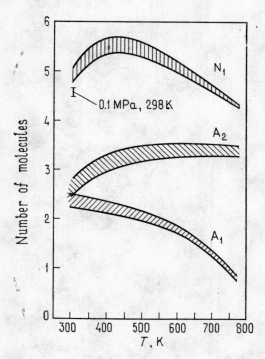

Fig. 3. Temperature dependence of the coordination number N. A_1 and A_2 are the contributions of structurally-distinguished molecules of the first and second types, respectively

However, such a ratio between two distinguished types of the nearest molecules cannot be the same for any randomly chosen point in liquid water, otherwise so a high correlation at 4.5 Å must not be observed. So, we ought to assume that the hydrogen bonded molecules are strongly correlated. In other words, there should be a tendency for these molecules to gather into fairly regular tetrahedral configurations, that is to say, to be preferably four or, at least, three-bonded. In turn it inevitably leads to the much lower concentration of the hydrogen bonds in the rest of the structure. Just such a structural unhomogeneity, being most likely a result of the fluctuations of entropy and density, causes a complicated shape of the first peak of $g(r)$.

These results are convenient to consider in terms of the percolation model developed by Stanley and Teixeira [3]. The model predicts, however, (as many other theories do) gathering of the tetrahedrally arranged molecules in small areas of liquid water. Assuming that the average number of bonds per molecule, $\bar{n}_{HB}$, is equal to ~ 2.4 we find that the probability for hydrogen bonding, P_b, at 298 K and 100 MPa is ~ 0.6. This value seems to be close yet to the percolation threshold below which the infinite cluster has not to exist. One can estimate temperature corresponding to the percolation threshold in water. For three-dimensional network of the diamond type the

threshold is known to be 0.39. As Fig. 3 shows this value is reached at about critical temperature.

Fig. 4 demonstrates structure functions and pair correlation functions of liquid water at pressures of 0.1 up to 200 MPa and at a constant temperature. A decrease of the correlation at 4.5 Å with increasing pressure may be noted. The shoulder of the first peak of $g(r)$ is seen even more clearly than in Fig. 1 and this feature obviously grows as pressure increases.

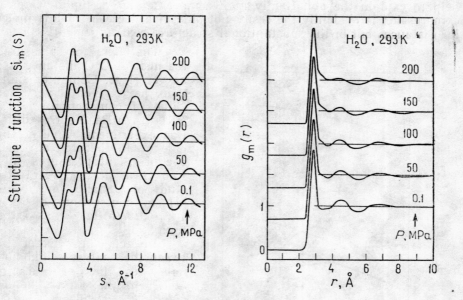

Fig. 4. Structure function and pair correlation functions of liquid at increasing pressure

Just such effects we have observed on heating. Redistribution of intensity between two components of the doublet main peak of the structure function takes place at isobaric heating as well. Thus, it turns out that both the temperature and the pressure alter the water structure in a similar way, namely, tetrahedral ordering becomes more weak. Some theoretical predictions of such a behaviour of the water structure have been made earlier, but all the same we were impressed by experimental evidence of this effect. Indeed, it is well known, for example, that the pressure and temperature influence the strength of hydrogen bonds in a quite opposite way.

To obtain a dependence of the coordination number on pressure the same procedure has been used as in the case of temperature. Fig. 5 shows an alteration of N as well as contributions of both structurally distinguished types of the molecules in liquid water.

N is about 4.4 at normal pressure and reaches ~4.9 at 200 MPa. However, the number of the first neighbours decreases unexpectedly from 2.4 to 2.2 molecules, whereas the number of the second neighbours increases from 2 up to 2.7 So, the decrease of the molecular density at ~4.5 Å is not merely a result of weakening of the orientational correlation. The reason may be simply a lessening of the "bonded" molecules. It is also clear that the increase of the coordination number occurs at the expense of the "unbonded" molecules only. Perhaps, this fact could help to explain the abnormally high compressibility of liquid water, for compression is evidently the result not only of a decrease of intermolecular distances but of the substantional structural rearrangement.

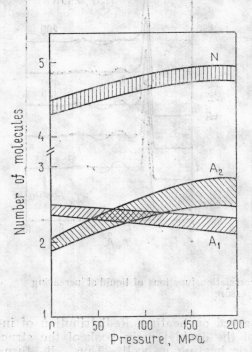

Fig. 5. Pressure dependence of N, A_1 and A_2

The effect of pressure on the nearest O-O distance is not as yet clear. Computer simulations have given a very small value of contraction, ~$1.5 \times \times 10^{-5}$ Å/MPa. Neutron diffraction experiments have not been able at all to catch any decrease of r_{oo} at pressures up to 1.56 GPa. In the recent X-ray study of Gaballa and Neilson [4] a contraction of about 1×10^{-4} Å/MPa has been found.

In fact, the distance between the nearest neighbouring molecules changes much greater. For the first "bonded" neighbours r_{oo} decreases from 2.83 to 2.77 Å under effect of a pressure of 200 MPa. Thus, the rate of the contraction is 2.6×10^{-4} Å/MPa. This value seems to be very high but at further increase of pressure it has most likely to decrease.

The data obtained in this study support an idea about structural inhomogeneity of liquid water undoubtedly arisen as a result of fluctuations. Indeed, a large variety of structural arrangements engender in water due to the fluctuations. However, some configurations may certainly be preferable, that is to say, they can arise more often or exist for a longer time. So, they should be more pronounced in the correlation functions.

It seems to us that inferences which can be drawn from this study allows for, at least, qualitative explanation of well-known anomalies of

water. These may be thought now as consequences of the only effect, namely, the decrease of a degree of tetrahedral ordering of the water structure.

REFERENCES

1. Gorbaty Yu. E. and Demianets Yu. N. (1983): *Chem. Phys. Lett.,* **100** (5), 450-454.
2. Gorbaty Yu. E. and Demianets Yu N. (1984): *Dokl. Akad. Nauk SSSR,* **275** (4), 903-907.
3. Stanley, H. E. and Teixeira, J. (1980): *J. Chem. Phys.,* **73** (7), 3404—4322.
4. Gaballa, G. A. and Neilson, G. W. (1983): *Mol. Phys.,* **50** (1), 97-104.

Complex Dielectric Constants of Light and Heavy Water Along the Coexistence Curve

Yu. A. LUBIMOV and O. A. NABOKOV

Moscow State University, Department of Physics, Moscow, USSR

1. INTRODUCTION

The measurements of the dielectric constant ε' and the dielectric loss ε'' of polar liquids at microwave frequencies allow information on the rate and mechanism of orientational motions of molecules in liquid to be obtained. The dielectric relaxation of water has been investigated in a limited temperature interval: systematic measurements were performed up to about 340 K.

A careful analysis of the most complete data on complex dielectric constant $\varepsilon^* = \varepsilon' - i\varepsilon''$ of water measured over a wide frequency range and at 268-333 K [1] reveals the absence of the dielectric relaxation time distribution at least within the limits of the accuracy of measurements (contrary to some early results). Thus, the dispersion of the dielectric permittivity of water can be represented by Debye equations with a single dielectric relaxation time τ_D:

$$\varepsilon' = \varepsilon_\infty + (\varepsilon_s - \varepsilon_\infty)/(1 + \omega^2 \tau^2_D), \quad \varepsilon'' = (\varepsilon_s - \varepsilon_\infty)\omega \tau_D/(1 + \omega^2 \tau^2_D) \qquad (1)$$

Here ω is the circular frequency, ε_s and ε_∞ are the static and the high-frequency dielectric constants, respectively, which as ε^* are temperature dependent. The dielectric relaxation of heavy water was systematically measured only by Collie et al. [2] at temperatures 278-333 K. The frequency dependence of ε^* was found to obey Debye equations also.

For polar liquids the dependence of $\ln \tau_D$ on inverse temperature is known to be linear. However, for water a deviation from the linearity is found at temperatures close to its melting point; like deviation has been noticed at 323-333 K [1]. The dielectric relaxation measurements at higher temperatures are necessary to investigate subsequent temperature variations of τ_D and to elucidate the mechanism of molecular orientational motions in liquid water. We report some results of the measurements of the complex dielectric constant of water and heavy water up to 533 K.

2. EXPERIMENTAL

The dielectric permittivity measurements were performed by the cavity resonator method [3] at frequencies 9.29-9.33 GHz (the resonant frequency variations are due to a thermal expansion of the cavity). The

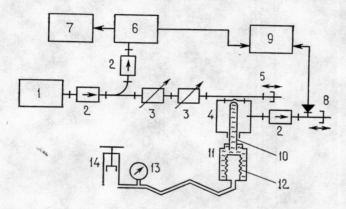

Fig. 1. Schematic diagram of the apparatus: *1* — microwave oscillator, *2* — ferrite isolator, *3* — precision attenuator, *4* — cavity, *5* — short circuit, *6* — heterodyne transfer oscillator, *7* — digital frequency meter, *8* — crystal detector, *9* — two-beam oscilloscope, *10* — quartz tube filled with water, *11* — pressure-transfer chamber, *12* — bellows, *13* — manometer, *14* — hydraulic press

cylindrical H_{011} cavity was used; a quartz tube filled with liquid under pressure equal to that of saturated vapor was placed at its axis. The tube was connected to the pressure-transfer chamber separated by bellows. The required pressure was produced by the hydraulic press and transferred to the tube through the flexible oil line and the bellows (Fig. 1). The resonant cavity temperature was controlled with an accuracy of ±0.03 K, the axial and radial temperature gradients in the cavity being the same.

The H_{011} mode was excited in the cavity by two windows at the top of the cavity, separated by a half-wavelength. Other modes were effec-

tively suppressed by this method of excitation and by the microwave absorber mounted behind the plunger at the other end of the cavity. The quality factor Q of the empty silvered cavity was about 10400 (at 313 K).

Values of ε' and ε'' were determined from the resonant frequency shift and the Q change of the cavity, containing the water-filled tube, relative to resonant frequency and the Q of the cavity with the empty tube. In calculating ε' and ε'' the dimensions of the cavity and the tube and also the influence of quartz tube walls were taken into account by the relations developed by Hallenga [4].

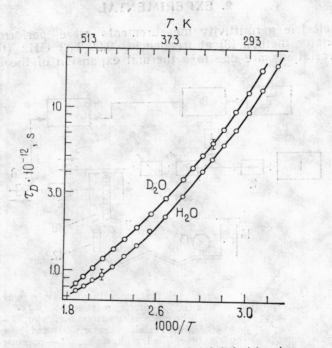

Fig. 2. Temperature dependence of dielectric relaxation time of H_2O and D_2O

The resonant frequency and Q of the cavity were measured by the dynamic method. The frequency modulated signal from a microwave oscillator was fed to the cavity input and also to the heterodyne transfer oscillator connected to a digital frequency meter (Fig. 2). The resonance curve received from the cavity output was registered by the two-beam oscilloscope. The second beam was used to observe the frequency mark received from the transfer oscillator. The mark was placed at the half-height of the resonance curve by a precision attenuator. Thus the mark permitted to determine the resonant frequency and the half-width of the resonance curve.

The samples of bidistilled water and distilled heavy water with D_2O content of 99.8% were used. Before measurements, the dc conductivity of the samples was about $(1-3) \times 10^{-4}$ S/m; after measurements, the dc conductivity increased by 2-3 times. The ε' and ε'' values for H_2O and D_2O at different temperatures are listed in the table. The accuracy of ε' and ε'' measurements was 2 and 4%, respectively. At 470-533 K there may be some contribution of dc conductivity to ε''. According to the available estimations [5, 6], this contribution may be about 4%.

3. RESULTS AND DISCUSSION

The relaxation time τ_D was calculated by the relations as follows from Debye equations

$$\tau_D = \varepsilon''/(\varepsilon' - \varepsilon_\infty)\omega \tag{2}$$

$$\tau_D = (\varepsilon_s - \varepsilon')/\varepsilon''\omega \tag{3}$$

The Eq. (3) may be used only when temperature is below 410 K because at higher temperatures the difference $\varepsilon_s - \varepsilon'$ becomes comparable with errors of measuring ε' and ε_s. At temperatures below 410 K, τ_D calculated by Eqs. (2) and (3) coincide within the limits of error. This confirms that the relaxation time distributions are negligible both in H_2O and D_2O.

The high-frequency dielectric constant ε_∞ appearing in Eq. (2) is unlikely to be measured directly. It can be determined by extrapolating the complex dielectric constants to infinite frequency. The average polarization of water calculated by the Clausius-Mossotti equation for ε_∞ values obtained by analyzing the ε^* data [1] equals 10.2 cm³ (per mol). Because of the insufficient accuracy of this method, which is due to the choice of dispersion relations and the probable contribution of orientational polarization, it is desirable to obtain ε_∞ independently.

ε_∞ was calculated by a method based on the assumption of additivity of distortion polarization of interatomic bonds of the molecule of the substance investigated and on the application of Clausius-Mossotti relation [7, 8]. This method takes into account both the electronic and the atomic polarizabilities of a molecule, because the bond distortion polarizations are calculated by the experimental values of ε_s and the polarization of corresponding nonpolar substances. The polarization of O—H bond was not determined [7, 8]. But the measurements of the dielectric constant ε_s and the polarization of nonpolar p-benzendiol $(C_6H_4(OH)_2)$ permit to calculate the distortion polarization of water [9]: $P_{H_2O} = 8.5$ cm³. It is quite close to the polarization of ice obtained from ε_∞ measurements [10, 11]. For instance, ice I has $P = 7.9$ cm³ at 10 K.

The difference between ε_∞ and n^2 (n is the refractive index of ice) is usually explained by the contribution of infrared bands mainly of those responsible for the translational and the rotational vibrations of water molecules [12]. The problem whether the contribution to ε_∞ of ice is due to the orientational polarization (i. e. the palorization of the

51

Table 1

T (K)	Water					Heavy water					Water		Heavy water	
	ε'	ε''	ε_∞	$\tau_D \cdot D$	$\tau_D \cdot T/\eta$	ε'	ε''	ε_∞	$\tau_D \cdot D$	$\tau_D \cdot T/\eta$	$\tau_D \cdot D$	$\tau_D \cdot T/\eta$	$\tau_D \cdot D$	$\tau_D \cdot T/\eta$
273	45.4	41.1	3.68			—	—	—			1.05	0.96	—	—
283	56.0	37.5	3.68			45.3	38.5	3.37			1.05	0.99	1.11	0.96
293	63.0	31.4	3.67			55.4	35.2	3.37			1.01	0.98	1.08	0.99
303	65.6	25.9	3.66			60.3	30.1	3.36			1.03	0.99	1.02	0.96
313	66.0	21.2	3.64			61.8	25.0	3.35			1.00	1.00	1.00	1.00
323	65.2	17.0	3.62			62.1	20.6	3.33			0.99	1.01	0.97	1.01
333	63.4	13.9	3.60			61.5	17.0	3.32			1.00	1.03	0.99	1.04
343	61.4	11.4	3.57			60.2	14.0	3.29			1.01	1.03	0.98	1.04
353	59.3	9.25	3.54			58.3	11.6	3.27			0.98	1.01	0.98	1.06
373	54.7	6.32	3.48			54.0	8.05	3.22			0.97	1.00	0.97	1.07
393	49.8	4.66	3.40			49.7	6.01	3.15			1.04	1.04	1.02	1.10
413	45.4	3.45	3.33			45.6	4.52	3.09			1.06	1.06	1.07	1.13
433	41.4	2.67	3.25			41.6	3.47	3.02			1.10	1.09	1.10	1.17
453	37.8	2.12	3.16			37.9	2.73	2.94			1.15	1.14	1.12	1.20
473	34.3	1.71	3.07			34.5	2.16	2.86			1.20	1.19	1.13	1.24
493	30.8	1.41	2.97			31.1	1.69	2.77			1.26	1.26	—	1.24
513	27.9	1.18	2.87			27.8	1.33	2.69			1.33	1.34	—	1.25
533	25.0	0.98	2.76			24.4	1.07	2.58			1.40	1.41	—	1.30

Note to the table: For H_2O below 333 K the values of ε', ε'' and τ_D have been taken from Ref. [1]; for D_2O below 303 K from Ref. [2]; the values of $\tau_D \cdot D$ and $\tau_D \cdot T/\eta$ have been normalized to the corresponding values at 313 K; the self-diffusion coefficients D have been taken from Ref. [14].

permanent dipoles of water molecules) is still undecided. As the polarization of ice coincides virtually with P_{H_2O} obtained from ε_s measurement of nonpolar p-benzendiol, the contribution of the orientational polarization may probably be neglected.

The ε_∞ of heavy water are also calculated by the Clausius-Mossotti equation with $P_{D_2O}=8.0$ cm³. As follows from the ε_∞ measurements of D_2O ice [11], the distortion polarization of D_2O should be almost similar to that of H_2O. Note that ε_∞ insignificantly affects the dielectric relaxation time of water: if ε_∞ varies by 15% the τ_D changes by about 1%. The ε_∞ values for H_2O and D_2O are listed in the table.

The τ_D are determined with an accuracy of 7%. The temperature dependences of the relaxation time of H_2O and D_2O are shown in Fig. 2; below 340 K for H_2O and below 310 K for D_2O the available data [1, 2] are used. The considerable curvature of the $\ln \tau_D = 1/T$ plot confirms the fact that the activation energies of the dynamic processes in water are strongly temperature dependent. The ratio of D_2O and H_2O relaxation times is 1.27 ± 0.04 at 280-480 K and decreases to 1.12 at 530 K. This decrease may be partly due to the unaccounted contribution of dc conductivity in ε'' of H_2O above 490 K.

No satisfactory model of liquid water has yet been developed. A "reduced" model was proposed to explain the dielectric relaxation of supercooled water [13]. The approximations used therein permit to describe quantitatively the τ_D temperature variations only close to the melting point. A considerable discrepancy between the calculated and experimentally determined values is noticed at 323 K.

The τ_D temperature dependence of water is known to be satisfactorily described by Debye-Stocks model in which the orientational motion of a molecule is considered as simple rotation of a sphere in viscous medium. The successful application of this model may probably be explained by the strong interrelation between translational and orientational motions of water molecules. The fact that $\tau_D \cdot T/\eta$ and $\tau_D \cdot D$ (where η is viscosity and D is self-diffusion coefficient) are temperature independent has been interpreted as evidence of the only mechanism of the molecular motions. As is seen from the table, these expressions both for H_2O and D_2O remain constant (within the error limits) only below 430 K. A steady increase in $\tau_D \cdot T/\eta$ and $\tau_D \cdot D$ with temperature indicates that a correlation between translational and orientational motions of water molecule is weakened as temperature rises.

4. ON THE TEMPERATURE DEPENDENCE OF KIRKWOOD CORRELATION FACTOR

The Kirkwood correlation factor g is the measure of the orientational arrangement of molecules in polar liquids. The g-factor of water is known to depend strongly on ε_∞ values. Ordinarily, only electronic polarization is taken into account, i. e. the g-factor of water is calculated assuming $\varepsilon_\infty = n^2$.

There is however another viewpoint according to which the distortion polarization of molecules in liquid water increases in contrast to the polarization of steam because of the formation of hydrogen bonds [15, 16]. Thus, the infrared and submillimeter absorption in liquid water is assumed to be caused by the induced dipole moments rather than the permanent molecular dipoles. Our result ($P_{H_2O} = 8.5$ cm³) calculated by the bond additive polarization method argues in favour of this approach.

At present it is not possible to solve the problem of distortion polarization of water, since another interpretation of the submillimeter absorption has been developed [17]. Nevertheless, it should be noted that the temperature dependences of g-factor, calculated with the assumption

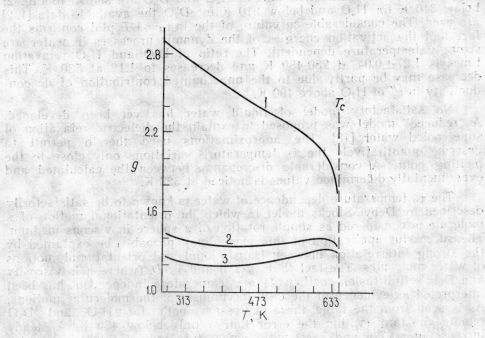

Fig. 3. Kirkwood correlation factor for water:
(1) $\varepsilon_\infty = n^2$; (2) $P_{H_2O} = 7.9$ cm³/mol; (3) $P_{H_2O} = 8.5$ cm³/mol

that $\varepsilon_\infty = n^2$, and with ε_∞ corresponding to $P = 8.5$ cm³, are essentially different (Fig. 3). With $\varepsilon_\infty = n^2$ the $(g-1)$ value due to the molecular environment contribution changes from 1.87 at 273 K to 0.94 at 633 K. This decrease should correspond to considerable variation in the molecular local arrangement of water. However, according to the X-ray scattering data [18], the number of molecules in the first coordination shell are in fact invariable with temperature.

Also, no experimental evidence confirms the appearance of long-existing free molecules in liquid water at high temperatures. Thus, the g-factor calculated from $P_{H_2O}=8.5$ cm^3 (or 7.9 cm^3) agrees better with the experimental data concerning temperature variations of the water structure since the $(g-1)$ value changes insignificantly with temperature in the whole region where water exists in the liquid state.

REFERENCES

1. Kaatze U. and Uhlendorf V. — *Z phys. Chem.*, 1981, Bd. 126, S. 151-165.
2. Collie C., Hasted J., and Ritson D. — *Proc. Phys. Soc.*, 1948, v. 60, p. 145-160.
3. Nabokov O. A. and Lubimov Yu. A. — Elektron. obrab. materialov, 1982, N 2, p. 45-48 (in Russian).
4. Hallenga K. — *Rev. Sci. Instrum.*, 1975, v. 46, p. 1691-1696.
5. Gancy A. B. — In: "Water and Aqueous Solutions. Structure, Thermodynamics and Transport Processes", N.-Y., 1972, p. 771-801.
6. Quist A. S. and Marshall W. L. — *J. Phys. Chem.*, 1965, v. 69, p. 2984-2987.
7. Boyer-Donzelot M. — *Bull. Soc. Chim. France,* 1970, N 2, p. 425-428.
8. Levin V. V. — In: "Phyzika i Phyzikochim. Zhidkostey", Issue 1, 1972, p. 176-190 (in Russian).
9. Lubimov Yu. A. and Nabokov O. A. — *Z. phizicheskoy khim.*, 1984, v. 58, No. 9 (in Russian).
10. Gough S. R. — *Canad. J. Chem.*, 1972, v. 50, p. 3046.
11. Johari G. P. — *J. Chem. Phys.*, 1976, v. 64, p. 3998.
12. Whalley E. — In: "Physics of Ice" N.-Y., 1969, p. 271-286.
13. Bertolini D. et al. — *Chem Phys. Lett.*, 1983, v. 98, p. 548-553.
14. Weingärtner H. — *Z. phys. chem.*, 1982, Bd. 132, S. 129-149.
15. Hill N. E. — *Trans. Faraday Soc.*, 1962, v. 59, p. 344.
16. Hasted J. B. — In: "Water — A Comprehensive Treatise", v. 1, N.-Y., 1972.
17. Apletalin V. N., Garin B. M., and Meriakri V. V. — Radiotechnika i Elektronika, 1983, v. 28, p. 1-16 (in Russian).
18. Eisenberg D. and Kauzmann W. "The Structure and Properties of Water", Oxford, 1969.

Near-Infrared Spectra of Water and Aqueous Electrolyte Solutions at High Pressures

K. SUZUKI

*Department of Chemistry, Ritsumeikan University,
Kita-ku, Kyoto 603, Japan*

1. INTRODUCTION

Pressure as well as temperature should be a useful tool to investigate physicochemical properties. Excellent spectroscopic investigations for compressed liquid water have been performed by Walrafen [1] at 28°C using laser Raman technique and by Franck and Roth [2] at 30-400°C using infrared technique. Valyashko et al. have recently observed infrared (IR) spectra of compressed aqueous solutions of LiCl [3] and NaClO$_4$ [4] at temperatures between 25 and 350°C. These investigations, however, do not cover the interesting temperature region below room temperature, and our preliminary study for compressed pure water is the only one that covers this region [5].

In this investigation, near-infrared (NIR) absorption spectra in the region of 9500-11 000 cm^{-1} (around 1.0 μ) for pure water and aqueous solutions containing various types of electrolytes were measured at various temperatures in the range from 10 to 55°C and pressures up to 500 MPa.

The main reason using NIR technique in the present investigation is as follows: It is rather difficult technically to measure high-pressure IR spectrum precisely by using the path length below 1 mm, and the greater separation between the absorption bands of the various types of water-molecular-species band is to be expected in NIR region than in IR region.

2. EXPERIMENTAL

Deionized water was distilled over alkaline permanganate before use. Alkali halides (LiX, NaX, and KX, where X is Cl, Br, and I), MgCl$_2$, NaClO$_4$, (n-C$_3$H$_7$)$_4$NBr, and (n-C$_4$H$_9$)$_4$NBr were supplied all in reagent-grade quality (Nakarai Chem. Co.), and were used without further purification.

A clamp-type high-pressure cell with saphire windows and a Hitachi 340 spectrophotometer were used to measure the NIR band, which is assigned to $2v_1+v_3$ [6]. Details of the apparatus and procedures were described elsewhere [7].

3. RESULTS AND DISCUSSION

3.1. BAND CONTOUR

Figure 1 shows the spectra of pure water and aqueous solutions of NaClO$_4$ and (n-C$_4$H$_9$)$_4$NBr at 0.1 MPa and 25°C. The spectra of alkali halides and MgCl$_2$ solutions are not shown, because they are not so much different from the spectrum of pure water. The spectrum of (n-C$_3$H$_7$)$_4$NBr solution is similar to that of (n-C$_4$H$_9$)$_4$NBr solution.

From these results, the effects of electrolytes on the structure of water can be classified into three types, that are of NaClO$_4$, of R$_4$NBr, and of others. By the addition of NaClO$_4$, the spectrum shifts towards a higher wavenumber and its contour sharpens. On the contrary, by the addition of R$_4$NBr, the spectrum shifts towards a lower wavenumber and its contour broadens. Other electrolytes investigated caused to shift the spectra to a slightly lower wavenumbers and exerted very little influence on the shape of spectra.

3.2. PURE WATER

Figure 2 shows the typical spectral change of pure water when pressure is changed. The spectra show an isobestic point at $10\,200\pm\pm100$ cm^{-1}. From this result, it is proposed that liquid water consists of a relatively small number of distinguishable molecular species.

The wavenumbers at absorption maxima, $\tilde{v}$ (max), of pure water are plotted against pressure at temperatures from 10-55°C in Fig. 3. A maximum exists at lower temperatures below room temperature, and the pressure at which the inversion appears increases as the temperature is decreased (i. e. 100 MPa at 25°C and 200 MPa at 10°C). Such an inversion phenomenon in the same temperature and pressure region has also been found in viscosity [8], proton NMR chemical shift [9], spin-lattice relaxation time [10], and self-diffusion coefficient [11] for liquid water or liquid heavy water.

The critical temperature where the inversion disappears is estimated to be about 38°C. At temperature above about 38°C, $\tilde{v}$ (max) decreases as the pressure is increased. It is also found from Fig. 3 that $\tilde{v}$ (max) under a constant pressure shifts to higher wavenumbers with increasing temperature.

The shift of the absorption spectrum toward higher wavenumbers reflects the rupture or loosening of hydrogen bond, because the O—H bond is free from the restraint, and *vice versa*. Assuming this fact, the following conclusion will be deduced.

When the pressure is increased at the lower temperatures of 10 and 25°C, the rupture of ice-I-like open structure, which is favorably ruptured by compression because of its bulkiness, competes with the formation of the packed structure, which is favorably formed by compression, and a maximum appears as the result at about 200 and 100 MPa, respectively. At high temperatures of 40 and 55°C, the remaining open structure is already very few at atmospheric pressure and the formation of the packed structure surpasses the rupture of the open structure from the beginning of compression. As a result, $\tilde{v}$(max) decreases as the pressure is increased.

3.3. AQUEOUS ELECTROLYTE SOLUTIONS

Figure 4 shows the typical effects of some electrolytes on $\tilde{v}$(max) *vs* pressure plots at 25 and 40°C. The maximum which appears for pure water at 25°C disappears by the addition of the electrolytes shown here ($NaClO_4$, NaCl, and $MgCl_2$) and other alkali halides investigated (NaBr, NaI, LiX, and KX, where X is Cl, Br, and I).

Figure 5 shows the effect of NaCl concentration on $\tilde{v}$(max) *vs* pressure plot at 25°C. The maximum shifts to a lower pressure and then disappears with an increase of concentration. The same trend was also found for other electrolytes mentioned above.

This fact suggests that these electrolytes rupture ice-I-like open structure. The addition of electrolytes except for $NaClO_4$ brings about the shift of $\tilde{v}$(max) towards lower wavenumbers. If a cation is adjacent to a water molecule, the oxygen atom of the water molecule will be attracted and the hydrogen atom will be somewhat repelled. This will result in a loosening of the O—H bond and its stretching frequency will shift to a longer wavelength. If an anion is adjacent to water molecule, the water will essentially be attracted through the hydrogen atom of the water molecule, which also causes a loosening of the O—H bond [12]. On the other hand, $\tilde{v}$(max) shifts to a higher wavenumber by the addition of $NaClO_4$. This fact together with the fact shown in Fig. 1 suggests that ClO_4^- ion remarkably destroys the open structure of water and the ion-dipole interaction with water is insignificant.

Figure 6 shows the dramatic effect of $(n\text{-}C_4H_9)_4NBr$ on $\tilde{v}$(max) vs pressure plot. The addition of $(n\text{-}C_4H_9)_4NBr$ brings about the appearance of a maximum even at 40°C where the maximum does not appear for pure water. This fact implies the formation of water structure which is similar to the clathrate open structure around $(n\text{-}C_4H_9)_4N^+$ ion. On the other hand the maximum did not appear by the addition of $(n\text{-}C_3H_7)_4NBr$ (up to 4.1 m) which unlike $(n\text{-}C_4H_9)_4NBr$ is unable to form a clathrate hydrate.

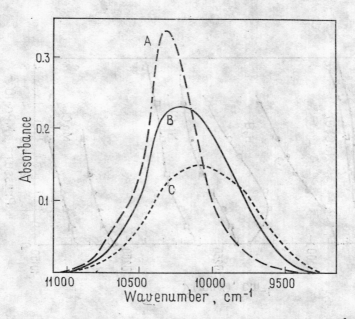

Fig. 1. NIR spectra of pure water (B), aqueous solutions of NaClO$_4$ $(A,$ saturated$)$, and $(n$-C$_4$H$_9)_4$NBr $(C,$ 1.0 m$)$ at 25° C and 0.1 MPa

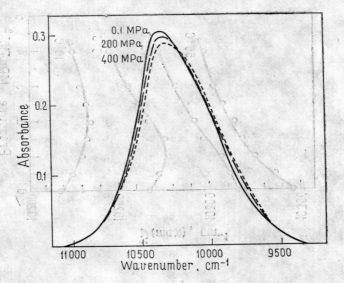

Fig. 2. Pressure effect on NIR spectra of H$_2$O at 55° C

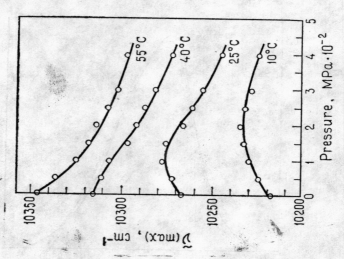

Fig. 4. Absorption maximum, $\tilde{v}(max)$, vs pressure for pure water ($\bigcirc$ $\bullet$) and saturated aqueous solutions of $MgCl_2$ ($\triangle$ $\blacktriangle$), $NaCl$ ($\square$ $\blacksquare$) and $NaClO_4$ ($\triangledown$ $\blacktriangledown$) at 25°C (opened) and 40°C (closed)

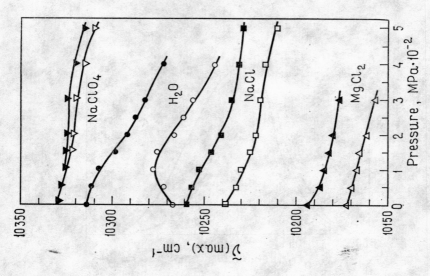

Fig. 3. Absorption maximum, $\tilde{v}(max)$, pressure for pure water at each temperature

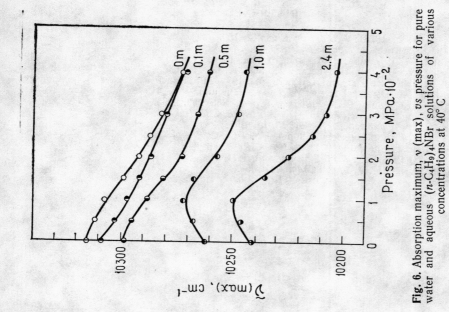

Fig. 6. Absorption maximum, ν (max), vs pressure for pure water and aqueous (n-C_4H_9)$_4$NBr solutions of various concentrations at 40°C

Fig. 5. Absorption maximum, ν (max), vs pressure for pure water and aqueous NaCl solutions of various concentrations at 25°C

REFERENCES

1. Walrafen, G. E. (1973): *J. Solution Chem.* **2**, 159.
2. Franck, E. U., Roth, K. (1967): *Disc. Faraday Soc.* **29**, 108.
3. Valyashko, V. M., Buback, M., and Franck, E. U. (1980): *Z. Naturforsch.* **35a**, 549.
4. Valyashko, V. M., Buback, M., and Franck, E. U. (1981): *Z. Naturforsch.* **36a**, 1169.
5. Suzuki, K., Taniguchi, Y., and Tsuchiya, M. (1979): in High Press. Sci. and Technology, Timmerhaus, K. D. and Barber, M. S. eds. (Plenum, New York and London) p. 548.
6. Buijs, K. and Choppin, G. R. (1963): *J. Chem. Phys.* **39**, 2035.
7. Suzuki, K. and Tsuchiya, M. (1975): *Bull. Chem. Soc. Jpn.* **48**, 1701.
8. Bett, K. E. and Cappi, J. B. (1965): *Nature* **207**, 620.
9. Linowsky, J. W., Liu Nan-I, and Jonas J. (1976): *J. Chem. Phys.* **65**, 3383.
10. DeFries, T. and Jonas, J. (1977): *J. Chem. Phys.* **66**, 896.
11. Woolf, L. A. (1975): *J. Chem. Soc. Faraday I* **71**, 784.
12. Bunzl, K. W. (1967): *J. Phys. Chem.* **71**, 1358.

Surface Tension of Heavy Water

N. B. VARGAFTIK, B. N. VOLKOV, and L. D. VOLJAK

*Moscow Aviation Institute, Department of Physics,
Moscow, USSR*

By now sufficient number of experimental studies have been made into the surface tension (σ) of heavy water (D_2O) over the entire range of liquid state. These have been analysed by Vargaftik [1]. The tables of recommended values of σ for D_2O were based on the most reliable works of Flood [2], Jones [3], Lachs [4], Takeuchi [5], Phibbs [6], and Vargaftik [7, 8]. On the basis of these tables a report containing the recommended values of σ for D_2O, their tolerances and also an equation to approximate these values was submitted at a meeting of the Executive Committee of IAPS (London, 1980).

In this equation the critical temperature of D_2O was taken equal to 644.16 K. Much later, studies [10] yielded a value of 643.89 K.

With consideration of the new critical temperature Straub recalculated the coefficients in the approximating equation describing surface tension of heavy water, and prepared a new report on surface tension of heavy water [11], that differed from the previous one only by coefficients in the approximating equation.

Lately the values of specific volumes of D_2O in liquid and gaseous phases were corrected [12]. As in calculating the surface tension use is made of specific volumes, it is advisable to recalculate σ by taking into account the new values of specific volumes.

In order to check the values of σ for D_2O in the near-critical range, we made experiments in this range. The procedure and the equipment used were similar to those employed in Ref. [8]. Measurements were taken by the differential capillary method. The same quartz capillaries

63

were used as in Ref. [8], their radii being $r_1 = (0.1933\pm0.0002)\times10^{-3}$ m and $r_2 = (0.5448\pm0.0004)\times10^{-3}$ m. The investigated heavy water was taken from the same lot as used in Ref. [8]. Its electrical conductivity was 3×10^{-8} Ohm^{-1}m^{-1}. The concentration of D_2O was 99.7%. The obtained results are given in Table 1, where ΔH is the difference of heavy water levels in capillaries and a^2 is the capillary constant.

The temperature dependence of the contact angle of D_2O on the capillaries was taken into account considering the work [13]. The densities of liquid and gaseous phases were taken from Ref. [12]. The uncertainty of the measured values of surface tension, given in Table 1, was within $\pm10^{-4}$ N/m.

Table 1

t	$\Delta H\times10^{-3}$	$a^2\times10^{-6}$	$\sigma\times10^{-3}$
°C	m	m²	N/m
368.88	0.578	0.165	0.19
369.04	0.545	0.155	0.18
369.56	0.472	0.134	0.14
369.62	0.475	0.135	0.14
369.94	0.391	0.111	0.10

The table of recommended values of σ for D_2O is based on the values of Table 1 and on the results of Refs. [2, 3, 4, 5, 6, 7] and [8], which were corrected according to the work [12]. These values of surface tension are given in column 2 of Table 2.

The tolerances $\Delta\sigma$ for recommended values of σ_{exp} are given in column 3 and they are the same as in Refs. [1, 9] and [11]. The grounds for these tolerances are represented in Ref. [1].

The recommended values of surface tension for heavy water are well described by the equation of the same type as the approximating equation for ordinary water:

$$\sigma = B\left(\frac{T_c-T}{T_c}\right)^{\mu} \cdot \left(1 - b\frac{T_c-T}{T_c}\right)$$

where $T_c = 643.89$ K

$B = 240.0\times10^{-3}$ N/m

$b = -0.647$

$\mu = 1.254$

The differences $\delta\sigma$ between the values of σ and σ_{exp} are given in column 5.

64

Table 2

t	$\sigma_{exp} \times 10^{-3}$	$\Delta\sigma \times 10^{-3}$	$\sigma \times 10^{-3}$	$\delta\sigma \times 10^{-3}$
°C	N/m	N/m	N/m	N/m
3.8	74.98	0.53	74.85	—0.13
10	74.10	0.51	73.99	—0.11
20	72.63	0.51	72.55	—0.08
30	71.10	0.50	71.05	—0.05
40	69.52	0.49	69.50	—0.02
50	67.87	0.48	67.89	0.02
60	66.18	0.47	66.21	0.03
70	64.43	0.45	64.48	0.05
80	62.63	0.44	62.69	0.06
90	60.79	0.43	60.85	0.06
100	58.90	0.41	58.95	0.05
110	56.96	0.40	57.01	0.05
120	54.97	0.39	55.02	0.05
130	52.94	0.37	52.98	0.04
140	50.88	0.36	50.90	0.02
150	48.78	0.34	48.78	0
160	46.64	0.33	46.62	—0.02
170	44.47	0.31	44.42	—0.05
180	42.26	0.30	42.19	—0.07
190	40.01	0.29	39.92	—0.09
200	37.72	0.26	37.63	—0.09
210	35.39	0.26	35.31	—0.08
220	33.04	0.26	32.96	—0.08
230	30.63	0.26	30.60	—0.03
240	28.20	0.26	28.23	0.03
250	25.80	0.26	25.82	0.02
260	23.41	0.25	23.45	0.04
270	21.02	0.23	21.06	0.04
280	18.65	0.22	18.68	0.03
290	16.28	0.20	16.31	0.03
300	13.95	0.18	13.94	0.01
310	11.67	0.17	11.67	0
320	9.44	0.15	9.50	0.06
330	7.25	0.14	7.22	—0.03
340	5.15	0.12	5.12	—0.03
350	3.18	0.10	3.16	—0.02
360	1.40	0.10	1.40	0
370	0.07	0.10	0.05	—0.02
T_c	0	—	—	—

REFERENCES

1. Vargaftik N. B., Volkov B. N., and Voljak L. D. (1979): Water and Steam. Pergamon, Oxford.
2. Flood R. and Tronsted L. (1936): *Z. Phys. Chem.* **A-175**, 347-352.
3. Jones G. and Ray W. A. (1937): *J. Chem. Phys.* 5, 505.
4. Lachs H. and Minkow J. (1937): *Roczniki Chemic.* **17**, 363-364.
5. Takeuchi T., Sagito T., and Inai T. (1937): *Proc. of Phys. — Mat. Soc. of Japan*, **18**, 552-554.
6. Phibbs M. K. and Guiger P. A. (1951): *J. Chem.* **29**, 173-181.
7. Vargaftik N. B., Voljak L. D., and Volkov N. B. (1970): In: "Thermophysical Properties of Fluids". Nauka, Moscow (in Russian).
8. Vargaftik N. B., Voljak L. D., and Volkov B. N. (1973): *Teploenergetika*, No. 8, 80-82.
9. Release on Surface Tension of Heavy Water Substance. (1980): IAPS.
10. Blank G. (1969): Wärme und Staffübertragung. 2, 53-59.
11. Release on Surface Tension of Heavy Water Substance (1983): IAPS.
12. Hill P. G., Chris MacMilan, and Lee V. (1982): *J. of Phys. and Chem. Ref. Data*, II (I), 1-14.
13. Voljak L. D., Stepanov V. G., Tarlakov J. V. (1975): *J. of Physical Chemistry.* **49**, 2931-2933.

Steam Absorptivity in the Range 6-125 μ m

M. A. STYRIKOVICH

Institute of High Temperatures, USSR Academy of Sciences, Moscow, USSR

E. G. KOKHANOVA and G. V. YUKHNEVICH,

N. S. Kurnakov Institute of General and Inorganic Chemistry, USSR Academy of Sciences, Moscow, USSR

Radiation heat exchange plays an important role in the heat transfer processes in various heat exchangers. Obviously, for calculating the radiation component of a heat transfer process it is necessary to know the optical constants of substances or materials between which heat exchange takes place. Therefore, one must study optical properties in the wavelength range (0-5000 cm^{-1}) corresponding to the radiation maximum at working temperatures (300-700°C) (Fig. 1a) of modern industrial plants. In this frequency range all metals and alloys used for making heat exchangers can be well approximated by the gray body model. Their coefficient of grayness is independent of temperature and pressure.

The optical properties of water which is the working substance of all heat plants are, on the contrary, strongly influenced by the above-mentioned factors (P and T). The water absorption spectra measurements at high pressures and temperatures in the region 0-5000 cm^{-1} are very complicated. The principal technical difficulties are caused by the necessity to machine the cell windows from the solid, transparent and insoluble in water material. Such properties of the material as transparency in the wavelength range 0-2500 cm^{-1} and insolubility in water are very seldom compatible.

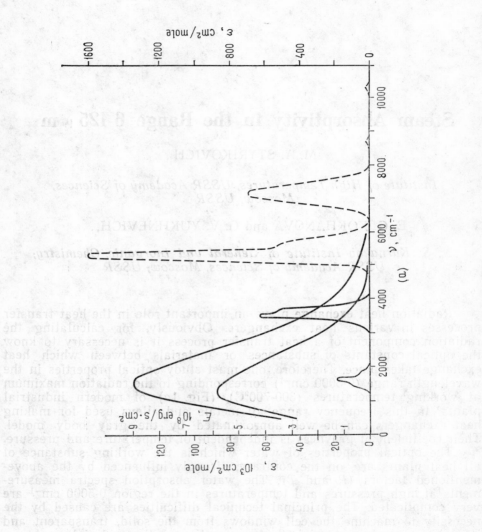

Steam Absorptivity in the Range 8-125 μm

N.A. STRIGUTSCH

Institute of High Temperatures, USSR Academy of Sciences,
USSR

O.B. LVOVA and G.P. YUKHNEVICH,

Nesmeyanov Institute of Element-Organic Chemistry,
Academy of Sciences, Moscow, USSR

Radiation heat transfer plays an important role in the heat transfer processes involving heat exchangers. Obviously, for calculating the radiation component of heat transfer process it is necessary to know the optical constants of substances or materials between which heat exchange takes place; therefore, to study optical properties in the wavelength range 80-1250 cm⁻¹ corresponding to the radiation maximum for high temperatures (300-1000°C). For most of modern industrial plastics in the frequency range ... are used for modelling their characteristics were approximated by the gray-body model. ... The values of temperature, concentration, and pressure ... behaviour of the substance was determined by the optical properties of the above-mentioned substance. ... The water absorption spectral measurement at high pressure and temperatures in the region 3000 cm⁻¹ are even complicated. The principal technical difficulties raised by the necessity to develop transparent in the solid, transparent and insoluble ... materials. Such properties of the materials are transparency in the spectral range 0.2500 cm⁻¹ and insolubility in water are too seldom compatible.

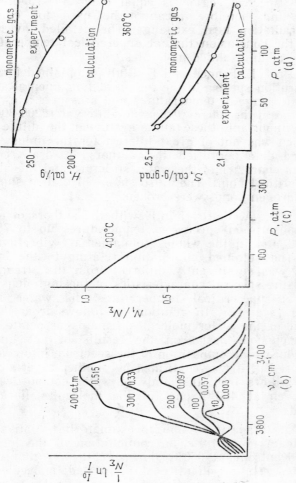

Fig. 1. Steam at 400° C: (*a*) The absorption spectra. The black body radiation at 600° C and 400° C; (*b*) The absorption spectra at pressures 10, 100, 200, 300 and 400 atm; (*c*) The dependence of relative quantity of the monomers on pressure; (*d*) The dependence of thermodynamic potentials on pressure at 360°C: *1* — enthalpy, *2* — entropy

A large number of studies of water absorption spectra were carried out by Prof. W. Luck and coworkers in 1963-1971. In these experiments, water of different isotopic composition was investigated including H_2O at temperatures up to 400°C [1-3]. These authors studied the overtone and composite vibration bands of water lying in the high-frequency range (5000-10 000 cm^{-1} (1-2 μm)). The radiation of the black body heated up to 400-600°C in this spectral region is equal only to 1/10 of its total radiation (Fig. 1a) and therefore it cannot be used for engineering calculation of the radiation heat exchange. The detailed investigations led the authors to the conclusion that at supercritical temperatures the steam consisted only of the isolated molecules.

Later, in 1967, Prof. E. Franck and K. Roth [4] obtained for the first time the water absorption spectra in the region of fundamental stretching vibrations in the wide range of temperatures and pressures. However, the spectra were measured in the narrow frequency interval (2200-3000 cm^{-1}) and, what is more important, these authors studied the dilute solution of D_2O in H_2O which was not of great interest for the industrial plant design. The obtained data permitted the authors to draw conclusion about the continuous character of the water structure and the absence of peculiarities in the critical point. They also supposed that at supercritical temperatures all hydrogen bonds were broken.

Four years later Dr. A. Vetrov jointly with the authors of Refs. [5-8] studied the steam absorption spectra at temperatures up to 700°C and pressures up to 500 atm in the whole fundamental vibration region (1500-5000 cm^{-1}). The obtained information relating to the interaction of water molecules was significantly different from the already known one. An analysis of the spectral characteristics of the high-density steam [5-8] showed that at supercritical temperatures the water molecules remained H-bonded (Fig. 1b). In addition, as follows from the experimental results, part of the water molecules remains associated even in dry steam. This conclusion, based on the results of not unequivocal mathematical procedure — decomposition of the spectrum into components, was not indisputable. But it was confirmed by another experiment of the authors [9, 10]. As follows from the study of water composite vibration band 2150 cm^{-1}, even at 400°C part of the molecules librated, i. e. remained associated.

The obtained data made it possible to estimate the relative quantity of monomers in steam (Fig. 1c). The computation of the high-density steam enthalpy and entropy at 360°C showed (Fig. 1c) that the consideration of estimated relative quantity of associated molecules led to good agreement with the experiment [8, 11] (Fig. 1d). Thus, we believe that the existence of H-bonded water molecule associations at supercritical temperatures was proved. However, the structure and size of the water clusters that form in high-density steam remained unknown. These structural parameters could not be determined because the measurements were carried out in the region of water intramolecular vibrations. This information can be obtained only from the intermolecular vibration

spectra lying in the middle and far infrared regions (from 6 to 100 μm) where no measurements have been carried out yet.

As was shown by quantum-mechanical calculations [12-14], there exists a large set of energetically equal clusters different in size and structure. Thus, a great number of diverse associations are expected in real high-density steam, the distribution of which is completely conditioned by temperature and pressure. The calculations of various configurations of water trimer and tetramer have shown that differences in position and intensity of the absorption bands of these associations can be detected only in the range 200-600 cm^{-1} [15, 16]. The results of the discussed computations showed that the nature of the water molecule cluster formation was rather complicated. Therefore the experimental investigation of the process will be very difficult.

The present work is the first attempt to solve this problem by studying the water cluster vibrational spectra in the middle and far infrared regions, which also yield information concerning the steam absorptivity in the frequency region below 1500 cm^{-1}. In this region, as it follows from Fig. 1a, the black body radiates almost half of the energy at temperatures of interest. Thus it becomes possible to completely calculate the radiation heat exchange.

To solve these problems, we developed a special high-temperature and high-pressure spectral cell. Its construction similar to those of the two-pass cells used by E. Franck [4] and A. Vetrov [5-8] with some differences. Of these, the most essential are: first, our cell has a diamond window, second, the cell aperture is only 4 mm in diameter. Its size was limited by the diamond plate dimensions (approximately 7 mm in diameter and 2 mm in thickness). Besides, some modifications have been made in the window sealing system. In our experiment the temperature was measured directly in the working cavity of the cell (volume ~3 cm^3). The junction of the thermocouple introduced into the cell was near the diamond (Fig. 2). The temperature was measured accurate to ±0.1°C and pressure to ±0.5 atm. The state of the system under investigation was completely determined by its temperature because the absorption spectra were registered at a constant volume.

The outer surface of the cell is in contact with an electrical heater, the current through which is regulated by a special electronic device. Thanks to the second thermocouple placed in the cell body (Fig. 2) the device continuously monitored the temperature in the working cavity. In our experiment the temperature remained constant at ±0.3°C. The corresponding pressure variations at 400°C were less than ±1 atm.

The spectra were measured with BRUKER IR Fourier Transform Spectrometer IFS-113 v. The transmittance of the empty cell was taken as a 100% line and the transmittance of the empty cell without a mirror — as a 0% line (Fig. 3). The diamond and mirror reflectivity was practically independent of pressure. Therefore the 0% line of the cell and its 100% line — the sum of diamond and mirror reflections — in our measurements were stable. The vibrational spectra of 105 μm thick layer of steam at 400°C and different densities (0.0125-0.5 g/cm^3 (pressures

36-380 atm. respectively)) are given in Fig. 3. Therein are shown the transmittance spectra in the 1750-100 cm^{-1} range where half of the vibrational-rotational bending band of water molecules and the all their rotational spectrum are located. It was difficult to obtain spectra at higher frequencies because of the diamond window absorption. The

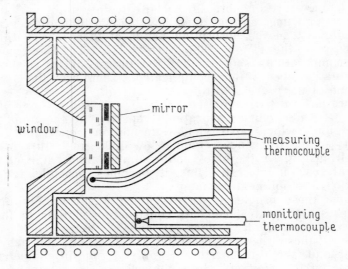

Fig. 2. The construction of spectral cell

rotational structure lines broaden at high pressure and thus it became possible to measure them without any distortion with a conventional spectrometer. In our experiment the resolution was 4 cm^{-1}.

In order to reveal the changes in the properties of molecule as functions of steam density it is convenient to make use of the absorption cross-section i. e. of Bugger absorption coefficient related to one molecule. The absorption cross-section spectra are shown in Fig. 4. As follows from the figure, the steam molecule cross-sections in absorption ranges 1500-1200 cm^{-1} and 600-400 cm^{-1} remain constant and are independent of density, which changes from 0 to 0.3 g/cm^3. At the same time in water vapor window (1200-600 cm^{-1}), the density increase results in the enhancement of the absorption cross-sections. This indicates that besides the isolated molecules there are some additional absorption centers in the water vapor. Note that relative number of the new centers increases with density.

The dependence of absorption cross-sections on density at different frequencies is plotted in Fig. 5. It is evident that at densities less than 0.1 g/cm^3 the cross-sections are in direct proportion to the water molecule concentration. Consequently, the number of new absorption centers changes as the square of the concentration. This unequivocally indicates

that the absorption in the frequency interval 1200-600 cm^{-1} is due to water dimers. The decrease in the slope of curves with increase in steam density shows that, in addition to dimers, clusters of more than two water molecules appear. An analysis of the curves also indicates that the cluster absorptivity in this spectral range is less than the dimer one. The absorption bands of the larger water molecule associations may occur in other frequency intervals.

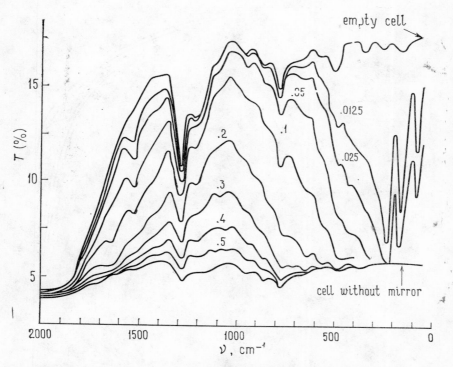

Fig. 3. The absorption spectra of different density (0.0125-0.5 g/cm³) steam at 400° C

The obtained data enabled us to draw the following conclusions concerning the steam absorptivity and the character of water molecule interactions.

(1) The steam absorptivity in the 10 μm water vapor window is mostly due to the formation of stable water molecule associations.

(2) The water molecules form clusters living more than 10^{-13} s at supercritical temperatures.

(3) These clusters exist even in a low-density vapor, where the average distance between the molecules equals 15 Å.

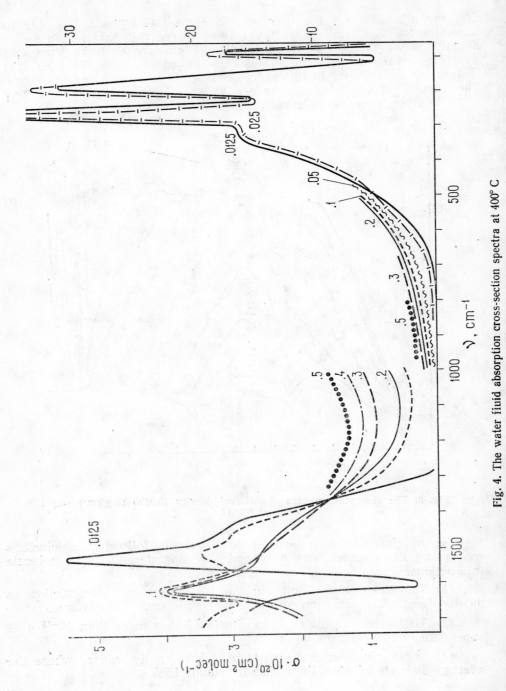

Fig. 4. The water fluid absorption cross-section spectra at 400° C

(4) The relative quantity of the clusters at steam densities less than 0.1 g/cm³ changes as square of density. This dependence is typical of dimers.

(5) The formation of clusters larger than dimers was observed at densities more than 0.1 g/cm³.

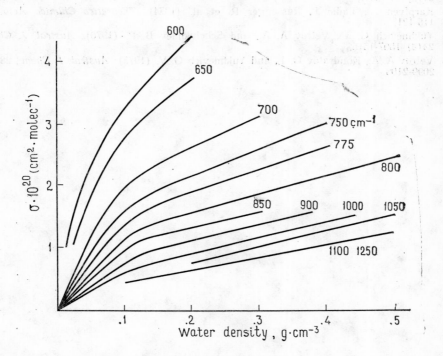

Fig. 5. The dependence of absorption cross-sections on density at different frequencies

REFERENCES

1. Luck W. A. P. (1965): *Ber. Bunsenges. Physic. Chem.* **69**, 65 and 626.
2. Luck W. A. P. and Ditter W. (1969): *Z. Naturforsch.* **24**, 482.
3. Luck W. A. P. and Ditter W. (1970): *J. Phys. Chem.* **74**(21) 3687.
4. Franck E. U. and Roth K. (1967): *Disc. Faraday Soc.,* **43**, 108-114.
5. Styrikovich M. A., Yukhnevich G. V., Vetrov A. A., and Vigasin A. A. (1973): *Dokl. Akad. Nauk SSSR,* **210**(21) 321-323.
6. Yukhnevich G. V. and Vetrov A. A. (1973): *Optika i spektroskopiya,* **34**(4) 672-677.
7. Vetrov A. A., and Yukhnevich G. V (1975): *Optika i spektroskopiya,* **39**(4) 488-492.
8. Styrikovich M. A., Vetrov A. A, and Yukhnevich G. V. (1976): *Dokl. Akad. Nauk SSSR,* **226**(1) 136-139.
9. Yukhnevich G. V. and Vetrov A. A. (1970): *Dokl. Akad. Nauk SSSR,* **194**(3) 557-559.

10. Yukhnevich G. V. "Infrakrasnaya Spektroskopiya Vodi", "Nauka", Moscow, 1973, 118 and 144-145.

11. Styrikovich M. A., Vigasin A. A., and Yukhnevich G. V. (1976): *Teplofizika vysokikh temperatur,* 14(4) 739-743.

12. Lentz B. R. and Scheraga H. A. (1973): *J. Chem. Phys.,* 58(12) 5296-5308.

13. Kistenmacher H., Lie G. C., Popkie H., and Clementy E. (1974): *J. Chem. Phys.,* 61(2) 546-551.

14. Karpfven A., Ladic J., Russegger P. et al. (1974): *Theoretica Chimica Acta,* 34, 115-121.

15. Yukhnevich G. V., Vetrov A. A., and Schelukhaev B. P. (1970): *Austral. J. Chem.,* 23(8) 1507-1515.

16. Vetrov A. A., Kondratov O. I., and Yukhnevich G. V. (1975): *Austral. J. Chem.,* 28(11) 2099-2106.

Rayleigh Light Scattering in the Supercritical Range of Fluids

Y. GARRABOS, R. TUFEU, and B. LE NEINDRE

Laboratoire des Interactions Moléculaires et des Hautes Pressions C.N.R.S., Université Paris-Nord, Villetaneuse, France

1. INTRODUCTION

In our laboratory we have performed several light-scattering experiments near the liquid-gas critical point of pure fluids [1-6]. These experiments allow one to determine the divergence of different static and dynamic properties [7, 8] at the approach of the critical point along the critical isochore in the one-phase region.

Static properties were obtained by various experimental methods. Measurements of the intensity of the scattered polarized light at 90° lead to the relative variation of the divergence of the isothermal compressibility. Measurements of the depolarization ratio of the double scattered light and turbidity lead to the absolute value of the scattering cross-section and the correlation length. A comparison with PVT measurements in the overlapping range ($T-T_c \simeq 1$ K) gives the value of the local dielectric constant and allows one to determine the isothermal compressibility.

Divergent behaviours of the isothermal compressibility k_T and the correlation length ξ, in the range $T-T_c \leq 1$ K, were described respectively by the following asymptotic laws [7]:

$$k_T = \Gamma_0^+ \tau^{-\gamma} \tag{1}$$

$$\xi = \xi_0^+ \tau^{-\nu} \tag{2}$$

where $\tau = (T-T_c)/T_c$; γ and ν are universal critical exponents. The values of γ and ν are fixed to the theoretical values ($\gamma = 1.24$ and $\nu = 0.63$) [7]. Consequently, from our experimental determination of k_T and ξ, the amplitude values Γ^+_0 and ξ^+_0 are obtained.

Using our values of Γ^+_0 and ξ^+_0, the universal amplitude ratio:

$$R = \xi_0^+ [B^2/(k_B T_c \Gamma_0^+)]^{1/d} \qquad (3)$$

was calculated [3-6]. We used the values of the amplitude B of the top of the coexistence curve available in the literature. The mean experimental value of $R = 0.71$ in close agreement with the theoretical one ($R_{Th} = 0.67$) [7]. The difference may be related to the non-analytical correction terms of the asymptotic laws defined by Eqs. (1) and (2).

Dynamic properties were determined from measurements of the time correlation function using a digital auto-correlator. The thermal diffusivity coefficient D_T is related to the decay rate of the exponential correlation function.

Far away from the critical point ($T > T_c + 1$ K) in the hydrodynamic region, the D_T values obtained from light scattering are compared to ($\lambda/\rho C_p$) ratios; λ is the thermal conductivity coefficient; C_p is the specific heat at constant pressure. In our laboratory λ is measured by the classical stationary method of coaxial cylinders [5, 6]. C_p is calculated from the thermodynamic relation:

$$C_p - C_v = \rho^{-1} \cdot (\partial P/\partial P)^2_\rho \cdot T \cdot k_T \qquad (4)$$

In Eq. (4), the divergence of k_T is determined by the methods previously described. Values of $(\partial P/\partial T)_{\rho_c}$ and C_v are obtained from PVT data and direct measurements available in the literature.

Near T_c in the critical region ($T < T_c + 1$ K), the singular part D_c of the thermal diffusivity coefficient obtained by light scattering is compared to the theoretical one in order to test the scaling relation [8]:

$$D_c = \Lambda \cdot [k_B \cdot T/(6\pi\eta\xi)] \qquad (5)$$

Λ is a universal constant; D_c is the singular part of D_T; η is the shear viscosity and ξ is the static correlation length. Using our determination of ξ following turbidity measurements and the small divergent values of η found in the literature, we obtained a value of Λ close to 1. More recent theoretical calculations yield $\Lambda_{Th} = 1.063$ [9] which is in agreement with our experimental value.

The standard arrangements for turbidity and correlation function measurements are shown in Figs. 1 and 2. Typical results of turbidity and thermal diffusivity behaviours for NH_3 case obtained from these experiments are shown in Figs. 3 and 4 [4-6].

2. TURBIDITY MEASUREMENTS

The turbidity $\mathscr{C}$ is the apparent attenuation of intensity of light passing through a scattering medium. $\mathscr{C}$ is obained by the relation:

$$I_T/I_0 = \exp(-\mathscr{C}/l) \tag{6}$$

where I_T is the intensity of light transmitted through the sample of length l and filled with a fluid; I_0 is the intensity of light transmitted through the empty cell. This practical definition is correct when the intensity I_R of the incident light is constant.

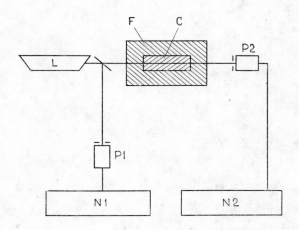

Fig. 1. Experimental set up for turbidity measurements

In the schematic experimental set up (Fig. 1), the deflected part of the incident 6328 Å He-Ne laser beam (L) is collected by a photodiode ($P1$) and its intensity I_R is measured by a nano-ammeter ($N1$). The intensities of the transmitted light I_0, I_T, are measured by similar devices ($P2$), ($N2$). The temperature of the cell (C) is controlled by a thermostated furnace (F). This temperature is measured by a quartz thermometer.

The results of NH_3 turbidity measurements (Fig. 2) show two typical ranges. For $T-T_c$ close to 1 K, the turbidity behaviour reflects the isothermal compressibility behaviour (dotted line in Fig. 2). The deviation observed with respect to this behavior when $T-T_c \leqslant 0.1$ K is related to the divergence of the correlation length. In this temperature range the value of the asymptotic amplitude ξ^+_0 is determined [4-6].

Analogous results were obtained in our laboratory for several pure fluids. In Table 1, the corresponding values of ξ^+_0 are shown and compared to values calculated by phenomenological methods [10, 11]. The

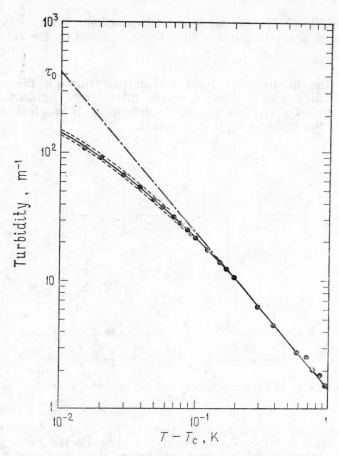

Fig. 2. Turbidity of ammonia along the critical isochore

Table 1. Amplitude of the correlation length of some pure fluids

Substances	Length of the cell (mm)	$\xi^+_0 (10^{-10}$ m)		
		Experiments	Sengers [10]	Garrabos [11]
Xe	10	1.85±0.15	1.89	1.76
CO$_2$	10	1.50±0.10	1.56	1.46
SF$_6$	10	1.90±0.10	1.96	1.84
C$_2$H$_6$	26	1.80±0.10	1.84	1.80
CClF$_3$	26	1.95±0.10		1.82
NH$_3$	30	1.40±0.08	1.44	1.35
H$_2$O			1.34	1.20

agreement between calculated and measured $\xi^+{}_0$ amplitudes is within the error bar of the experimental values.

3. TIME CORRELATION FUNCTION MEASUREMENTS

The characteristic time of the decay rate of entropy fluctuations is:

$$t_D = (D_T \cdot q^2)^{-1} \tag{7}$$

where D_T is the thermal diffusivity coefficient and q is the transfer wave vector.

This time is obtained by homodyne detection of the light scattered at 90° by the sample cell (C) (Fig. 3). The incident light is provided by the He-Ne laser (L) and focussed in the cell by the lens $(L1)$. The scattered light is collected by the lens $(L2)$ which conjugates the scattering volume and the pinhole (T) in front of the photomultiplier (P). The analysis of the photo-current is made by a digital correlator (D) through the standard amplifier — discriminator (A).

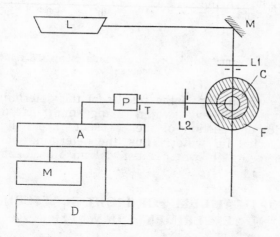

Fig. 3. Experimental set up for time correlation function measurements

Results obtained from time correlation function measurements yield the thermal diffusivity coefficient [4-6]. The variation of this coefficient in terms of T-T_c is shown in Fig. 4 for NH_3. These optical measurements allow one to extend much closer to T_c (from $T = T_c + 2$ K down to $T = T_c + 0.01$ K) the thermodynamic determination of D_T above $T = T_c + 2$ K. In this range the thermal diffusivity was calculated from the measurement of the thermal conductivity λ, using the relation:

$$D_T = \lambda / \rho C_p \tag{8}$$

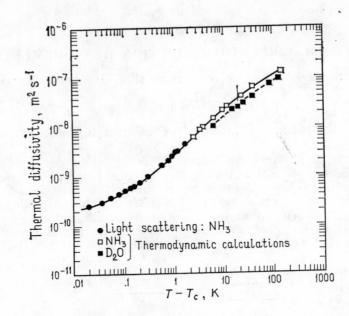

Fig. 4. Thermal diffusivity of ammonia along the critical isochore

We have shown in Fig. 4 the results of the determination of D_T by Eq. (8) in the case of heavy water using our determination of λ, presented in another section of the present proceeding. A similar behaviour is expected for water. Thus, the determination of D_T by light scattering experiments in water must extend this behaviour to lower values of T-T_c.

4. OPTICAL CELL FOR LIGHT SCATTERING EXPERIMENTS IN WATER

In Fig. 5 is shown a longitudinal section of the optical cell used near the critical point of water. The cell is provided with four quartz windows (W), located at 90°. Windows are attached on four conical flanges (E) which ensure the tightening of the cell. The cell is at constant volume and is closed by a needle valve set up in the wall. The cell is put in a cylindrical furnace made of aluminium. The heating of the furnace is provided by eight longitudinal resistances. The temperature is measured by a platinum resistance set up in the wall of the cell. The temperature control is realized by classical devices which ensure a temperature stability at the level of the cell better than 2×10^{-4} K.

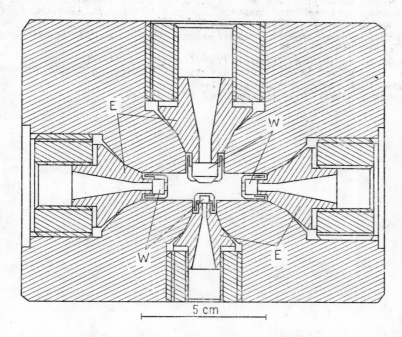

Fig. 5. Optical cell for light scattering experiments in critical water

5. CONCLUSION

Measurements of turbidity and time correlation can be performed to provide a determination of the correlation length and isothermal compressibility very close to the critical point of water. These optical methods associated with thermodynamic measurements will cover a large temperature range.

REFERENCES

1. Garrabos, Y., Tufeu, R., and Le Neindre, B. (1979): *J. Chem. Phys.* **68**, 495.
2. Garrabos, Y., Tufeu, R., Le Neindre, B., Zalczer, R., and Beysens, D. (1980): *J. Chem. Phys.* **72**, 4637.
3. Beysens, D., Tufeu, R., and Garrabos, Y. (1979): *J. Phys. Lett.* (Paris) 40, L-623.
4. Tufeu, R., Le Taief, A., and Le Neindre, B. (1981): In "Proceedings of the 8th Symposium on Thermophysical Properties", Ed. Sengers, J. V. (A.S.M.E., N. Y., 1982), p. 451.
5. Tufeu, R., Ivanov, D. Y., Garrabos, Y., and Le Neindre, B. (1984): *Ber. Bunsenges. Phys. Chem.* 88, 422-427.
6. Tufeu, R., Garrabos, Y., and Le Neindre, B. (1979): in "Thermal Conductivity 16", Ed. Larsen, D. C. (Plenum Press, N. Y., 1982), p. 605-612.

7. For critical static properties, see for example: "Phase Transitions, status of the experimental and theoretical situation", Cargèse (1980): Eds. Lévy, M., Le Guillou, J. C., and Zinn-Justin, J. (Plenum Press, N. Y., 1982).

8. For critical dynamic properties, see for example, Hohenberg, P. C. and Halperin, B. I. (1977): *Rev. Mod. Phys.* **49**, 435.

9. Paladin, G. and Peliti, L. (1982): *J. Phys. Lett.* (Paris), **43**, L-15.

10. Sengers, J. V. and Levelt-Sengers (1978): in "Progress in Liquid Physics", Ed Croxton, C. A. (Wiley, N. Y., 1978), p. 144.

11. Garrabos, Y. (1984): To be published in *J. Phys.* (Paris).

AQUEOUS SOLUTIONS

AQUEOUS SOLUTIONS

Investigation and Systematization
of Physico-Chemical Properties of Water-Steam
Contaminants with the Aim to Optimize
Water-Chemistry Conditions of Power Plants

O. I. MARTYNOVA

Moscow Power Engineering Institute, Moscow, USSR

The dependability of any component of power plants — condensers, preheaters, steam generators, and the whole setting system — is closely related to the water-chemistry conditions (regimes) in which the equipment is operated (i. e. it is related to the environment, or medium, the equipment is in contact with). This environment is never, as well known now, absolutely pure water and steam, whose properties are established rather complete to-day, but it is always impure water and steam, contaminated by salts, corrosion products, gases, organic matters, chemical additives, etc. The chemical term for impure water is solution.

To control the allowable — from safe power plant operation point of view — contamination level, water and steam chemistry specifications had (and still have) to be set up, as well as the means to maintain these specifications. This required, in the time when power engineering passed over to high pressures, the creation of a new engineering science, the so-called water regime, specific for different types of power plants and the whole circuit's materials of construction.

The theoretical basis of this science in a large complex of physico-chemical and thermo-physical (thermodynamic, kinetic, electrochemical, etc.) characteristics of specific for modern power plant's cycles media at its particular high temperature parameters. So, having in reality always

to deal with *solutions* of all possible impurities, that may be present in water and steam, these impurities' own properties have to be investigated and reference data accumulated for them, according to a special list of priority set up by IAPS.

With regard to these obtained characteristics, interaction processes of various impurities with materials of construction, in particular, equipment elements with their specific thermophysical features and design, including heat and mass exchange, hydrodynamic conditions, etc. must be established with a special emphasis to electrochemical corrosion.

Of course, to understand and control all these very complicated processes, one has to start with a meticulous compilation of the already stored in some other areas of science and technology experimental data on properties of solutions, typical for power plants steam and water circuits. This has to be followed by critical evaluation and mathematical processing of the experimental data. Most representative equations for correlating these data are obtained and verified making them available as much as possible for industrial use. But a further broad development of experimental and theoretical studies of not yet established data, their reliability evaluation is necessary. This includes, first of all, thermodynamic characteristics, velocity constants of chemical reactions and other properties of all contaminants, additives and interaction products in the whole region of steam-power plants cycles parameters P-V-T-X. To provide the power engineering industry with reliable numerical data, the accumulation and systematization of the whole available information in automated data banks is necessary and is already in progress. This involves the optimization of already existing and the development of new water-chemistry regimes, new selection of materials of construction, corrosion prevention measures, etc. Many relevant data already exist, more are being produced and processed, particularly under guidance of IAPS, mainly in the field of *thermodynamic properties* of systems important for water chemistry technologies of power plants cycles and design of new equipment.

So, comparatively investigated is, as far as establishing reliable phase diagrams and mathematical equations, the influence of temperature, pressure and concentration on the behaviour of very important, from the point of view of power plants operation system, solutions of silica in water and steam. Silica is one of the main deposition-forming (in turbines), though not corrosion-aggressive agent. A lot of data is now available about sodium chloride and hydroxide solutions, these being most dangerous from the local corrosion's point of view, and even, to a certain extent, about such extremely complicated compounds as corrosion products -- iron oxides (hydrocomplexes) and to a less extent, but nevertheless there is aluminum, cobalt and copper oxides. Since recently, great attention has been paid to solubility of certain gaseous substances, particularly to non-polar oxygen and hydrogen, in water and steam.

Of great significance are reliable data on concentrated solutions, peculiarities of compounds with high solubility in water like sodium chloride and hydroxide, activity coefficients and, of course, possible con-

centration phenomena in various power plant's circuit equipment parts and specific locations of different design. These data are very important for geothermal power plants, of course.

Taking into account the great number of solutions' properties, in comparison with pure water or steam, and not having yet the long term experience of investigating and evaluating these properties like the excellent work having been done by IAPS with these pure substances, it is of foremost importance to establish those solutions' characteristics, most critical from their damage-causing point of view. So, for comparatively sparingly soluble silica, one of the main in turbine deposit-forming but non-agressive contaminants, such critical parameter obviously is just its solubility in water and steam as well as distribution coefficient, now having been estimated with adequate accuracy for most conditions. On the contrary, contaminants like sodium hydroxide, highly soluble in water but not in steam, do not form solid deposits in turbines but their concentrated solutions are corrosion-agressive, especially, as now known rather well, in the first portions of moisture near the so-called Wilson (transition) zone of steam turbines. For compounds like sodium hydroxide, at complete evaporation forming only a liquid melt, not solid deposits, even comprehensive, coordinated, thermodynamically consistent transport data in accordance with steady-state distribution between water and steam are insufficient because they can determine only the flow bulk concentration but not most important possible local condensed concentrations, in, say, crevices, porous deposits, gaps, etc. This concerns substances like NaCl, too, but most of all NaOH.

The task of obtaining new data for power industry and science is aggravated by the fact that it is still *insufficient* — even though perhaps inevitably, for the beginning — to compile data characterizing water-steam mono- and multicomponent systems, typical for various power plant conditions, but isolated from materials of construction, the solutions operate in contact with. That is why, it is necessary to compile, get experimentally and handle also data concerning the interaction of solution's processes going on in specific locations with these materials, taking into account the various electrochemical potentials, transport properties and surface-active properties of the yielded systems, as well.

With regard to the fact that most water-chemistry regimes are connected with a so-called correction of the working medium by various chemical additives, naturally one more task arises — the investigation, again in a broad parameter range, of not only these isolated additives, first of all ammonia and its derivatives, including filming amines, sodium phosphate and its derivatives, boric acid, potassium and lithium hydroxide, various polymeres, chelates, surface active species, etc. but of processes and products of their interaction with inevitable water and steam impurities, too. Of course, a lot of information about these systems is already available, but it has to be systematized, evaluated and made available for industrial use.

And what is more, the majority of the some very important, from the point of view of water chemistry's optimization, contaminants,

including silica and practically all corrosion products (metal oxides and their hydroxocomplexes) are inclined to form colloidal systems and even, to a certain extent, at concentrations high enough, stereostructures. This means that the scope of IAPS W. G. IV problems is extremely complicated by the necessity to characterize these systems quantitatively. First of all, surface and electrophoretic properties of colloidal particles have to be determined, including adsorption ability, surface electrokinetic potentials as function of pH, this being of great importance for a variety of processes in power plants.

As some now used water regimes are either oxidizing or reducing, the majority of metal oxides (hydroxocomplexes) contain metal ions of different valency. This makes the development of redoxmetric measurements, carried out in these systems, be of great importance. One of the most important challenges in so doing, especially at high temperatures, is the estimation of a quantitative relationship between the metal's various corrosion potentials and the redoxpotential (measured by an inert electrode) of the solution, the metal is operating in.

Summing up, it has to be underlined, once more, that most accumulated data on solubility of various components in steam and water, their distribution coefficients, etc. have been obtained for thermodynamically equilibrium conditions and, what is even more important, for monocomponent but not for multicomponent systems. We hope to hear from Prof. Pitzer, whether, for all properties, equations can be generalized to multicomponent systems and even to complex brines.

At the same time, in power plant cycles equilibrium state is an exception (perhaps — steam generation processes) as well as an exception is the existence of monosolutions. That is why, in future work main attention has to be paid not only to thermodynamics but also to kinetics of processes, as near as possible approaching reality conditions.

The given, not at all being complete, enumeration of questions, whose solution is of vital importance for the understanding of the complicated physical and chemical processes, taking place in thermal power plants working media circuits, indicates the broad scope and level of problems IAPS W. G. IV has to solve to provide a scientific basis to increase the reliability of steam power plants equipment. This goal is alike important for all countries.

Thermodynamic Properties of Aqueous NaCl from 273 to 823 K with Estimates for Higher Temperatures

KENNETH S. PITZER

*Department of Chemistry and Lawrence Berkeley Laboratory,
University of California, Berkeley, California 94720, USA*

ABSTRACT

Equations based on theory, to the degree feasible, are developed which represent the various thermodynamic properties of aqueous NaCl. In the range 273-573 K, 0-1 kbar, and to saturation molality, the experimental data base is excellent. An equation based on a virial expansion and a Debye-Hückel term represents the data accurately. This form of equation has been generalized to multicomponent systems; hence these results are readily extended to complex brines as the properties for other components become known.

In the range above 573 K the data are more limited and of less accuracy, but a comprehensive treatment is still possible. The equation is one used before for ionic systems miscible to a fused salt. The supercooled liquid NaCl is taken as a reference state. The equation is valid from 423 to 823 K and thus has an extensive overlap with the lower-temperature equation. Estimates are also presented for temperatures above 823 K.

INTRODUCTION

The thermodynamic properties of aqueous sodium chloride are important in many areas including geochemistry, oceanography, and many

91

industrial processes. In practical steam power generation the water is not perfectly pure and the most common impurity is NaCl. For many purposes the properties of seawater and other natural waters may be approximated by sodium chloride solutions of appropriate composition. Since aqueous NaCl is the electrolyte which has been most widely investigated, it is of theoretical interest as the prototype of an ionic system including a polar molecular solvent.

There are various empirical compilations of the volumetric and thermodynamic properties of $NaCl-H_2O$, including a comprehensive one by Khaibullin (1980). The primary purposes of the present investigations have been, (a) to represent the experimental data by analytical equations, and (b) to incorporate very recent measurements of high accuracy. An equation of state has the advantage that the various properties interrelated by thermodynamics are all given by the appropriate derivatives of the parent equation. This makes available convenient computation of any desired property and assures consistency with all thermodynamic relationships. It was possible to use equations where the composition dependence has a theoretical basis in molecular-statistical terms and where this compositional dependence can be generalized to multicomponent systems.

Although the properties of aqueous NaCl are known over a very wide range of temperature, pressure, and composition, the accuracy and completeness of the measurements varies greatly for different ranges of temperature and pressure. Also the character of water changes greatly and therewith the extent to which water modifies the interionic forces. For both sets of reasons, it is difficult to use a single equation to express the properties over the entire range. Fortunately, there is an equation of state for pure water valid in good approximation over the entire range of interest and having a theoretically based form so that extrapolation to higher temperature is probably reliable. We use the equation of Haar, et al. (1981) without correction for the region immediately around the critical point (Kestin, et al. 1984). For the dielectric constant (relative permittivity) of water the empirical equation of Uematsu and Franck (1980) is valid from 273 to 823 K, but its form does not allow extrapolation to higher temperature. Since one of our objectives is to develop equations valid at still higher temperatures, we have chosen to use a theoretically based equation (Pitzer, 1983a) for the dielectric constant which should be reasonably accurate to very high temperature above the range of experimental measurement; it fits the experimental measurements for liquid-like densities above 573 K, which extend to 823 K, and is valid for steam at any temperature. At lower temperatures the relatively simple equation of Bradley and Pitzer (1979) fits the data for the liquid and agrees very precisely with the values recommended by the International Union of Pure and Applied Chemistry (Kienitz and Marsh, 1981).

Recently my colleagues and I have developed two equations for the thermodynamic properties of aqueous NaCl. The first, by Pitzer, Peiper, and Busey (1984), depends on an earlier equation of Rogers and Pitzer

(1982) for the volumetric properties as well as a large array of activity, enthalpy and heat capacity measurements extending to 573 K, 1 kbar, and saturation molality. This equation, designated PPB hereafter, follows the pattern used for room-temperature electrolyte thermodynamics in using molality as the measure of composition and in adopting a form known to be accurate for many aqueous electrolytes, including mixtures, near 298 K. Since the PPB equation has been described rather comprehensively, the present paper will include only a general description and, in the *Appendix*, a concise statement of the exact mathematical expressions and parameters.

The second equation, by Pitzer and Li (1983) was designed for application over a much higher temperature range extending even to pure fused NaCl. While presently available experimental data allowed evaluation of parameters only to 823 K, there is every reason to expect the form of the equation to continue to be useful to much higher temperature provided the density remains fairly high. While there are many differences between this second, PL, equation and the PPB equation, both assume that the NaCl is present primarily as ions. At low densities, such as the steam side of the two-phase region, there is extensive ion pairing which must be recognized in an accurate treatment. The extent to which the PL equation gives useful estimates at intermediate densities and concentrations is discussed in the latter part of this paper.

The original publication of Pitzer and Li (1983) presented the development of the PL equation for activities of both H_2O and NaCl but gave only very brief comments about other properties. In this paper the development of the PL equation is reviewed; the differences from the PPB equation are clearly described, and then new calculations are presented of the enthalpy and entropy from the PL equation for the range above 573 K.

Finally, there is a discussion of the few data and of estimates of the properties of $NaCl-H_2O$ above 823 K.

TREATMENT VALID TO 573 K (PPB EQUATION)

There are extensive measurements of several thermodynamic properties for $NaCl-H_2O$ in the range to 573 K or in some cases a little higher temperature. A set of equations representing these properties to nearly full experimental accuracy has been developed and published very recently (Pitzer, et al., 1984). Extensive tables are included as well as references to the sources of experimental data. It would be inappropriate to repeat that presentation in full detail, hence, I present only a general description and include in the *Appendix* a concise statement of the exact equations with a listing of the numerical parameters.

The composition dependence at any given temperature and pressure is represented by a simple, theoretically-based equation (Pitzer, 1973) which has the advantage that it has been successfully extended to multicomponent systems (Pitzer and Kim, 1974). The pure electrolyte parameters are the dominant terms for the more complex systems. The PPB

equation uses molality which is the most widely used composition variable for aqueous electrolytes.

For a single solute with singly charged ions one may define the excess Gibbs energy in the molality system as

$$G^{EX} = G - n_1 G°_1 - n_2 \bar{G}°_2 + 2n_2 RT (1 - \ln m) \tag{1}$$

where n_1 and n_2 give the content in moles of solvent and salt of molar Gibbs energies $G°_1$ and $\bar{G}°_2$ in their reference states, respectively, and m is the molality. Then it is assumed that

$$G^{EX}/n_w RT = f(I) + 2m^2 B(I) + 2m^3 C \tag{2}$$

where n_w is the number of kg of solvent ($n_1/n_w \cong 55.5$ mol·kg^{-1}) and $f(I)$ is a function of a Debye-Hückel type representing the effect of long-range electrostatic forces. B and C are second and third virial coefficients representing the effect of short-range, solvent-modulated forces between pairs and triplets of solute particles respectively. Higher virial terms could be included but are not needed for NaCl or for most solutes. Although for the 1 : 1 electrolyte the ionic strength I is equal to the molality m, the separate symbol I is used to emphasize a dependency on ionic strength in a multicomponent solution. For a different salt $f(I)$ remains the same, but B and C are solute specific. The ionic strength dependence of B, however, is given by a general expression which has a theoretical basis (Pitzer, 1973). The detailed forms of $f(I)$ and $B(I)$ are given in the *Appendix*.

The chemical potential and the activity for either component are given by the usual definitions

$$\mu_i = (\partial G/\partial n_i)_{T,P,n_j} \tag{3}$$

$$\ln a_i = (\mu_i - G°_i)/RT. \tag{4}$$

The molality, $m = 1000 \, n_2/M_1 n_1$ with M_1 the molecular mass of the solvent, is unsymmetrically related to n_1 and n_2. Hence, there are different expressions for a_1 and a_2

$$\ln a_1 = \frac{M_1}{1000} \left[-2m + \frac{G^{EX}}{n_w RT} - m \left(\frac{\partial (G^{EX}/n_w RT)}{\partial m} \right)_{P,T} \right] \tag{5}$$

$$\ln a_2 = 2 \ln m + [\partial (G^{EX}/n_w RT)/\partial m]_{P,T} \tag{6}$$

In the molality system the activity coefficient of a symmetrical ionic solute is

$$\gamma_\pm = a_2^{1/2}/m \tag{7}$$

$$\ln \gamma_\pm = (1/2) [\partial (G^{EX}/n_w RT)/\partial m]_{P,T} \tag{8}$$

while the osmotic coefficient is defined to represent solvent activity as

$$\varphi = -(1000/2M_1 m) \ln a_1 \tag{9a}$$

$$\varphi - 1 = \frac{1}{2} \left[\left(\frac{\partial (G^{EX}/n_w RT)}{\partial m} \right)_{P,T} - \frac{G^{EX}}{n_w RT m} \right]. \tag{9b}$$

94

The detailed expressions for $\gamma_\pm$ and φ with the forms adopted for $f(I)$ and $B(I)$ are given in the *Appendix*.

The excess enthalpy, entropy, heat capacity, volume, and related quantities are all given by appropriate derivatives of the excess Gibbs energy. If Y^{EX} represents any of these excess quantities, it has the same composition dependence as G^{EX} since the molality is independent of P or T. Thus

$$Y^{EX}/n_w = f^Y(I) + 2m^2B^Y(I) + 2m^3C^Y \qquad (10)$$

where the ionic strength dependence of $f^Y(I)$ is the same regardless of Y and the same is true for $B^Y(I)$.

The only temperature and pressure dependent quantity in $f(I)$ is the Debye-Hückel parameter which is fully determined by the properties of water. Both B and C contain parameters dependent on temperature and pressure which are expressed empirically. The appropriate derivatives then give B^H for enthalpy, B^S for entropy, B^V for volume, etc. The full array of parameters are given in the *Appendix*.

The full list of experimental data which were considered is given in the detailed paper. I only comment here about a few very important investigations including the precise osmotic coefficient measurements of Liu and Lindsay (1972) and the very recent heat of dilution measurements of Busey, et al. (1984). These two sets of data are consistent to high accuracy and determine the composition dependency at the higher temperatures. The absolute enthalpy and entropy values depend primarily on the heat capacity measurements at 3 mol·kg^{-1} by White and Wood (1982). Most important for the volumetric data base are the measurements of Hilbert (1979). These comments all relate to the higher temperatures; there are many accurate measurements at lower temperatures.

The final PPB equation gives the Gibbs energy as a function of temperature, pressure, and molality. Appropriate derivatives yield activity and osmotic coefficients, as well as the enthalpy, entropy, heat capacity, volume (or density), expansivity and compressibility. Tables are included giving the various parameters for various values of T and P. From these parameters it is a very simple calculation to insert the dependency on molality and to obtain any of the properties at a given value of m. In addition, there are tables of the various thermodynamic properties at appropriately spaced intervals of T, P, m for linear interpolation.

The pressure dependency of the Gibbs energy, which yields the volumetric properties, was developed by Rogers and Pitzer (1982) and was retained without change in the more complete PPB equation. Likewise, the earlier 1982 paper should be consulted for the details of the experimental volumetric data and for the tables of volumetric parameters and explicit values of the specific volume, expansivity, and compressibility. In order to maintain optimum accuracy for the low temperature range where there are very precise experimental data, a "low temperature" fit

95

(valid to 358 K) was made in addition to the general fit valid from 273 to 573 K with lower precision.

Table 1 gives a list of the tables of parameters as functions of T and P and a list of tables of properties as functions of T, P, and m which are available in one or the other paper. As samples of the property tables and for use in comparison with results for higher temperatures, Tables 2 and 3 give the specific enthalpy and the specific entropy for 1000 bar.

Table 1. Tavles of parameters as functions of T and P and of thermodynamic properties as functions of T, P, and m based on the PPB equation for NaCl-H_2O

Parameter tables: all given by Pitzer, et al. (1984);
the last three also given by Rogers and Pitzer (1982).

1. Standard Gibbs energies, Debye-Hückel A_φ parameter, and virial coefficients for NaCl(aq)
2. Standard entropies, Debye-Hückel A_S parameter, and virial coefficients for the NaCl(aq) entropy
3. Standard enthalpies, Debye-Hückel A_L parameter, and virial coefficients for the NaCl(aq) enthalpy
4. Standard heat capacities, Debye-Hückel A_J parameter, and virial coefficients for the NaCl(aq) heat capacity
5. Standard volumes, Debye-Hückel A_v parameter, and virial coefficients for the NaCl(aq) volume
6. Standard expansivities, Debye-Hückel A_x parameter, and virial coefficients for the NaCl(aq) expansivity
7. Standard compressibilities, Debye-Hückel A_K parameter, and virial coefficients for the NaCl(aq) compressibility

Property Tables

A. Tables given by Pitzer, et al. (1984)

1. Activity coefficient of NaCl(aq)
2. Osmotic coefficient of NaCl(aq)
3. Entropy of solution of NaCl(aq)
4. Enthalpy of solution of NaCl(aq)
5. Relative entropy of NaCl(aq)
6. Relative enthalpy of NaCl(aq)
7. Relative heat capacity of NaCl(aq)
8. Density of NaCl(aq)
9. Specific entropy of NaCl(aq)
10. Specific enthalpy of NaCl(aq)

B. Tables given by Rogers and Pitzer (1982)

11. Specific volumes of NaCl(aq)
12. Expansivities of NaCl(aq)
13. Compressibilities of NaCl(aq)

The absolute reference state of energy or enthalpy for H_2O is the gas at zero Kelvin while for NaCl it is the infinitely dilute solute standard state at 298.15 K and 1 atm. Another set of reference states in common use is pure liquid water and infinitely dilute NaCl, both at 273.15 K (0°C). For conversion of our values in Table 2 to the latter reference states, add 1997.65 $J \cdot g^{-1}$ for H_2O and subtract 52.1 $J \cdot g^{-1}$ for

Table 2. The specific enthalpy (J·g⁻¹) of NaCl(aq) at 1000 bar (100 MPa)

t °C	$m=0.1$ mol/kg	$m=0.25$ mol/kg	$m=0.5$ mol/kg	$m=0.75$ mol/kg	$m=1.0$ mol/kg	$m=2.0$ mol/kg	$m=3.0$ mol/kg	$m=4.0$ mol/kg	$m=5.0$ mol/kg	$m=6.0$ mol/kg
0.0	—1891.2	—1874.8	—1848.2	—1822.4	—1797.4	—1704.8	—1622.1	—1547.9	—1480.6	—1419.2
10.0	—1852.2	—1836.2	—1810.0	—1784.7	—1760.2	—1669.0	—1587.6	—1514.3	—1447.9	—1387.1
20.0	—1812.9	—1797.6	—1771.6	—1746.8	—1722.7	—1633.1	—1552.9	—1480.6	—1414.9	—1354.9
25.0	—1793.2	—1777.6	—1752.3	—1727.7	—1703.9	—1615.0	—1535.5	—1463.7	—1398.5	—1338.7
30.0	—1773.4	—1758.0	—1732.9	—1708.6	—1685.0	—1597.0	—1518.1	—1446.8	—1382.0	—1322.5
40.0	—1733.6	—1718.6	—1694.0	—1670.2	—1647.1	—1560.7	—1483.1	—1412.9	—1348.9	—1290.2
50.0	—1693.8	—1679.1	—1655.1	—1631.8	—1609.1	—1524.4	—1448.2	—1379.1	—1315.9	—1257.9
60.0	—1654.0	—1639.6	—1616.1	—1593.3	—1571.1	—1488.0	—1413.2	—1345.2	—1283.0	—1225.6
70.0	—1614.0	—1600.0	—1577.1	—1554.7	—1533.0	—1451.7	—1378.2	—1311.3	—1249.9	—1193.3
80.0	—1574.1	—1560.4	—1538.0	—1516.2	—1494.9	—1415.2	—1343.1	—1277.4	—1217.0	—1161.2
90.0	—1534.0	—1520.7	—1498.8	—1477.5	—1456.8	—1378.8	—1308.1	—1243.5	—1184.2	—1129.1
100.0	—1493.9	—1480.9	—1459.6	—1438.9	—1418.6	—1342.4	—1273.1	—1209.7	—1151.3	—1097.2
110.0	—1453.8	—1441.1	—1420.4	—1400.1	—1380.4	—1305.9	—1238.1	—1175.9	—1118.6	—1065.3
120.0	—1413.5	—1401.2	—1381.1	—1361.3	—1342.1	—1269.4	—1203.0	—1142.1	—1085.8	—1033.5
130.0	—1373.1	—1361.2	—1341.7	—1322.5	—1303.8	—1232.9	—1168.0	—1108.4	—1053.2	—1001.8
140.0	—1332.7	—1321.2	—1302.2	—1283.6	—1265.4	—1196.4	—1133.0	—1074.6	—1020.5	—970.1
150.0	—1292.2	—1281.1	—1262.7	—1244.7	—1227.0	—1159.9	—1098.1	—1040.9	—987.9	—938.5
160.0	—1251.6	—1240.9	—1223.2	—1205.7	—1188.6	—1123.4	—1063.1	—1007.3	—955.4	—906.9
170.0	—1210.9	—1200.6	—1183.5	—1166.7	—1150.2	—1086.9	—1028.2	—973.7	—922.8	—875.3
180.0	—1170.1	—1160.3	—1143.9	—1127.7	—1111.7	—1050.4	—993.3	—940.1	—890.4	—843.8
190.0	—1129.2	—1119.8	—1104.1	—1088.6	—1073.2	—1014.0	—958.5	—906.5	—857.9	—812.3
200.0	—1088.2	—1079.3	—1064.3	—1049.4	—1034.7	—977.6	—923.7	—873.0	—825.5	—780.7
210.0	—1047.1	—1038.6	—1024.4	—1010.2	—996.1	—941.2	—888.9	—839.6	—793.0	—749.2
220.0	—1005.8	—997.9	—984.5	—971.0	—957.5	—904.8	—854.2	—806.2	—760.6	—717.7
230.0	—964.4	—957.0	—944.4	—931.7	—918.9	—868.4	—819.6	—772.8	—728.3	—686.1
240.0	—922.8	—915.9	—904.2	—892.3	—880.2	—832.1	—785.0	—739.5	—695.9	—654.4
250.0	—881.0	—874.7	—863.9	—852.8	—841.5	—795.9	—750.5	—706.2	—663.5	—622.7
260.0	—838.9	—833.3	—823.5	—813.2	—802.8	—759.7	—716.0	—673.0	—631.1	—590.9
270.0	—796.7	—791.7	—782.9	—773.6	—764.0	—723.6	—681.7	—639.8	—598.7	—559.0
280.0	—754.1	—749.9	—742.2	—733.9	—725.2	—687.6	—647.5	—606.8	—566.3	—526.9
290.0	—711.3	—707.9	—701.4	—694.2	—686.4	—651.7	—613.5	—573.8	—533.7	—494.4
300.0	—668.2	—665.7	—660.5	—654.4	—647.7	—616.2	—579.8	—540.8	—501.0	—461.5

Table 3. The specific entropy (J·K⁻¹·g⁻¹) of NaCl(aq) at 1000 bar (100 MPa)

t °C	$m=0.1$ mol/kg	$m=0.25$ mol/kg	$m=0.5$ mol/kg	$m=0.75$ mol/kg	$m=1.0$ mol/kg	$m=2.0$ mol/kg	$m=3.0$ mol/kg	$m=4.0$ mol/kg	$m=5.0$ mol/kg	$m=6.0$ mol/kg
0.0	3.504	3.495	3.477	3.458	3.437	3.350	3.260	3.171	3.085	3.002
10.0	3.644	3.634	3.614	3.593	3.571	3.478	3.384	3.292	3.202	3.117
20.0	3.780	3.769	3.748	3.725	3.701	3.603	3.505	3.409	3.317	3.229
25.0	3.847	3.835	3.813	3.789	3.765	3.664	3.564	3.466	3.373	3.284
30.0	3.913	3.900	3.877	3.853	3.828	3.724	3.622	3.522	3.427	3.337
40.0	4.042	4.028	4.004	3.977	3.951	3.842	3.735	3.632	3.535	3.442
50.0	4.167	4.152	4.126	4.098	4.070	3.956	3.845	3.739	3.638	3.544
60.0	4.288	4.273	4.245	4.216	4.186	4.067	3.951	3.842	3.739	3.642
70.0	4.406	4.390	4.360	4.329	4.298	4.174	4.055	3.942	3.836	3.738
80.0	4.521	4.504	4.472	4.440	4.408	4.279	4.156	4.040	3.931	3.830
90.0	4.633	4.615	4.582	4.548	4.514	4.381	4.253	4.134	4.023	3.920
100.0	4.742	4.723	4.688	4.653	4.618	4.480	4.349	4.226	4.112	4.006
110.0	4.848	4.828	4.792	4.756	4.719	4.576	4.441	4.315	4.199	4.091
120.0	4.952	4.931	4.893	4.856	4.818	4.670	4.531	4.402	4.283	4.173
130.0	5.053	5.031	4.992	4.953	4.914	4.762	4.619	4.487	4.365	4.252
140.0	5.152	5.129	5.089	5.048	5.008	4.851	4.705	4.570	4.445	4.330
150.0	5.249	5.225	5.183	5.141	5.100	4.939	4.789	4.650	4.523	4.406
160.0	5.344	5.319	5.276	5.233	5.189	5.024	4.870	4.729	4.599	4.479
170.0	5.437	5.411	5.366	5.322	5.277	5.107	4.950	4.806	4.673	4.551
180.0	5.528	5.501	5.455	5.409	5.363	5.188	5.028	4.881	4.746	4.622
190.0	5.617	5.589	5.542	5.494	5.447	5.268	5.104	4.954	4.817	4.691
200.0	5.705	5.676	5.627	5.578	5.529	5.346	5.178	5.025	4.886	4.758
210.0	5.791	5.761	5.710	5.660	5.610	5.422	5.251	5.095	4.954	4.824
220.0	5.875	5.844	5.792	5.740	5.689	5.497	5.322	5.164	5.020	4.889
230.0	5.959	5.926	5.872	5.819	5.767	5.569	5.392	5.231	5.085	4.952
240.0	6.041	6.007	5.951	5.897	5.843	5.641	5.460	5.296	5.149	5.014
250.0	6.121	6.087	6.029	5.973	5.917	5.711	5.526	5.361	5.211	5.075
260.0	6.201	6.165	6.106	6.048	5.991	5.779	5.591	5.424	5.273	5.136
270.0	6.279	6.242	6.181	6.121	6.063	5.847	5.655	5.485	5.333	5.195
280.0	6.357	6.319	6.255	6.194	6.134	5.912	5.718	5.546	5.392	5.254
290.0	6.434	6.394	6.329	6.265	6.203	5.976	5.779	5.605	5.450	5.312
300.0	6.510	6.468	6.401	6.335	6.271	6.039	5.838	5.663	5.508	5.370

Table 4. Uncertainty estimates for the PPB equation

Part A. Volumetric Properties   Uncertainties

Property	$t/°C$	p/bar	m	low T fit	high T fit
Volume	0-25	1.01	0-5.5	120 ppm	150 ppm
	0-25	1-1000	0-2.0	120 ppm	150 ppm
	25-85	1-1000	0-5.5	70 ppm	150 ppm
	85-300	1-1000	0-5.5	—	700 ppm
Expansivity	0-25	1.01	0-4.0	1%	—[a]
	0-25	1-1000	0-2.0	1%	—[a]
	25-85	1-1000	0-4.0	1%	5%
	85-300	1-1000	0-4.0	—	5%
Compressibility	0-25	1-1000	0-2.0	.5%	—[a]
	25-85	1-1000	0-5.0	.5%	5%
	85-300	1-1000	0-5.0	—	5%

[a] Not recommended.

Part B. Thermal and Activity Properties.

Property	m	25°C		200°C		300°C	
p/bar		200	1000	200	1000	200	1000
φ	0-6	0.002	0.004	0.004	0.008	0.006	0.02
$\ln \gamma_\pm$	1.0	0.002	0.006	0.005	0.013	0.015	0.05
	3.0	0.003	0.010	0.008	0.02	0.018	0.06
	6.0	0.004	0.012	0.01	0.03	0.020	0.08
$\Delta H°_s/RT$	—	0.06	0.22	0.08	0.30	0.8	2.5
$\varphi L/RT$	1.0	0.01	0.03	0.03	0.09	0.4	1.5
	3.0	0.02	0.10	0.03	0.12	0.5	2.0
	6.0	0.03	0.12	0.04	0.15	0.8	3.0
$\overline{C}°_{p,2}/R$	—	0.6	1.8	1.5	4.0	20.	40.
$\varphi C_p/R$	1-4	0.3	1.2	0.6	2.5	10.	20.

NaCl. Our entropies are absolute values, i. e., they are referenced to states of H_2O and NaCl which approach zero entropy as $T \rightarrow 0$ K.

Estimates of uncertainty are given in Table 4 for calculated values of various thermodynamic quantities as a function of temperature, pressure, and molality. The uncertainties are very small over most of the range of conditions but increase considerably near 573 K and especially above 200 bar and near 573 K. The solubility of solid NaCl was not used in the evaluation of the PPB equation, but it is shown that the final results for the activity coefficient are consistent with the measured solubility to 1% at the extremes of temperature and to 0.2% over most of the range.

TREATMENT VALID FROM 373 TO 823 K (PL EQUATION)

As the temperature increases substantially above 573 K there are several changes in the $NaCl$-H_2O system which are so great that a difference in general formulation of treatment is indicated. Most obviously, pure water at moderate pressure is gas-like rather than liquid-like above the critical temperature. The solute properties in the infinitely dilute reference state display extreme behavior near the critical point of water. Hence this infinitely dilute state is not a useful reference state for the NaCl. The solubility of solid NaCl increases rapidly. Indeed at its melting point, 1074 K, the system becomes a fluid miscible in all proportions. The molality becomes infinite for a pure fused salt; thus it is not a convenient measure of composition for very soluble systems. Even at 823 K the saturated solution is 63 wt. % NaCl. At temperatures above 700 K and below the critical pressure, the more concentrated of the co-existing phases is liquid-like. The dilute phase is essentially steam; only very near the critical pressure does the concentration of NaCl become substantial. For some purposes, however, the presence of even a very small concentration of NaCl can be very important. The NaCl in the dilute phase exists primarily as ion pairs (Quist and Marshall, 1968; Pitzer, 1983b) whereas in the concentrated phase it is ionized (Pitzer, 1984a). Thus, at these higher temperatures, there is a wide range of composition in which the solvent is liquid-like and the solute is ionized. It is for this range of ionized, liquid-like nature that an equation was developed by Pitzer and Li (1983).

The original publication emphasized the selection of the form of the PL equation and the data for the activity of both H_2O and NaCl to which it was fitted. After reviewing these aspects, the present paper will emphasize the calculation of enthalpy and entropy. Then there is a discussion of the extent to which the PL equation may yield less accurate but still useful estimates for more dilute solutions or at still higher temperatures or pressures.

Simple 1-1 nitrate salts have much lower melting points than NaCl. They form aqueous solutions of very large solubility at temperatures near 373 K. A very simple equation was found to be adequate (Pitzer, 1981) to describe the behavior of some of these nitrate solutions that are similar in many respects to the NaCl solutions at much higher temperatures. Thus Dr. Li and I first explored the adequacy of this simple equation for $NaCl$-H_2O in the range below 573 K where the properties are well established. We found remarkably good agreement for a very simple equation from 373 to 573 K, to 1 kbar, and over the entire range of composition. As expected, the accuracy of agreement is less than that of the more complex PPB equation, but the accuracy is high enough to be very useful in a range of less precise experimental data. Consequently, the same equation was fitted to the more limited data in the range to 823 K and good agreement was obtained.

Several differences between this PL treatment and the PPB equation should be emphasized.

1. The measure of composition is mole fraction on an ionized basis,.

$$x_1 = n_1/(n_1 + 2n_2) \tag{11}$$
$$x_2 = 2n_2/(n_1 + 2n_2) \tag{12}$$

with n_1 and n_2 the numbers of moles of H_2O and $NaCl$, respectively. The terms for ideal behavior are different on a mole fraction basis from those on a molality basis. This yields a different definition of the excess Gibbs energy as follows:

$$G_x^{EX} = G - n_1 G^\circ{}_1 - n_2 G^\circ{}_2 - RT(n_1 \ln x_1 + 2n_2 \ln x_2) \tag{13}$$

where the subscript x is a reminder of the mole fraction basis.

2. The solute reference state, at any given temperature and pressure, is supercooled liquid NaCl. At temperatures below 600 K this is evaluated from solution data just as is the case for the infinitely dilute reference state. But one can use the heat of fusion of NaCl, and the differences in heat capacity and volume between solid and liquid, to extrapolate properties of liquid NaCl well below the melting point. A smooth curve was adopted interpolating properties of supercooled liquid NaCl between the values from 373 to 573 K and the extrapolated curve valid at higher temperature. This is expressed as a Gibbs energy of solution (or of fusion) $\Delta_s G^\circ$ and is shown in Fig. 1 as $\Delta_s G^\circ/RT$ as a function of T. Now the molar reference-state Gibbs energy of NaCl becomes.

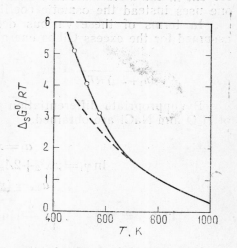

Fig. 1. The Gibbs energy of fusion to the supercooled liquid. See text for explanation of the curves

$$G^\circ{}_2 = G^\circ(NaCl, c) + \Delta_s G^\circ \tag{14}$$

where $G^\circ(NaCl, c)$ is for crystalline NaCl and all terms are on a molar basis.

3. The Debye-Hückel term in the PL equation contains the limiting law, and is equivalent to the $f(I)$ of the PPB equation in the limit at low concentration. But the two expressions, although similar in form, differ at any finite concentration and differ substantially at high concentration. The ionic strength is now defined as

$$I_x = \frac{1}{2} \sum_i z_i^2 x_i \tag{15}$$

with z_i the charge and x_i the mole fraction of the ith ion. For NaCl this becomes $I_x = x_2/2$.

4. The single Margules term in mole fraction for short-range forces in the PL equation is completely different from the power series in molality of the PPB equation. At finite concentration differences in the expressions for short-range forces compensate for differences in the ideal-solution terms and the Debye-Hückel term. The theoretical rationale for the use of a Margules term for an ionic system has been discussed elsewhere (Pitzer, 1981).

5. While the activity of water is the same in either equation and the activity of NaCl differs only from the difference of reference states, the activity coefficients are different. For the solute in the mole fraction system $\gamma_{\pm,x} = a_2^{1/2}/x_2$ whereas in the molality system $\gamma_\pm = a_2^{1/2}/m$. For the solvent in the mole fraction system $\gamma_1 = a_1/x_1$, but in the molality system one uses instead the osmotic coefficient defined in equation (9a).

In terms of these various definitions the following expression is assumed for the excess Gibbs energy:

$$\frac{G_x^{EX}}{(n_1+2n_2)\,RT} = wx_1\,x_2 - \left(\frac{4A_r\,I_r}{\rho}\right)\ln\left(\frac{1+\rho I_x^{1/2}}{1+\rho 2^{-1/2}}\right) \qquad (16)$$

By appropriate differentiation the activities and activity coefficients of H_2O and NaCl are obtained.

$$a_1 = x_1\gamma_1 \qquad (17)$$

$$\ln \gamma_1 = wx_2^2 + 2A_x I_x^{3/2}/(1+\rho I_x^{1/2}) \qquad (18)$$

$$a_{2,x} = (x_2\gamma_{\pm,x})^2 \qquad (19)$$

$$\ln \gamma_{\pm,x} = wx_1^2 - A_x\left\{\left(\frac{2}{\rho}\right)\ln\left(\frac{1+\rho I_x^{1/2}}{1+\rho 2_x^{-1\,2}}\right) + \frac{I_x^{1/2}-2I_x^{3\,2}}{1+\rho I^{1/2}}\right\} \qquad (20)$$

In these expressions A_x is the Debye-Hückel parameter on a mole-fraction basis

$$A_x = (1/3)\,(2\pi N_A d_1/M_1)^{1/2}(e^2/4\pi\varepsilon_0 DkT)^{3/2} \qquad (21)$$

with d_1, and D the density and dielectric constant (relative permittivity) of water. The permittivity of free space is ε_0, however, the factor $4\pi\varepsilon_0$ is unity in the e.s.u. system of units in which much electrolyte research is reported. The density is obtained from the equation of Haar, et al. (1981) while the dielectric constant is obtained from the equation of Bradley and Pitzer (1979) below 600 K and the theoretically based equation of Pitzer (1983a) above 600°K. As a measure of the uncertainty related to the joning of the two equations one can compare the results in the temperature range 573-623 K where both equations yield reasonable agreement with experimental dielectric constant data.

The parameter ρ is related to the hard-core, interionic repulsive distance a in Debye-Hückel theory,

$$\rho = a\,(2e^2N_A d_1/M_1\varepsilon_0 DkT)^{1/2}. \qquad (22)$$

The hard-core model is only an approximation hence it seems best to evaluate ρ empirically provided the corresponding a-value is reasonable. Thus we adopt only the factor $(d_1/DT)^{1/2}$ from equation (22) and evaluate a numerical coefficient to best fit the data for the activity of water in NaCl solutions from 373 to 573 K with the result

$$\rho = 2150\,(d_1/DT)^{1/2} \tag{23}$$

with d_1 in $g \cdot cm^{-3}$ and T in K. The numerical factor corresponds to $a = 5.7$ Å which is a very reasonable value.

Calculations for 373-573 K: In this fitting of water-activity data the Margules parameter w is freely adjusted at each temperature while A_x is determined by the properties of pure water. The osmotic coefficients from the recent comprehensive treatment (Pitzer et al., 1984) were adopted up to 4 $mol \cdot kg^{-1}$ while above that molality the directly measured values of Liu and Lindsay (1972) were used. Typical fits of $\ln a_1$ for the saturation pressure at several temperatures are shown in Fig. 2. The values of w are given in Table 5 and shown on Fig. 3.

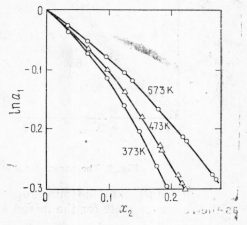

For the saturated solution at any temperature, the activity of the NaCl equals that of the solid. If $\Delta_s G°$ is the Gibbs energy of

Fig. 2. Comparison of curves for water activity calculated from the PL equation with experimental values

Table 5. Pressure and temperature dependence of $-w$ and $\Delta_s\,G°/RT$ values with various pressures in bars

T, K	P_{sat}	4 0	600	800	1,000	1	P_{sat}	1,000
	$-w$					$\Delta_s G°/RT$		
373	3.14	3.23	3.27	3.30	3.34	7.008	7.008	7.133
423	2.72	2.79	2.82	2.85	2.88	6.207	6.217	6.325
473	2.11	2.15	2.18	2.20	2.22	5.147	5.149	5.261
523	1.48	1.52	1.54	1.57	1.59	4.084	4.088	4.193
573	0.89	0.91	0.93	0.95	0.98	3.075	3.084	3.180
623		0.5	0.5_3	0.5_6	0.5_9	2.35		2.45_5
673		0.3_7	0.2_5	0.2_5	0.3	1.81_5		1.91_5
723			0.2_5	0.1_5	0.1	1.43		1.53
773				0.2	0.0_5	1.14		1.24
823				0.3	0.0_5	0.91		1.00_5

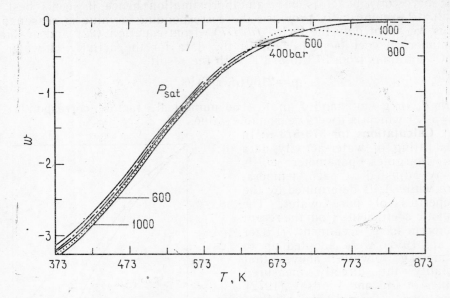

Fig. 3. The parameter w as a function of T and P

fusion or solution to the pure supercooled liquid NaCl, which is chosen as the reference state for the mixed system, then

$$\Delta_s G^\circ/RT = -2 \ln (x_2 \gamma_{\pm,x})_{sat}. \tag{24}$$

Solubility values were given by Liu and Lindsay (1972), and the resulting values of $\Delta_s G^\circ/RT$ are given in Table 5 and are plotted in Fig. 1 together with a calculated, extrapolated curve based on the heat of fusion and other properties of liquid and solid NaCl. The latter curve is taken from the JANAF tables (1971) where the pertinent data are assembled. Our values below 600 K are not expected to fit the extrapolated curve, but it is apparent that a very reasonable curve can be interpolated between the values below 600 K and the calculated curve above 700 K.

Calculations were also made for higher pressures. The change with pressure of an activity is given by

$$(\partial \ln a_i/\partial P)_T = (\overline{V}_i - V^\circ_i)/RT \tag{25}$$

where $\overline{V}_i$ is the partial molar volume at a given composition while V_i° is the molar volume of component i in its reference state. Similarly, the Gibbs energy of solution of the solid changes with pressure according to

$$(\partial (\Delta_s G^\circ)/\partial P)_T = V^\circ_2 (liq) - V^\circ_2 (cry). \tag{26}$$

The molar volumes of liquid and crystalline NaCl are available (Kirshenbaum, et al., 1962; A. I. P. Handbook, 1972) and the former was

extrapolated to lower temperature to allow integration of equation (26). The data for the activity of water in solutions under high pressure are given in the general treatment up to 573 K, 1 kbar, and 6 mol·kg⁻¹. The volumetric data of Hilbert (1979) were used to convert to higher pressure the data of Liu and Lindsay (1972) for solutions of high molality.

Calculations for 573-823 K: In the temperature range above 573 K, the data for water activity are of lower precision but the two-phase, steam-liquid equilibrium is well established through 823 K (Khaibullin and Borisov, 1966; Sourirajan and Kennedy, 1962; Urusova, 1974a, b; Urusova and Ravich, 1971; Khaibullin, 1980; Parisod and Plattner, 1981; Wood et al., 1984). Volumetric measurements as well as phase equilibria are reported by Urusova (1975), Khaibullin (1980), and by Gehrig, et al. (1983) as well as by earlier investigators. Although the steam phase dissolves more NaCl with increasing temperature, the activity a_1 is close to unity except very near the critical point. We take pure steam as the reference state for water. Then for the saturated liquid at P, $\gamma_1(sat, liq) = a_1(vap)/x_1(sat, liq)$ with $a_1(vap) \cong 1.0$.

To interpret these γ_1 values, they must be converted to a constant pressure by integration of equation (25). For pure steam the value of $V°_1$ is obtained from the equation of Haar, et al. (1981). When $V°_1$ becomes very large, however, it is more convenient to obtain the integral of that term from the difference in Gibbs energy with pressure

$$G°_1(P_2) - G°_1(P_1) = \int_{P_1}^{P_2} V°_1 dP \tag{27}$$

which is also given from the equation of state. For the solution $\overline{V}_1$ and $\overline{V}_2$ were obtained from the volumetric data of Urusova (1975) by finite difference between values for adjacent mole fractions. The results were smoothed and interpolated graphically and inserted in equation (25), which was integrated graphically.

The available experimental information concerning the solute is the pressure and composition of the liquid at the triple point where steam, liquid solution, and solid NaCl are in equilibrium. Measurements of the solid solubility are reported by Keevil (1942) in addition to Urusova and others cited above. Thus, equation (24) can be applied under these conditions with $\Delta_s G°$ taken from Table 5 or Fig. 1. Then the resulting $\gamma_\pm$ can be converted to higher pressure by integration of the equation

$$\partial \ln \gamma_{\pm,x}/\partial P = (\overline{V}_2 - V°_2)/2RT \tag{28}$$

where $V_2°$ is the molar volume of the supercooled liquid NaCl.

While the accuracy of these $\overline{V}_1$ and $\overline{V}_2$ values is not high, it is sufficient to maintain reasonable accuracy in the values of γ_1 and $\gamma_{\pm, x}$ up to 1000 bar at 723, 773, and 823 K. At 623 K and 673 K the data of Urusova (1975) extend only to 225 bar and 600 bar, respectively.

We expect equations (16), (18), and (20) to apply at pressures high enough that pure water has a liquid-like density. Thus we first tested

equations (18) and (20) on values of γ_1 and $\gamma_{\pm,x}$ converted to the highest pressure of the volumetric data of Urusova (1000 bar at 723, 773, 823 K, 600 bar at 673 K, 225 bar at 623 K). A good fit was obtained in all cases. Reasonably good fits were obtained at somewhat lower pressures varying from 215 bar at 623 K to 685 bar at 823 K. The resulting values of w are summarized in Table 5 and are shown on Figs. 3 and 4. Above 573 K the uncertainty in w is about 0.05 at best and often 0.1 or more. Even so $\ln \gamma_1$ is determined to about 0.03 and sometimes much better throughout the range of solubility. The uncertainty in $\ln \gamma_{\pm,x}$ is, likewise, about 0.03 or less for the saturated solution but becomes larger for more dilute solutions. Since the uncertainties in individual points for w are as large as the difference between curves for adjacent pressures, the points are omitted on Figs. 3 and 4. Figure 5 shows typical fits of experimental values of $\ln \gamma_1$.

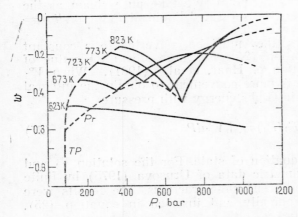

Fig. 4. The parameter P_r (dashed) and w (solid unlabeled curves)

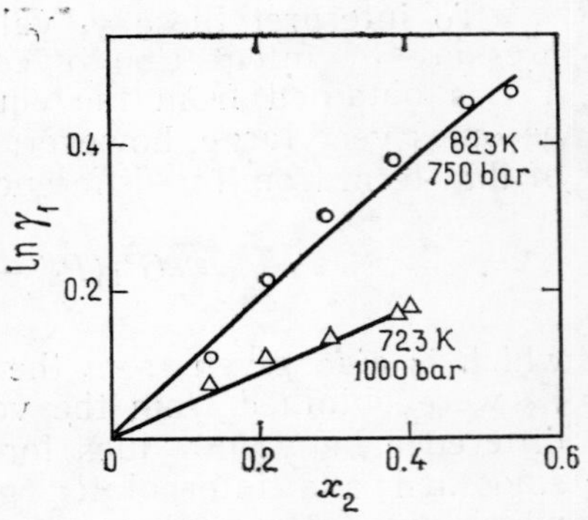

Fig. 5. Comparison of calculated curves for the activity coefficient of water with measured values

At 623 and 673 K the values of w were estimated for higher pressures (above 225 and 600 bar, respectively). Some guidance was obtained from the volumetric data of Hilbert (1979) and Gehrig, et al. (1983), but their measurements do not extend near saturation composition. Thus the choice of these w values was guided primarily by the well-determined values at lower and higher temperatures. From Fig. 3 it is apparent that the w-values follow a reasonable pattern.

In much of the range of very high temperature, this representation of the properties of $NaCl-H_2O$ is valid only at x_2 values somewhat above that of the critical point at a given temperature. Consequently, one may question the appropriateness of the Debye-Hückel term. It was found that alternate equations without a Debye-Hückel term required additional empirical parameters to provide an equally good representation of the

data. Thus one can justify the retention of the Debye-Hückel term on a purely empirical basis. Also at pressures very substantially above the critical pressure the ion-pairing effect will become small and the present type of equation will then be valid down to zero x_2.

As the pressure is further reduced, however, the Debye-Hückel term based on the real dielectric constant of steam does become unsatisfactory. The concentrated solution is liquid-like even at the lower pressure and a "liquid-like" Debye-Hückel term is needed. While other methods were tested, our choice of procedure is to adopt at each temperature a reference pressure P_r and to use the solvent properties for that P_r for the Debye-Hückel term at all lower pressures. Correspondingly, the activity coefficients of water are modified to refer to a hypothetical "liquid" H_2O retaining the volume for P_r at the lower pressures. This change in γ_1 is given by

$$\ln \gamma_1{}^* - \ln \gamma_1 = \int_P^{P_r} [V^\circ{}_1(P_r) - V^\circ{}_1(P)] dP/RT \tag{29}$$

where $\gamma_1{}^*$ is the modified activity coefficient. Since the hypothetical liquid water has higher activity than real steam at the same pressure, the activity coefficient is lowered by the correction. The modified $\gamma_1{}^*$ values were fitted satisfactorily by equation (18) while the $\gamma_{\pm, x}$ values were fitted by equation (20). Curves for w as a function of P for various temperatures are shown in Fig. 4 as the solid curves above and to the left of the dotted curve which shows P_r. Table 6 gives values of P_r and of d_1 at P_r.

Table 6. Values of P_r at various temperatures and of d_1 at various pressures in bars

T, K	P_r, bar	p_r	200	400	600	800	1,000
					d_1		
623	215	0.6096	0.6008	0.6721	0.7109	0.7393	0.7622
673	353	0.4785	0.1005	0.5237	0.6125	0.6593	0.6926
723	490	0.3918	0.07873	0.2709	0.4799	0.5637	0.6138
773	627	0.3585	0.06771	0.1780	0.3384	0.4570	0.5282
823	686	0.3021	0.06043	0.1434	0.2529	0.3619	0.4444

While it is certainly possible to develop an empirical equation for w, it would have to be quite complicated to represent the complex behavior. Thus it seems best to use graphical methods or simple interpolation formulas for w. Such methods are also adequate for P_r. In conclusion of this section, previous equations are rearranged and combined to give explicitly the Gibbs energy of the solution

$$G = n_1 G^\circ{}_1 + n_2 [G^\circ{}_2(c) + \Delta_s G^\circ] + RT(n_1 \ln x_1 + 2n_2 \ln x_2)$$
$$+ (n_1 + 2n_2) RT \left[w x_1 x_2 - \left(\frac{4A_x I_x}{\rho}\right) \ln \left(\frac{1 + \rho I_x}{1 + \rho 2^{-1/2}}\right) \right]. \tag{30}$$

Enthalpy and Entropy from the PL Equation. Since the Gibbs energy is known over a range of temperature, the enthalpy and entropy can be obtained from the appropriate temperature derivatives. However, because w is known only as a table of values at a series of temperatures and for other reasons, it is more convenient to use finite differences than analytical derivatives for the term in excess Gibbs energy. Both specific enthalpy and specific entropy were calculated from (G/T) and G, respectively, with 50° differences in T for 1000 bar and for 500 bar.

For the calculations above 723 K, 500 bar is less than P_r as defined above. Here one must calculate the Gibbs energy G_1^* of the hypothetical water at 500 bar but with the density corresponding to P_r.

$$G^*_1(P, T) = G°_1(P_r) - (P_r - P) V°_1(P_r). \tag{31}$$

Then equation (30) is modified by the substitution of G_1^* for $G_1°$ and the use of A_x at P_r, T instead of A_x at P, T. Then H_1 and S_1 are calculated by finite differences.

The resulting specific enthalpies and entropies are given in Tables 7 through 10 and shown in Figs. 6 through 9. The absolute reference states are the same as those for the PPB equation as stated above. The results are quite smooth through 648 K but then show some small but unreasonable oscillation at higher temperature and especially at 500 bar. This is to be expected in view of the uncertainties in w and other parameters at very high temperature. The values of H and S above 648 K were smoothed and in some cases rounded. Since the two-phase region extends below 798 K for most of the composition range, Tables 8 and 10 are terminated at 748 K.

The values in Tables 7 and 9 can be compared with those in Tables 2 and 3 for 548 K; the agreement is excellent. Indeed, the H and S values

Table 7. Specific enthalpy of NaCl(aq) at 1000 bar in $J \cdot g^{-1}$

			T/K					
m	x_2	$wt \%$	548	598	648	698	748	798
0.1	0.0036	0.58	—775	—556	—328	—85	174	445
0.25	0.0089	1.44	—769	—552	—328	—89	167	430
0.5	0.0177	2.84	—760	—545	—326	—93	158	410
0.75	0.0263	4.20	—750	—538	—324	—95	150	395
1.0	0.0348	5.52	—742	—530	—320	—95	140	380
2.0	0.0672	10.47	—701	—497	—302	—95	130	350
3.0	0.0976	14.92	—661	—464	—282	—85	130	330
4.0	0.126	18.95	—621	—430	—259	—76	130	320
5.0	0.153	22.62	—580	—400	—240	—60	130	320
6.0	0.178	25.96	—540	—370	—210	—40	150	320
10.0	0.265	36.9	—400	—250	—120	30	190	330
15.0	0.351	46.7	—260	—130	— 1	100	250	380
20.0	0.419	53.9	—	— 15	70	170	300	410
30.0	0.519	63.7	—	—	200	280	390	480
40.0	0.590	70.0	—	—	—	350	450	540

from the PL equation can be compared with those from the PPB equation over the temperature range 448-573 K and the full ranges of pressure and molality and the agreement is very good throughout.

A comparison was also made of enthalpies of dilution calculated from our equations with those measured at approximately 400 bar and 623 and 673 K by Busey, et al. (1984). At 673 K the PL equation yields values within 5% of the measurements which is excellent agreement under these circumstances. At 623 K the calculated values are too small by about 15%. The behavior of w is particularly uncertain near this temperature. Also this is near the respective upper and lower temperature limits of validity of the two equations for dielectric constant. Hence, this agreement is probably as good as could be expected. At each temperature the composition dependence of the heat of dilution is given quite accurately by the PL equation.

Most of the curves on Figs. 6-9 are nearly straight lines with small and slowly changing curvature. The marked exceptions are the enthalpy and entropy curves for pure H_2O at 500 bar. These show S-shapes with steep regions near 700 K which represent the vestige of the vaporization process. Consistent with this picture, the curves for the more dilute solutions intersect the two-phase boundary about 750 K. At 1000 bar this residual of the vaporization effect has diminished and moved near 800 K. It is barely noticeable for pure water and no longer apparent for the solutions. The two-phase region is near 900 K at 1000 bar.

The PL equation assumes an ionized salt and is fitted above the critical temperature of water only to data for solutions more concentrated than the critical composition. It is known that NaCl in steam is largely ion-paired and that this ion-pairing extends to the critical region.

Table 8. Specific enthalpy of NaCl(aq) at 500 bar in $J \cdot g^{-1}$

					T/K		
m	x_2	$wt \%$	548	598	648	698	748
0.1	0.0036	0.58	—789	—552	—286	(45)	(409)[a]
0.25	0.0089	1.44	—784	—550	—291	(23)	(396)
0.5	0.0177	2.84	—776	—546	—295	(—6)	(377)
0.75	0.0263	4.20	—768	—542	—298	(—27)	(362)
1.0	0.0348	5.52	—759	—536	—300	—44	349
2.0	0.0672	10.47	—721	—511	—296	—86	311
3.0	0.0976	14.92	—683	—483	—285	—103	287
4.0	0.126	18.95	—645	—454	—269	—107	270
5.0	0.153	22.62	—607	—425	—251	—103	260
6.0	0.178	25.96	—570	—395	—231	—93	250
10.0	0.265	36.9	—433	—282	—147	—33	250
15.0	0.351	46.7	—286	—159	—48	70	270
20.0	0.419	53.9	—	— 56	38	130	310
30.0	0.519	63.7	—	—	175	260	370
40.0	0.590	70.0	—	—	—	350	440

[a] See text for the uncertainty of values at low molality and high temperature.

m	x_2	wt %	548	598	648	698	748	798
					T/K			
0.1	0.0036	0.58	6.32	6.70	7.07	7.43	7.78	8.14
0.25	0.0089	1.44	6.28	6.66	7.02	7.38	7.73	8.07
0.5	0.0177	2.84	6.22	6.59	6.95	7.30	7.64	7.97
0.75	0.0263	4.20	6.16	6.53	6.87	7.22	7.55	7.87
1.0	0.0348	5.52	6.10	6.47	6.80	7.14	7.47	7.78
2.0	0.0672	10.47	5.88	6.24	6.55	6.86	7.1_5	7.4_5
3.0	0.0976	14.92	5.69	6.03	6.31	6.62	6.9_0	7.1_7
4.0	0.126	18.95	5.51	5.85	6.11	6.4_0	6.6_5	6.9_0
5.0	0.153	22.62	5.36	5.68	5.9_5	6.2_0	6.4_5	6.7_0
6.0	0.178	25.96	5.22	5.53	5.7_5	6.0_0	6.2_5	6.5_0
10.0	0.265	36.9	4.77	5.04	5.2_5	5.4_5	5.7_0	5.9
15.0	0.351	46.7	4.4	4.6	4.8_0	5.0_0	5.2_0	5.3_5
20.0	0.419	53.9	—	4.3_5	4.5_0	4.6_5	4.80	4.9_5
30.0	0.519	63.7	—	—	4.0_0	4.1_5	4.3_0	4.4_5
40.0	0.590	70.0	—	—	—	3.9	4.0	4.1

At 1000 bar and below 823 K the ion pairing is probably not extensive enough to seriously affect the enthalpy values in Table 7 or the entropies in Table 9. The effect will be much greater at 500 bar and can be evaluated from very recent measurements for 3.2% (0.546 mol·kg⁻¹) solution by Bischoff and Rosenbauer (1984). They measured the density over a range of temperature and pressure and calculated the difference from 1000 bar to lower pressures for both H and S. In Table 11 our results for this difference are compared with their values. There is essentially perfect agreement at 548, 598, and 648 K. There is a definite difference at 698 K and a larger difference at 743 K which is the maximum temperature of the experimental measurements. The differences are in the expected direction with the apparent molal enthalpy or entropy of ions smaller than that for ion pairs at a given T and P. The dominant cause is the stronger hydration of ions as compared to ion-pairs.

In Tables 8 and 10 the values at low m and high T are enclosed in parentheses to indicate uncertainty from this ion pairing effect. From Table 11 one can estimate the magnitude of the error at 500 bar from the differences in a given function at a particular temperature. The error at 1000 bar should be much smaller and is probably negligible. At molality lower than 0.5 the error per mole of NaCl may be somewhat larger but the net effect on the specific enthalpy or entropy should be no larger than that for the 0.5 mol·kg⁻¹ solution.

Discussion of the PL Equation: From Fig. 3 it is apparent that w is large and negative at low temperature. This represents a dominance of ion hydration over the other interparticle interactions. But with increase in temperature w appears to approach a small and relatively constant value. Then the departure from ideal behavior arises primarily from the Debye-Hückel term which arises from the long-range aspect of interionic

Table 10. Specific entropy of NaCl(aq) at 500 bar in $J \cdot K^{-1} g^{-1}$

					T/K		
m	x_2	$wt \%$	548	598	648	698	748
0.1	0.0036	0.58	6.40	6.82	(7.24)	(7.73)	(8.24)[a]
0.25	0.0089	1.44	6.36	6.77	(7.19)	(7.65)	(8.17)
0.5	0.0177	2.84	6.30	6.70	7.10	(7.53)	(8.06)
0.75	0.0263	4.20	6.24	6.63	7.02	7.42	(7.96)
1.0	0.0348	5.52	6.18	6.57	6.94	7.32	(7.8)
2.0	0.0672	10.47	5.95	6.32	6.66	7.0	7.5
3.0	0.0976	14.92	5.75	6.10	6.41	6.7	7.2
4.0	0.126	18.95	5.56	5.90	6.19	6.5	6.9
5.0	0.153	22.62	5.40	5.72	6.00	6.3	6.7
6.0	0.178	25.96	5.26	5.57	5.82	6.0	6.5
10.0	0.265	36.9	4.80	5.07	5.28	5.5	5.8
15.0	0.351	46.7	4.41	4.64	4.81	5.0	5.3
20.0	0.419	53.9	—	4.34	4.48	4.7	4.9
30.0	0.519	63.7	—	—	4.05	4.2	4.4
40.0	0.590	70.0	—	—	—	3.9	4.1

[a] See text for the uncertainty of values at low molality and high temperature.

Table 11. Change in enthalpy and entropy with pressure from 500 to 1000 bar for 0.546 molal NaCl

T/K	548	598	648	698	743	Source
$\Delta H / J \cdot g^{-1}$	18	0	−31	−120	−273	B.R.[a]
$\Delta H / J \cdot g^{-1}$	16	1	−30	−85	−205	P.[b]
$\Delta S / J \cdot K^{-1} g^{-1}$	−0.08	−0.10	−0.17	−0.29	−0.50	B.R.[a]
$\Delta S / J \cdot K^{-1} g^{-1}$	−0.08	−0.11	−0.15	−0.23	−0.38	P.[b]

[a] Bischoff and Rosenbauer, 1984.
[b] This research.

forces. It appears possible that w may remain small at still higher temperature, but that will remain a speculation until further evidence becomes available.

The equations and parameters described above provide a full representation of various thermodynamic properties of NaCl-H$_2$O for the range 373-823 K, 0-1 kbar, and composition to saturation except for dilute solutions above 573 K where ion-pairing is substantial. With one exception, all of the equations for G_x^{EX}, ln γ_1 and ln $\gamma_{\pm, x}$ are so simple that there seems to be no need for tabulation of numerical values of these quantities. A simple programmable calculator is adequate for the calculations. The exception is the complex equation of Haar et al. (1981) for the density of pure H$_2$O. For the convenience of users, Table 6 gives densities above 573 K for several pressures, including P_r. The equations

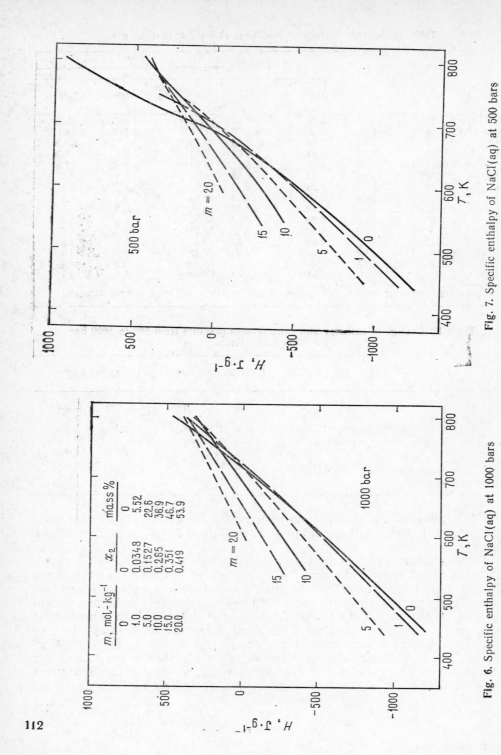

Fig. 7. Specific enthalpy of NaCl(aq) at 500 bars

Fig. 6. Specific enthalpy of NaCl(aq) at 1000 bars

m, mol·kg^{-1}	x_2	mass %
0	0	0
1.0	0.0348	5.52
5.0	0.1527	22.6
10.0	0.265	36.9
15.0	0.351	46.7
20.0	0.419	53.9

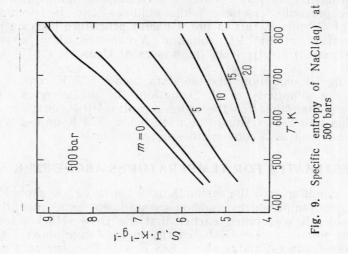

Fig. 9. Specific entropy of NaCl(aq) at 500 bars

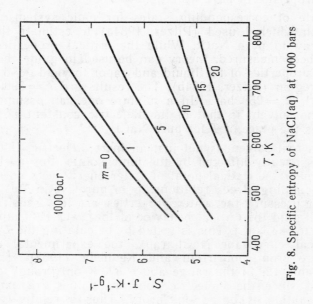

Fig. 8. Specific entropy of NaCl(aq) at 1000 bars

of this paper, together with those of Bradley and Pitzer (1979) or Pitzer (1983a) for the dielectric constant, then yield the Gibbs energy and the activity coefficients directly. Tables of enthalpies and entropies were given in the preceding section. While volumes can be obtained from differences of G with P, this is not a useful procedure because of limitations in precision of the PL equation. Volumetric data above 573 K should be obtained directly from the papers of Urusova (1975) or Gehrig et al. (1983).

While the PL equation fits the data below 573 K remarkably well in view of the simplicity of the equation, the precision of fit is substantially less than that for the more complex PPB equation. Thus the PPB equation and tables should be used below 573 K unless the simpler equation is adopted as a convenient approximation.

ESTIMATES FOR TEMPERATURES ABOVE 823 K

The PL equation and the methods used below 823 K should be applicable at higher temperatures as soon as sufficient experimental data become available. It was noted earlier that the Debye-Hückel term is the primary term for the excess Gibbs energy in the range above 700 K. Thus one can make crude estimates above 823 K with w either zero or a small negative value extrapolated from Fig. 3. The Debye-Hückel parameter is available from properties of pure water.

The postulate of corresponding states for ionic systems provides another approach which I used (Pitzer, 1984a) to estimate the critical curve all the way to pure NaCl. While the critical properties of pure NaCl have not been measured, theory can be used to guide the extrapolation of known properties of the liquid and vapor to yield good estimates for the critical region (Pitzer, 1984b). The results are $T_c = 3900$ K, $V_c = 490$ cm$^3 \cdot$mol^{-1}, $P_c = 258$ bar. In a mixture one can assume that the only effect of the solvent is that of its dielectric constant D. Then the product DT takes the place of T for pure NaCl.

Of course, a real solvent is not just an inert dielectric. If the nature of the solvent is greatly affected by the ionic solute, this postulate has no value. Thus, near the critical point of water one expects no agreement with this corresponding states relationship to pure NaCl. But at higher temperature and pressure the water properties are less sensitive to the presence of NaCl, and the critical behavior of the NaCl-H$_2$O mixture may relate primarily to the NaCl. This is tested by calculating the DT product along the critical curve for NaCl (aq); the experimental P_c values above 823 K are from Souririjan and Kennedy (1962). The results are shown in Table 12. In the range above 823 K only rough agreement can be expected since the dielectric constant values are extrapolated, although the equation is based on theory. Thus a relatively constant discrepancy of about 10% offers support for this very approximate model. At lower temperature, as expected, there is an increasingly rapid change in pattern as the critical point of water is approached.

Table 12. Test of a corresponding states model for the critical curve for NaCl(aq)

T_c/K	P_c/bar	DT/K exp'l	DT/K model
973	1237	4360	3900
923	1082	4290	3900
873	922	4220	3900
823	754	4070	3900
723	422	3470	3900

The critical curve for NaCl (aq) above 973 K can now be estimated on the basis that at any temperature the critical dielectric constant $D_c = 3900/T$. The density of steam is then calculated to fit this D_c, and, in turn, the steam pressure may be calculated to fit this density.

Except near the critical point of pure NaCl, the pressure contribution from NaCl can be neglected and the partial pressure of H_2O taken as the total pressure. Figure 10 is constructed on that basis except that a smooth curve is extended from 3500 K to the critical pressure of pure NaCl at 3900 K. It should be emphasized that the pressures above 1000 K on Fig. 10 are uncertain by at least 20% even if the corresponding states model is generally valid in this range.

A second check on this model is the critical volume which should be the same per mole of NaCl for the mixture as for pure NaCl. The experimental data for critical volumes extend only to 823 K where Urusova (1975) reports a density equivalent to about 480 cm³·mol⁻¹ in remarkable agreement with the 490 value for pure NaCl. Given a constant critical volume of about 490 cm³·mol⁻¹, one can calculate from the density of H_2O the critical composition. The results are shown in Fig. 11. This

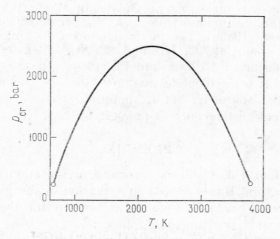

Fig. 10. Estimated critical pressure of NaCl(aq)

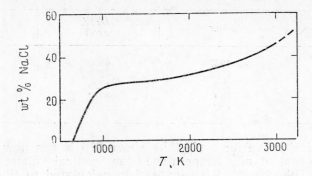

Fig. 11. Estimated critical composition of NaCl(aq)

curve is terminated near 3000 K in view of the increasing uncertainties, although it must extend to 100% at 3900 K.

The relatively flat portion of the critical composition curve near 30% by weight implies that there are about seven water molecules of hydration per NaCl in this range. This corresponds to a substantial inner shell around either separated ions or an ion pair. The rapid rise in critical pressure allows this relatively constant hydration in the 1000 to 2000 K range. At higher temperatures the degree of hydration decreases as expected.

Finally, one notes that there is need for a better treatment of dilute NaCl in steam. The solubility of solid NaCl in steam has been measured by a number of investigators with less than satisfactory agreement. The ionic dissociation was measured by Quist and Marshall (1968) and this was related to the information from mass spectrometry for hydrated Na^+ and Cl^- by Pitzer (1983b). The thermodynamics of dilute aqueous NaCl at the critical pressure at 723, 773, and 823 K was treated, including the ionization equilibrium, by Pitzer and Li (1984). These initial investigations fall far short of a complete treatment such as is now available for more concentrated solutions, but it should be possible to build such a complete treatment on the concepts now established.

APPENDIX

PPB Equation: There follows a concise presentation of the equation of Pitzer, Peiper, and Busey (1984) in extension of the general but brief description above. The basic equation is written for the Gibbs energy which is equation (1) rearranged as follows:

$$G = n_1 G°_1 + n_2 \overline{G}°_2 - 2n_2RT(1 - \ln m) + G^{EX} \tag{A-1}$$

with m the molality and G^{EX} the excess Gibbs energy for which the molality dependence is given by equation (2). The forms assumed for the functions in equation (2) are

$$f(I) = -A_\varphi (4I/b) \ln(1+bI^{1/2}) \tag{A-2}$$

$$B(I) = \beta^{(0)} + 2\beta^{(1)}[1-(1+\alpha I^{1/2})\exp(-\alpha I^{1/2})]/\alpha^2 I. \tag{A-3}$$

The two parameters b and α are assigned the constant values 1.2 and 2.0 $kg^{1/2} \cdot mol^{-1/2}$, respectively, independent of temperature or pressure. Indeed, $f(I)$ is a universal function. While $\beta^{(0)}$ and $\beta^{(1)}$ are specific to the solute, NaCl in this case, the ionic strength function multiplying $\beta^{(1)}$ is unchanged for most electrolytes. The Debye-Hückel parameter is

$$A_\varphi = (1/3)(2\pi N_A d_1/1000)^{1/2}(e^2/4\pi\varepsilon_0 DkT)^{3/2} \tag{A-4}$$

where the various quantities were defined in connection with equation (21).

The osmotic and activity coefficients are obtained from equations (8) and (9b) which yield

$$\varphi - 1 = -A_\varphi I^{1/2}/(1+bI^{1/2}) + m[\beta^{(0)}+\beta^{(1)}\exp(-\alpha I^{1/2})] + m^2 C^\varphi \tag{A-5}$$

$$\ln \gamma_\pm = -A_\varphi \left\{ \frac{I^{1/2}}{1+bI^{1/2}} + \frac{2}{b}\ln(1+bI^{1/2}) \right\} + m \cdot \left\{ 2\beta^{(0)} + \right.$$

$$\left. \frac{2\beta^{(1)}}{\alpha^2 I}\left[1-\left(1+\alpha I^{1/2}-\frac{\alpha^2 I}{2}\right)e^{-\alpha I^{1/2}}\right] + \frac{3m}{2}C^\varphi \right\} \tag{A-6}$$

The form for the ionic strength dependence of the second virial coefficient was first chosen for the osmotic coefficient, where it is very simple (Pitzer, 1973). The forms for the excess Gibbs energy and the activity coefficient are derived therefrom. C^φ is $2C$ for a $1:1$ salt.

A practical reference state for NaCl was chosen to be the mixture $NaCl \cdot 10H_2O$ or a molality $m_r = 5.5508$ $mol \cdot kg^{-1}$. This avoids a direct expression the extreme behavior of the infinitely dilute reference state as the critical point of water is approached. The Gibbs energy of NaCl in the infinitely dilute is obtained instead from the indirect expression

$$\frac{\bar{G}_2^\circ(T,P)-\bar{H}_2^\circ(298\ K, 1\ bar)}{RT} = (z_1 + z_2 P + z_3 P^2 + z_4 P^3)/T$$

$$+ z_5 + z_6 P + z_7 P^2 + z_8 P^3 + z_9 \ln T + (z_{10}+z_{11}P+z_{12}P^2)T$$

$$+ (z_{13}+z_{14}P)T^2 + z_{15}/T(T-227) + z_{16}/T(680-T)^3$$

$$- \frac{10[G_1^\circ(T, P-H_1^\circ(g, 0\ K)]}{RT} - \frac{G^{EX}(T, P, m_r)}{n_2 RT}. \tag{A-7}$$

Note that the reference state for NaCl is the enthalpy in the infinitely dilute standard state at 298.15 K while that for H_2O is the ideal gas at 0 K. Note also that the final term is the excess Gibbs energy of the solu-

117

tion at $m_r = 5.5508$ mol·kg^{-1}. Although the excess Gibbs energy is defined above, it is rewritten in detail as follows for any molality of NaCl

$$\frac{G^{EX}(T, P, m)}{n_2 RT} = -(4A_\varphi/b)\ln(1 + bm^{1/2}) + 2m\beta^{(0)}$$

$$+ (4\beta^{(1)}/\alpha^2)[1 - (1 + \alpha m^{1/2})\exp(-\alpha m^{1/2})] + m^2 C^\varphi. \tag{A-8}$$

The virial coefficient parameters are functions of temperature and pressure as follows:

$$\beta^{(0)} = z_{17}/T + z_{18} + z_{19}P + z_{20}P^2 + z_{21}P^3 + z_{22}\ln T$$

$$+ (z_{23} + z_{24}P + z_{25}P^2 + z_{26}P^3)T + (z_{27} + z_{28}P + z_{29}P^2)T^2$$

$$+ (z_{30} + z_{31}P + z_{32}P^2 + z_{33}P^3)/(T - 227)$$

$$+ (z_{34} + z_{35}P + z_{36}P^2 + z_{37}P^3)/(680 - T), \tag{A-9}$$

$$\beta^{(1)} = z_{38}/T + z_{39} + z_{40}T + z_{41}/(T - 227), \tag{A-10}$$

$$C^\varphi = z_{42}/T + z_{43} + z_{44}P + z_{45}\ln T + (z_{46} + z_{47}P)T$$

$$+ (z_{48} + z_{49}P)T^2 + (z_{50} + z_{51}P)/(T - 227)$$

$$+ (z_{52} + z_{53}P)/(680 - T). \tag{A-11}$$

In addition to a general fit for the range 273-573 K a limited fit was generated for the temperature range 273-358 K which requires fewer parameters but yields higher precision at low temperature. The two sets of parameters are given in Table A-1.

It should be noted that parameter z_5 involves the entropy of aqueous NaCl which was taken to be $13.88_6 R$ at 298.15 K based on the absolute entropy of solid NaCl. The equation for H_2O also yields the entropy on an absolute basis.

One may now calculate volumes, enthalpies, entropies, heat capacities or other quantities by appropriate derivatives with respect to pressure or temperature. Equations for the osmotic and activity coefficients were given above. These equations, with the parameters in Table A-1 and with the properties of pure water, give a very general and accurate expression of the properties of aqueous NaCl from 273-573 K, 1-1000 bar, and 0-6 mol·kg^{-1}. With high accuracy at saturation pressure but reduced accuracy at higher pressure, the molality range may be extended to saturation. Pitzer, et al., (1984) also note particular increases of uncertainty as the temperature approaches 573 K at certain pressures and compositions. They give extensive tables of the various quantities likely to be of interest as listed in Table 1 above.

Table A-1. Parameters z_i for equations A-6 to A-10

i	low T value	high T value
1	−71659.531	−71637.203
2	2.3483335	2.2209012
3	−8.3668484 × 10⁻⁵	−7.7991396 × 10⁻⁵
4	2.4018168 × 10⁻⁹	−4.8099272 × 10⁻⁹
5	624.88208	624.68125
6	−5.3697119 × 10⁻⁴	6.0159787 × 10⁻⁴
7	3.5126966 × 10⁻⁷	3.4069074 × 10⁻⁷
8	0	2.1962044 × 10⁻¹¹
9	−110.74702	−110.74702
10	0.038900801	0.039494473
11	2.6973456 × 10⁻⁶	−6.5313475 × 10⁻⁷
12	−6.2746876 × 10⁻¹⁰	−6.4781894 × 10⁻¹⁰
13	−1.5267612 × 10⁻⁵	−1.5842012 × 10⁻⁵
14	0	3.2452006 × 10⁻⁹
15	516.99706	516.99706
16	−5.9960301 × 10⁺⁶	−5.9960301 × 10⁺⁶
17	−656.81518	−656.81518
18	24.879183	24.869130
19	−2.1552731 × 10⁻⁵	5.3812753 × 10⁻⁵
20	5.0166855 × 10⁻⁸	−5.5887470 × 10⁻⁸
21	0	6.5893263 × 10⁻¹²
22	−4.4640952	−4.4640952
23	0.011087099	0.011109914
24	−6.4479761 × 10⁻⁸	−2.6573399 × 10⁻⁷
25	−2.3234032 × 10⁻¹⁰	1.7460070 × 10⁻¹⁰
26	0	1.0462619 × 10⁻¹⁴
27	−5.2194871 × 10⁻⁶	−5.3070129 × 10⁻⁶
28	2.4445210 × 10⁻¹⁰	8.6340233 × 10⁻¹⁰
29	2.8527066 × 10⁻¹³	−4.1785962 × 10⁻¹³
30	−1.5696231	−1.5793660
31	2.2337864 × 10⁻³	2.2022821 × 10⁻³
32	−6.3933891 × 10⁻⁷	−1.3105503 × 10⁻⁷
33	4.5270573 × 10⁻¹¹	−6.3813683 × 10⁻¹¹
34	5.4151933	9.7065780
35	0	−2.6860396 × 10⁻²
36	0	1.5344744 × 10⁻⁵
37	0	−3.2153983 × 10⁻⁹
38	119.31966	119.31966
39	−0.48309327	−0.48309327
40	1.4068095 × 10⁻³	1.4068095 × 10⁻³
41	−4.2345814	−4.2345814
42	−6.1084589	−6.1084589
43	0.40743803	0.40217793
44	−6.8152430 × 10⁻⁶	2.2902837 × 10⁻⁵
45	−0.075354649	−0.75354649
46	1.2609014 × 10⁻⁴	1.5317673 × 10⁻⁴
47	6.2480692 × 10⁻⁸	−9.0550901 × 10⁻⁸
48	1.8994373 × 10⁻⁸	−1.5386008 × 10⁻⁸
49	−1.0731284 × 10⁻¹⁰	8.6926600 × 10⁻¹¹
50	0.32136572	0.35310414
51	−2.5382945 × 10⁻⁴	−4.3314252 × 10⁻⁴
52	0	−0.091871455
53	0	5.1904777 × 10⁻⁴

ACKNOWLEDGEMENT

This work was supported by the Director, Office of Energy Research, Office of Basic Energy Sciences, Division of Engineering, Mathematics, and Geosciences of the U. S. Department of Energy under Contract No. DE-AC03-76SF00098.

REFERENCES

1. American Inst. Phys. Handbook (1972): McGraw-Hill Pub. Co., New York, 3rd Ed., Sect. 4, p. 139.
2. Bischoff, J. L. and Rosenbauer, R. J. (1984): to be published.
3. Bradley, D. J. and Pitzer, K. S. (1979): *J. Phys. Chem.* 83, 1599—1603.
4. Busey, R. H., Holmes, H. F., and Mesmer, R. E. (1984): *J. Chem. Thermodynamics,* 16, 343—372.
5. Gehrig, M., Lentz, H., and Franck, E. U. (1983): *Ber. Bunsenges. Phys. Chem.* 87, 597—600.
6. Haas, L., Gallagher, J. S., and Kell, G. S. (1981): in Proceedings of the Eighth Symposium on Thermophysical Properties, ed. Sengers, J. V., (Am. Soc. Mech. Engr., New York), vol. 2, pp. 298—302.
7. Hilbert, R. (1979): Dissertation (University of Karlsruhe, FRG).
8. JANAF Thermochemical Tables (1971): 2nd ed., Stull, D. R. and Prophet, H., Ed., NSRDS—NBS 37, U. S. Gov. Printing Office, Washington.
9. Keevil, N. B. (1942): *J. Am. Chem. Soc.* 64, 841—850.
10. Kestin, J., Sengers, J. V., Kamgar-Parsi, B., and Levelt Sengers, J.M.H. (1984): *J. Phys. Chem. Ref. Data,* 13, 175—183.
11. Khaibullin, I. Kh. (1980): Tables of Thermodynamic Properties of Gases and Liquids, No. 6: Aqueous and Vapor-Phase Solutions. Water-Sodium Chloride System. (Izdatel'stvo Standartov, Moscow, U.S.S.R.).
12. Khaibullin, Kh., and Borisov, N. M. (1966): *Teplofizika Vysokikh Temperatur,* 4, 518 (English translation 489—494).
13. Kienitz, H. and Marsh, K. N. (1981): *Pure and Appl. Chem.* 53, 1847—1858.
14. Kirshenbaum, A. D., Cahill, J. A., McGonigal, P. J. and Grosse, A. V. (1962): *J. Inorg. Nucl. Chem.* 24, 1287—1296.
15. Liu, C. and Lindsay, W. T., Jr., (1972): *J. Solution Chem.* 1, 45—69.
16. Parisod, C. J. and Plattner, E. (1981): *J. Chem. Eng. Data,* 26, 16—20.
17. Pitzer, K. S. (1973): *J. Phys. Chem.* 77, 268—277.
18. Pitzer, K. S. (1981): *Ber. Bunsenges. Phys. Chem.* 85, 952—959.
19. Pitzer, K. S. (1983a): *Proc. Nat. Acad. Sci. U.S.A.,* 80, 4575—4576.
20. Pitzer, K. S. (1983b): *J. Phys. Chem.* 87, 1120—1125.
21. Pitzer, K. S. (1984a): *J. Phys. Chem.* 88, 2689—2697.
22. Pitzer, K. S. (1984b): *Chem. Phys. Letters,* 105, 484—488.
23. Pitzer, K. S. and Li, Y.—g. (1983): *Proc. Nat. Acad. Sci. U.S.A.,* 80, 7689—7693.
24. Pitzer, K. S. and Li, Y.—g. (1984): *Proc. Nat. Acad. Sci. U.S.A.,* 81, 1268-1271.
25. Pitzer, K. S. and Kim, J. J. (1974): *J. Am. Chem. Soc.* 96, 5701—5707.
26. Pitzer, K. S., Peiper, J. C., and Busey, R. H. (1984): *J. Phys. Chem. Ref. Data,* 13, 1—102.
27. Quist, A. S. and Marshall, W. L. (1968): *J. Phys. Chem.* 72, 684-703.
28. Rogers, P. S. Z. and Pitzer, K. S. (1982): *J. Phys. Chem. Ref. Data,* 11, 15-81.
29. Sourirajan, S. and Kennedy, G. C. (1962): *Am. J. Sci.* 260, 115-141.
30. Uematsu, M. and Franck, E. U. (1980): *J. Phys. Chem. Ref. Data,* 9, 1291-1306.
31. Urusova, M. A. (1974a): *Russ. J. Inorg. Chem.* (English translation) 19, 450-454.
32. Urusova, M. A. (1974b): *Geochem. Int'l.* 9, 944-950.
33. Urusova, M. A. (1975): *Russ. J. Inorg. Chem.* (English translation) 20, 1716-1721.
34. Urusova, M. A. and Ravich, M. I. (1971): *Russ. J. Inorg. Chem.* (English translation), 16, 1534-1535.
35. White, D. E. and Wood, R. H. (1982): *J. Solution Chem.* 11, 223-236.
36. Wood, S. A., Crerar, D. A., Brantley, S. L., and Borcsik, M. (1984): *Am. J. Sci.* 284, 668-705.

120

The Solubility of Magnetite. A Critical Evaluation of the Compiled Data

G. BOHNSACK

Bayer AG, D 5090 Leverkusen 1, FRG

1. IMPORTANCE OF THE SOLUBILITY OF MAGNETITE IN GEOLOGY AND PARTICULARLY IN POWER STATION CHEMISTRY

The solubility of magnetite is of interest to geologists in connection with the formation of iron ore deposits in the form of natural magnetite. This concentration of iron at variance with the mean distribution of iron in the Earth's crust presupposes transportation processes and a considerable magnetite solubility. Accordingly, natural systems which could lead to corresponding solubility have been sought; and the dependence of the solubility of magnetite, mostly under acidic conditions, on the influence of dissolved substances in natural waters, and on temperature has been investigated.

In power station chemistry, as in geochemistry, hydrothermal conditions for the formation of minerals, for which the same transport and accumulation phenomena apply, are known. The main difference is that in modern power stations very pure water is used.

Magnetite is the thermodynamically stable final product of the reaction between iron and water. The iron/water system and the various solubilities in it are of central importance in the water chemistry of power stations. Magnetite, which forms spontaneously on carbon steel in contact with hot water at temperatures exceeding 200°C, provides a protective layer and is thus responsible for the chemical resistance of the steel to water and steam. It is only able to do so because its solubility

is still lower than that of the primary corrosion product, iron (II) hydroxide. If locally, through concentration of acid or alkali, present in the boiler water, the magnetite solubility is increased, destruction of the protective layer, severe corrosion, and serious damage must be expected. At low temperatures, at which the spontaneous formation of magnetite is blocked, water conditioning is needed to reduce the rate at which iron is dissolved.

The minimum of magnetite solubility represents the lowest solubility of iron in the oxygen-free iron/water system. It cannot be lowered unless oxygen is permitted to be present, in which case it falls to the still lower level of iron (III) oxide.

The solubility of magnetite is decisively important in closed hot water systems. This may be seen in the circulation system of a drum boiler and the primary circuit of a nuclear power station with pressurized water reactor. In such systems, which are in solubility equilibrium, the familiar transport phenomena, together with the formation of deposits, are controlled by possible supersaturation, whether in consequence of steam formation on the heated wall or in consequence of temperature changes. In nuclear power stations they are closely related to radioactive contamination by the ^{60}Co isotope. All that the power station chemist can do in such a case is to adjust the pH value to the level at which the solubility of magnetite is lowest in the temperature range concerned.

Owing to the importance, outlined above, of the solubility of corrosion products in power stations the IAPS's Working Group IV, which is concerned with the thermodynamic properties of solutions, has extended its programme to include the solubility of metal oxides and hydroxides as corrosion products, and has begun work on magnetite and its solubility.

During the last five years the published data have been collected. At the Working Group's annual meetings these data, together with all the problems connected with the solubility of magnetite, have been discussed at length. To give a visual impression of the various measured values and indicate how little agreement exists, I have attempted to plot the values and their scatter simultaneously.

Figures 1, 2, and 3 show the dependence of the solubility on the pH value, with temperature as the parameter, Figure 3 being the combination of Figures 1 and 2. Figure 4 shows the change in solubility with temperature and pH value.

It is evident from the figures that the range of variation of the various measured values is exceptionally wide. Where there is so little agreement — a lack of agreement which is both widespread and typical of solubility data for metal oxides of low solubility — a critical evaluation of the collected data is needed in order to show the state of knowledge at the present time.

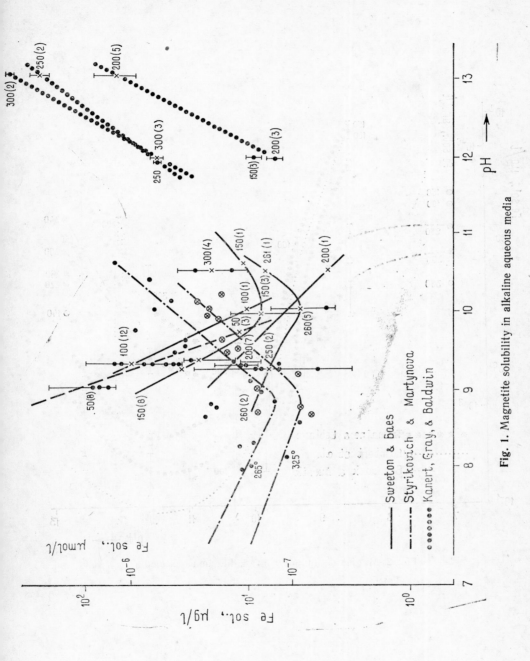

Fig. 1. Magnetite solubility in alkaline aqueous media

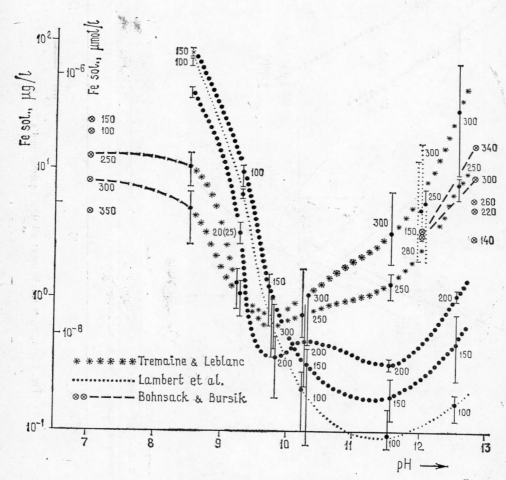

Fig. 2. Magnetite solubility in alkaline aqueous media

124

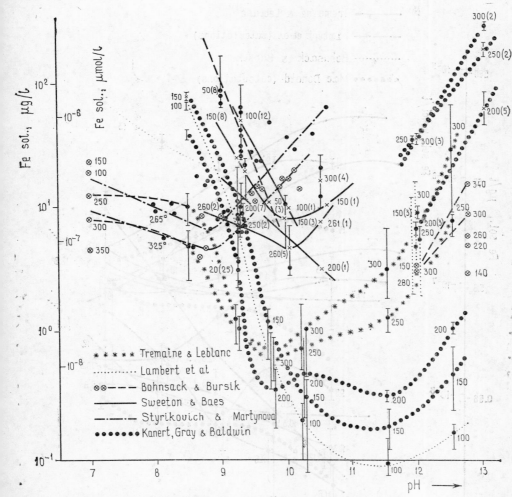

Fig. 3. Magnetite solubility in alkaline aqueous media

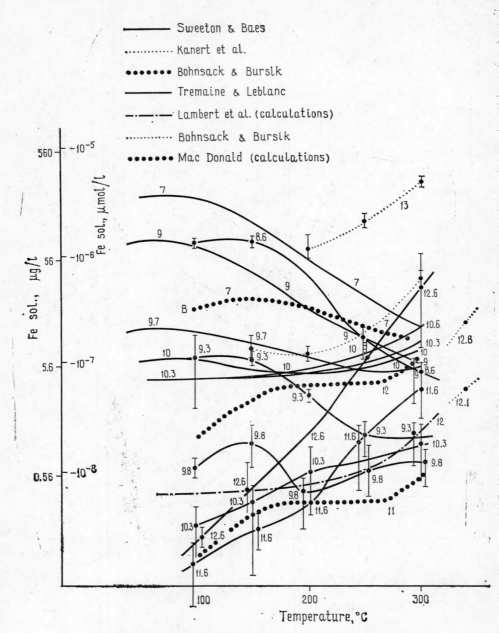

Fig. 4. Magnetite solubility in alkaline aqueous media

2. CRITICAL EVALUATION OF THE PUBLISHED DATA ON THE SOLUBILITY OF MAGNETITE

Not only experimentally determined magnetite solubility data, but also thermodynamically calculated magnetite solubilities, are given in a number of publications.

The measured and thermodynamically calculated solubility data must be considered separately.

2.1. CRITICAL EVALUATION OF THE MEASURED VALUES

The measured results must be evaluated according to all the parameters that have been shown to influence these results in the various systems. These parameters necessarily determine the structure of this chapter.

2.1.1. Surface Properties and Crystallinity of the Magnetite

In the dissolution of a solid, that is to say in the reaction which takes place at the interface, the surface properties of the solid are important. In general terms it may be said that the solubility of a solid depends on the activity of the solid phase. This, in turn, is determined by the following factors:
— The lattice energy
— The free energy of the surface
— The formation of a solid solution
— The degree of hydration of the surface

The lattice energy is specific to the substance. The free energy of the surface increases with the division of the solid or the increase in its specific surface area. It is also increased by lattice faults, such as dislocations and other heterogeneities of the surface. The solubility of very small and rapidly formed crystallites may therefore be considerably greater under some circumstances. As compared with larger and well formed crystals, they are metastable and in the state of recrystallization. The solubilities decrease with time, and under these conditions the lowest measured solubilities correspond best to a stable state of equilibrium. In accordance with the typical structure of the protective layer, the hot water at power stations is in contact with the hydrothermally formed epitactic upper layer, which consists of relatively large magnetite crystals.

Magnetite, which is a mixed condensate of the hydroxides of bivalent and trivalent iron, can also be considered to be a homogeneous solid solution of FeO in Fe_2O_3. In the formation of a solid solution the activities of the pure solid phases are changed. The dissolution takes place in accordance with the quotient of the solubility products, which is defined as the distribution coefficient D, provided the activities of the solids can be set at 1. If the solubilities of the components differ greatly, the less soluble phase necessarily accumulates at the surface. This alters

the activity of the solid solution considerably. If in this way or in consequence of hydrolytic precipitation during the dissolution, for example, a new solid phase is formed, it is also customary to speak of incongruent solubility.

As the solubility of iron(III) oxide hydrate is below the limit of analytical detectability, magnetite may be said to be in solubility equilibrium with iron(II) hydroxide. As the solubility of magnetite is very low, except in strong bases, it is appropriate to speak of a solubility equilibrium when a large surface (a sufficient quantity of magnetite) is in equilibrium with a limited amount of water. It is necessary to be wary if a limited amount of magnetite is treated with large amounts of solvent flowing at a high rate.

The aforementioned change in the surface of the dissolving magnetite does not occur, of course, if the solubility equilibrium is reached because magnetite is precipitated from an oversaturated iron(II) hydroxide solution. This process is described by the Schikorr reaction, in which, however, there is a redox step in which water acts as an oxidizing agent and releases hydrogen. The irreversibility of this redox step, which occurs spontaneously only at temperatures exceeding 200°C, is evident from the blocking which is seen at low temperatures.

The degree of hydration on the surface may influence the kinetics of the dissolution reaction.

In experiments aimed at determining the solubility of natural magnetite samples in acidic aqueous solutions Holser and Schneer [1] observed substantial changes, indicated by a change of colour from black to red, in the surfaces of the samples. They discuss the possibility of the formation of a passive layer, consisting of an oxide of trivalent iron of uncertain composition, at the surface of the magnetite.

Sweeton and Baes [2] considered the composition of the magnetite, which in their case was iron powder oxidized with steam, and the possibility that this may change during the experiments. They thought it possible that the magnetite may be oxidized to $H Fe_5O_8$ or γFe_2O_3. They determined the chemical composition with reference to iron, oxygen, and impurities, and also the lattice constant before and after the experiment, but were unable to detect a change. These investigation methods are too insensitive to detect changes in surface composition.

Tremaine and Leblanc [3] investigated the suface of their magnetite by ESCA. For this purpose they chose two trial runs in the alkaline range at pH values of 11.5 and 12.5 and at 573 K. Their findings did not agree with the theoretical expectation. The surface of the magnetite was covered by about four layers of iron(II) hydroxide over the whole length of the column at pH 12.5, but only in the upper part of the feed section at pH 11.5. In the alkaline range the solubility increases with increasing temperature and the gradient of this change becomes steeper as the alkali concentration rises. Therefore the hydroxide must have been precipitated from the oversaturated solution while the column was cooling. A corresponding investigation in the acidic or neutral range would undoubtedly be more informative because there would be less

likelihood of the solutions becoming oversaturated as they cooled; such an investigation has not yet been carried out.

Experimental findings relating to the influence of the crystallite size on the measured solubility of magnetite or to the time-related alteration of this solubility in consequence of crystal growth have been obtained by Irma Lambert [4] and co-workers, especially for the alkaline range. Their findings are confirmed by measurements made by the author of this article in the neutral range [5] and by those of Linnenbom [6].

The problems posed by the influence of the state of crystallization are encountered in connection with the solubility of solids in general. For example, they are particularly apparent in the difference between the solubilities of amorphous silicic acid and of quartz.

To some extent there is also a connection between the state of crystallization of the solid used for the solubility determination and the formation of suspended material. Where the crystallites are very small, fairly intensive dissolving-out of the soluble iron(II) component of the magnetite lattice may be expected to cause instability and facilitate the breakdown of the lattice. As the next parameter we must therefore define the nature of the solution.

2.1.2. The Aqueous Magnetite Solution

Metal oxides are in solubility equilibrium with aqueous solutions of their hydroxides. If the formation of the oxide from a hydroxide is regarded as a polycondensation, the dissolution reaction of oxides in water may be described as a hydrolysis. In the hydrolysis of a polymeric substance there is a continuous transition via oligomers to the monomers. Relatively large particles that can be separated with membrane filters are determined as suspended matter. Their presence causes wide variation of the measured values, especially at low solubilities at the detection limits of the analytical methods. Spindler (7) has termed this part of the solubility range of iron(III) oxide hydrates the range of colloidal solubility for which exact measurements cannot be given. The differential magnetite dissolution makes the relationships more complex, and, as explained above, it may intensify the formation of suspended matter. This applies particularly to the relatively alkaline range, in which the oxide bridges of the trivalent iron may also be attacked hydrolytically, with a resultant increase in the tendency of the lattice to break down.

All researchers by whom the solubility of magnetite has been measured have attributed the wide scatter of the iron concentration observed by them to the presence of suspended matter and have considered the influence of this factor in detail. As shown by Styrikovich and Martynova [8], by Sweeton and Baes, and, for example, by Helz [9], attempts to remove the suspended matter with suitable filters in the

form of metal frits or membranes reduced, but did not eliminate, the deviations.

The assumption of Sweeton and Baes [2] that a filter residue represented originally dissolved and later deposited material, and should therefore be considered as belonging to the dissolved iron, has led to excessively high solubility data. Helz showed that the suspended iron existed as an oxide and not as a hydroxide.

Tremaine and Leblanc [3] observed, when measurements were taken in a flowing medium, that stationary, lower, and considerably less widely varying readings were obtained as the throughput, and therefore the time, increased. Their way of arriving at reliable concentration measurements was more successful. It must therefore be said that the best and most reliable measured values at present available for a wide pH range are those of these authors. Their values are confirmed by those obtained in the alkaline range by Lambert [4] and co-workers, who considered the lowest measurements to be least falsified by suspended matter and used them to fix thermodynamic data.

Our own measurements [10] in the neutral and weakly alkaline ranges, obtained by differential analysis, likewise agree well.

As practical experience has shown, suspended corrosion products must be expected in power plant installations, particularly in loops. In systems of this kind, in which metallic iron controls the redox conditions, via the pores of the protective layer, solubility equilibrium results from magnetite deposition from oversaturated solutions. Suspended particles, acting as an inner surface, as it were, therefore accelerate the reaching of equilibrium at which they themselves grow.

But this differential analysis, by which an idea of the amount and nature of the suspended matter can be gained also, additionally compels one to seek an exact definition of the solution that is in equilibrium with magnetite. Owing to the already mentioned continuous transition from the polymer to the monomer, for which an almost arbitrary number of possible oligomers can be given, only the monomeric species Fe^{2+}, $FeOH^+$, $Fe(OH)_2$, $Fe(OH)^-_3$, and FeO^{2-}_2 can be clearly separated. The undissociated hydroxide $Fe(OH)_2$ then holds the key position in the precipitation process. It must have the lowest concentration, and an increase in its concentration means oversaturation. The corresponding species of trivalent iron are of theoretical importance only because, having poor solubility, except in the highly acidic and very highly alkaline ranges, they cannot be analytically detected.

Compliance with the principle that only monomers should be analysed is readily permitted by the colorimetric methods because the coloured complex with iron (II) as the central atom is formed only with monomers in the solution. But the exploitation of this elegant method presupposes an analytical process of such sensitivity that direct measurement is possible without the solution having to be concentrated. At the solubility minimum at pH values of 10 to 11 the sensitivity is not quite adequate.

2.1.3. Influence of the Analysis Method on the Measured Values

Reliable measured values are only possible if the analytical method is sufficiently sensitive. In general, colorimetry of the phenanthroline complex in the form in which it has been used to determine total iron has been employed. At a detection limit of 20 µg/l for iron, Holser and Schneer [1] were unable to determine the solubility of magnetite in pure water or in weakly alkaline solution. The same applies to Helz [8], who used atom absorption and gave only a detection limit of 1 mg/l. Sweeton and Baes [2], and also Tremaine and Leblanc [3] enriched the dissolved iron with the aid of cation exchangers. From the use of ion exchangers in water treatment it is known that their surfaces separate suspended iron oxide hydrates exceptionally well. Particularly if the sum of bivalent and trivalent iron is determined, the influence of the suspended matter is not eliminated. This explains the high and widely varying values of Sweeton and Baes. In Tremaine's investigation the reduction in the amount of suspended matter at high throughputs led to lower values with a narrower range of variation.

2.1.4. Influence of the Redox Conditions

The importance of the redox conditions in the determination of the solubility of magnetite has been discussed frequently and a definition of these conditions and a way of controlling them have been demanded. That is understandable, for the dissolved product, iron(II) hydroxide, is generally known to be sensitive to oxidation.

According to a widely used practice, the oxygen in the used water expelled with hydrogen, after which the water was saturated with hydrogen against the hydrogen partial pressure of 1 bar. This hydrogen is intended to control the redox conditions. If the Schikorr reaction is used as the dissolving reaction, then, theoretically, it influences the solubility of magnetite in accordance with $p^{1/3}$.

This influence of the hydrogen on the solubility has so far not been detected in iron concentration measurements.

In the Schikorr reaction water acts as an oxidizing agent for the dissolved iron. The reaction equilibrium therefore develops, at least at the surface of the magnetite, when iron(III) hydroxide is absent from the solution.

It is important, though, to exclude dissolved oxygen from the system, since otherwise the existence of the bivalent iron in the solution cannot be guaranteed. If, in accordance with power station operating practice, metallic iron takes control over the redox conditions, traces of oxygen can be tolerated, as we have been able to demonstrate.

The dissolution of the magnetite in water should be described as a purely hydrolytic reaction without a redox step. This is of considerable importance in connection with the solubility equilibrium, since reversibility must be demanded. As typically illustrated by the Schikorr reaction,

reversibility is not generally given for redox reactions or redox steps of composite reactions.

The proposal of Helz [9] that iron(III) oxide be used to control the oxidation-reduction — as a redox buffer, so to speak — may be appropriate for geologically relevant systems at very high temperatures at which the decomposition-oxygen pressures play a part. In our case, that of pure water and very dilute solutions at up to 350°C, this would alter the stoichiometric ratio between the bivalent and trivalent iron of the magnetite over the whole solid surface.

2.2. EVALUATION OF THE THERMODYNAMICALLY CALCULATED SOLUBILITIES

Evaluation of the calculated values is basically very simple, for the thermodynamically calculated solubilities are as good, but no better, than the thermodynamic data from which they are derived. Where magnetite is concerned, particularly the heat of formation of the dissolved ion species and their protolysis constants are not exactly known and, where they have been determined, they are unreliable. The solubility itself, its dependence on the pH value, and its dependence on temperature have been stated approximately correctly. Although the calculated values are presented in different ways, the set-up and calculation process have been fundamentally the same. It is not surprising that attempts have been made to derive better thermodynamic data from the measured solubilities with a view to securing close agreement between measured and calculated values.

2.3. PROSPECTS

The solubilities of magnetite obtained by Tremaine and Leblanc [3] as functions of pH value and temperature, which in the alkaline and neutral ranges agree well with the measured values of other authors, must be regarded as the best and most reliable measured values available. The mean values could be presented in skeleton tables. But it would be simpler to state, in accordance with the chemical thermodynamics, the equation for calculating the solubility equilibrium and the temperature-dependence of the heat of formation values. It has not yet been decided what form the chemical equation describing this equilibrium should take. A decision could be made easier by an investigation of the change in the surface properties of the dissolving magnetite as evidence of the differential dissolution.

REFERENCES

1. Holser, W. T. and Schneer, C. J.: Hydrothermal Magnetite, *Geological Soc. of America Bull.*, v. 72 (1961) 3, pp. 369-386.
2. Sweeton, F. H. and Baes Jr., C. F.: The Solubility of Magnetite and Hydrolysis of Ferrous Ion in Aqueous Solutions at Elevated Temperatures, *J. Chem. Thermodyn.* (1970) 2, pp. 479-500.
3. Tremaine, P. R. and Leblanc, J. C.: The Solubility of Magnetite and the Hydrolysis and Oxidation of Fe^{2+} in Water to 300°C, *J. of Solution Chem.* 9 (1980) 6, pp. 415-442

4. Lambert, I.; Beslu, P.; Lalet, A. and Montel, J.: Thermodynamique de Solubilisation de la Magnetite en Milieu Basique, Symposium Thermodynamics of Nuclear Materials (1979); Proceedings: Intern. Atomic Energ. Agency, Vienna, Paper IAEA-SM-236/12.
5. Bohnsack, G.: Das Verhalten von Eisen(II)—hydroxid bei höheren Temperaturen, *VGB Mitt.* **51** (1971) 4, pp. 328-338.
6. Linnenbom, V. J.: The Reaction between Iron and Water in the Absence of Oxygen, *J. Electrochem. Soc.* **96** (1958) 6, pp. 322-324.
7. Schindler, P.; Michaelis, W. and Feitknecht, W.: Löslichkeitsprodukte von Metalloxiden und —hydroxiden, 8. Mitteilung: Die Löslichkeit gealterter Eisen(III)-hydroxid-Fällungen, *Helvet. Chim. Acta* **46** II (1963) 46, pp. 444-449.
8. Styrikovich, M. A.; Martynova, O. J.; Kobjakov, J. F.; Men'shikova, V. L. and Reznikov, M. J.: Solubility of Magnetite in Boiling Water of High Temperature, *Teploenergetica* **18** (1971) 7, pp. 82-84.
9. Bohnsack, G. and Bursik, A.: Messungen zur Magnetitlöslichkeit in reinem Wasser unpublished.
10. Helz, G. R.: Hydrothermal Solubility of Magnetite Diss. The Pennsylvania State University, PH. D. 1971 Mineralogy, Microfilm No. 9475 University Microfilm AXEROX Comp., Ann Arbor/Michigan (1972).

Phase Equilibria and Properties of Aqueous Solutions at High Temperatures and Pressures

V. M. VALYASHKO

*N. S. Kurnakov Institute of General and Inorganic Chemistry,
USSR Academy of Sciences, Moscow, USSR*

The data on physico-chemical properties of aqueous systems at high temperatures and pressures are necessary for the development of a number of trends in chemistry, geochemistry, power industry, hydrothermal crystal growth and materials synthesis, hydrometallurgy, and electrochemistry.

These wide and differing requirements for information concerning hydrothermal systems assume, together with obtaining of new experimental data, special importance in establishing general regularities of behaviour of various properties, creating qualitative and quantitative models of high-temperature equilibria, allowing the number of complicated high-temperature measurements to be reduced, and, in some cases, evaluating the properties of interest.

HYDROTHERMAL SYSTEMS AND THE STATE OF THEIR STUDY

A comparison of the properties of water and aqueous solutions, measured at different temperatures by various investigators [1-5], shows that beginning from 200-250°C the properties of aqueous systems differ so much from those at low temperatures that one can say about some specificity of hydrothermal systems, connected first of all with the thermal change of molecular structure of water.

At present, special equipment and techniques have been developed to study aqueous systems at high parameters, which make it possible

Table 1

	Li	Na	K	Rb	Cs	NH₄	Tl	Pb	Mg	Ca	Sr	Ba	Cu	Ag	Zn	Cd	Hg	Mn	Co	Ni	UO₂
F	448	370	632 / 200	225		237	250			421		395		250							376
Cl	556 / 700 / 1000	700 / 700 / 800	650 / 700 / 1000	200 / 280	200 / 800 / 1000	417 / 600	381	407	350 / 2 0 / 600	350 / 388 / 600	412 / 200	524 / 200 / 750	360	359	250	481	275	430	720 / 240 / 500	240	
Br	350 / 350 / 600	678 / 350 / 800	421 / 350 / 300			462 / 800	421	307			383	415	330	349	210	419	237				
I	350 / 350	600 / 350 / 800	602 / 350	380	346			360	248				340	365		385	255				
O/OH	473	550 / 415 / 360	460 / 400 / 700			700		200	200	500 / 340	475	360 / 750	200	260	200				200	200	200
CNS	350 / 300	350 / 300	336 / 340 / 800	200						350		417 / 310		209 / 222 / 222							200
NO₃	370	540 / 315	450 / 340	200						350		225									
CO₃		275	450		292																
SO₄	422 / 315	500 / 367 / 300	500 / 500 / 800		292	410	380		350 / 300 / 350	450	600	280			300	200		200	205	350	365 / 375
PO₄	300	500				200				300											
MoO₄		567								300											
WO₄		550								500											
BₓOy	300	350 / 300 / 280	260				262														

Explanation to Table

556	↑	Phase equilibria
700	↑	Density
1000	↑	Electroconductivity

to determine thermodynamic, transport, spectral, relaxation and diffraction properties of hydrothermal solutions. However, the volume of available information on the properties of hydrothermal systems is not sufficient both as to the number of systems and the quality of data.

Table I lists temperatures (°C), up to which the phase equilibria (the upper figure in the square of the Table), density (middle figure) and electrical conductivity (lower figure) of the water-salt systems indicated thereat have been studied. The dash or empty square means that numerical data above 200°C are not available. The table contains not all the properties and systems studied at high temperatures and pressures; it gives only a general idea of the state of study of hydrothermal equilibria. It is obvious that maximum information has been obtained on phase equilibria, primarily on solubility and vapour pressure. There is considerably less information on density and electrical conductivity of high-temperature solutions. Still less experimental data are available on other enumerated properties of hydrothermal solutions. Usually these are isolated works concerning the study of alkali metals' salts solutions.

COMPLETE PHASE DIAGRAMS OF BINARY AND TERNARY SYSTEMS

In order to generalize the available data on phase equilibria over a wide range of parameters of state it is convenient to construct phase diagrams. A set of heterogeneous equilibria and sequence of their realization in many systems are similar and may be described by the same topological scheme of phase diagrams (by a qualitative phase diagram plotted in dimensionless coordinates).

When we exclude from consideration such phase transformations as polymorphism, formation of solid solutions and compounds and assume that immiscibility of liquids at high pressures is always bounded by a critical curve (l_1-l_2), then all the variety of qualitative phase diagrams of binary aqueous-salt systems may be depicted in all by 6 topological schemes (main types) of *complete phase diagrams*.

By complete phase diagrams we mean those describing all possible in the system equilibria with the participation of gaseous (g), liquid (l), and/or solid (s) phases over the entire range of parameters at which the noncrystalline phases remain heterogeneous.

The structure of topological schemes of complete phase diagrams is uniform for all the systems consisting of components with different volatility; among those systems, besides water-salt ones are organic, water-organic, and gases systems. Consideration of equilibrium data for all these systems increases the number of main types up to 9. At the same time it is obvious that these 9 types are characteristic only of the present-day state of our knowledge of these systems, but do not exhaust all the conceivable variants of complete phase diagrams.

Recently, we have proposed a method of theoretical derivation of topological schemes of the phase diagrams, based on the principle of their continuous topological transformation [6].

To explain the idea of this principle, we consider the results of analytical investigation of an equation of state. The continuous variation of equation's parameters (including the molecular parameters of components) results in a number of continuously changing types of phase diagrams (see, for example, [7]). Continuous transformation of one type into another proceeds via the appearance of the boundary version of phase diagram possessing simultaneously the properties of both types. The boundary variants of diagrams occurring on solving the equation of state or upon deriving phase diagrams with the help of statistical physics [7, 8] cannot be realized since they, in violation of the Gibbs Rule, contain equilibria which are possible only in more complicated (ternary) systems.

Nevertheless, this approach, considered all diagrams as the stages of continuous process of topological transformation, points towards the possibility of deriving new phase diagrams, if the regularities of topological transformation of individual elements of diagrams are known. To state these regularities, we have analyzed the sequence of phase equilibria transformation in numerous systems studied experimentally and theoretically [6, 9].

The result of these studies is a classification of complete phase diagrams of binary systems (Fig. 1) including 8 already known types (1a, 2a, 1b, 1b′, 1c, 1c′, 2c″, 2c′) and 4 new types (1b″, 2b′, 2b″, 1c″) of complete phase diagrams. The diagrams are arranged in the sequence of their topological transformation, this being demonstrated by the boundary versions situated between them (in the frames).

This classification is complete within the framework of the above mentioned and the new limitations: only one region of immiscibility can exist in a system; all geometrical elements of phase diagrams, their phase reaction and shape (but not the combination of these elements) are confirmed by experimental investigations of real systems.

The method of topological transformation makes it possible by cancelling the restrictions to theoretically derive all conceivable variants of the binary systems' diagrams [9].

The classification given in Fig. 1 may be used to construct (within the same restrictions) complete phase diagrams of ternary systems, if the phase diagrams of the boundary binary systems are known [6]. On passing from one boundary system to another through the three-component region of composition, the phase diagrams of the boundary systems should have the continuous topological transformations. All steps of these transformations are shown in Fig. 1, including the nonvariant points of ternary systems, described by the boundary versions of binary phase diagrams.

Thus, wide opportunities appear for theoretical derivation of any phase diagrams of two- and three-component systems, and this, in turn, enables one to plan the experiment in a certain direction and choose the variant of phase diagram, corresponding to the system in question, using a limited number of experiments.

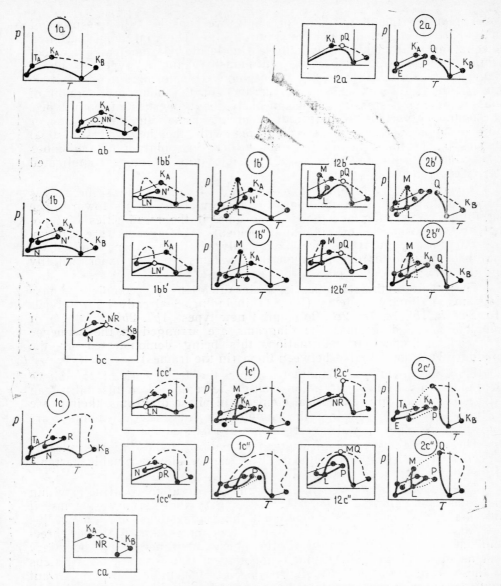

Fig. 1. Classification of main types of complete phase diagrams (*P-T* projection) of binary systems, consisting of volatile (*A*) and nonvolatile (*B*) components, characterized by complete immiscibility in a solid state and possibility of appearance of limited region of liquid phase immiscibility. The diagrams in frames are the boundary versions of phase diagrams, complicated by nonvariant equilibria of highest orders (empty points pQ, MQ, pR and others). Filled points are the nonvariant equilibria of one-(points T_A, T_B, K_A, K_B), and two-component (points *p*, *Q*, *R* and others) systems. Thin lines are the monovariant curves of one-component systems. Thick lines are the monovariant curves of binary systems. Dashed lines are the monovariant critical curves of binary systems. Dotted lines are the metastable parts of monovariant curves, bounding the region of immiscibility.

EFFECT OF TEMPERATURE ON CONCENTRATIONS OF HYDROTHERMAL SOLUTIONS UNDER CONDITIONS OF MONOVARIANT EQUILIBRIA

So far we have considered the data on phase equilibria within the frameworks of qualitative topological schemes, while the compositions of equilibrium phases and their changes were beyond the scope of our study. At the same time, an analysis of numerical data on temperature dependence of composition of hydrothermal electrolyte solutions makes it possible to establish a general regularity for all the systems under study [10].

In high-temperature monovariant equilibria of solubility and immiscibility, a negative temperature coefficient of variation in liquid phase composition always becomes positive with increasing amount of electrolyte in the solution. The sign of the temperature coefficient changes in a relatively narrow range of concentration, which is constant under the conditions of different equilibria, slightly changes in the solutions of similar electrolytes, but regularly decreases with the increase of ion charge (Table 2).

Table 2

System	mol. %	Phase equil.	System	mol. %	Phase equil.
Na_2SO_4-NaOH-H_2O-NaCl	10.2	l-g-s	Na_2CO_3-NaOH-H_2O	10.4	l-g-s
Na_2SO_4-NaCl-H_2O	10.4	l-g-s	K_2SO_4-KCl-H_2O	7.9	l-g-s
Li_2SO_4-H_2O	6.5	$l(fl)$-s	$KLiSO_4$-H_2O	5.1	$l(fl)$-s
	6.55	l_1-l_2-s		4.8-5.1	l_1-l_2-s
	6.55	l_1-l_2		4.9-5.1	l_1-l_2
Na_2SO_4-H_2O	7.6	$l(fl)$-s	Na_2WO_4-H_2O	2.5-3.2	l_1-l_2
	7.4-7.6	l_1-l_2-s		2.3	l_1-l_2-g
	7.4-7.6	l_1-l_2	Na_2MoO_4-H_2O	2.1	l_1-l_2-g
Na_2CO_3-H_2O	6.7	$l(fl)$-s		2.2	l_1-l_2
	7.4-8.3	l_1-l_2-s	$BaCl_2$-H_2O	1.2-2.2	l_1-l_2-g
	7.4-8.3	l_1-l_2		1.5-5.4	l_1-l_2-s
K_2SO_4-H_2O	4.6	$l(fl)$-s		4.5-5.5	l_1-l_2
	4.3-4.8	l_1-l_2-s	UO_2SO_4-H_2O	1.5-2	l_1-l_2-g
	4.3-4.5	l_1-l_2		1.5-2	l_1-l_2
			UO_2F_2-H_2O	5.1-6.1	l_1-l_2-g

TEMPERATURE DEPENDENCE OF ELECTRICAL
CONDUCTIVITY AND MOLAR
VOLUMES OF DIFFERENT CONCENTRATION
HYDROTHERMAL SOLUTIONS

As seen from Table 1, the list of hydrothermal systems, for which volumetric and conductometric data are available, is rather short. This list is even shorter for the behaviour of these properties over a wide concentration interval at elevated temperatures.

The main part of high-temperature conductometric measurements is related to the determination of dissociation constants and is therefore limited by the range of low concentrations. Only the data of Ref. [11] make it possible to have an idea about temperature dependence of electrical conductivity of aqueous LiCl, LiBr, and CsCl solutions, concentrations up to 20-30 mol%, over a wide range of parameters. An analysis of these data shows [10] that the maximum of equivalent electrical conductivity, typical of the curves of temperature dependence of dilute solutions, disappears on increasing concentrations of 1:1 electrolytes up to 10-15 mol%. At higher concentrations the conductivity of aqueous solutions increases linearly with temperature, similarly to that of electrolyte melt.

High-temperature volumetric measurements, having a relative error less than $\pm 1\%$, carried out for strong hydrothermal solutions, are known only for the $NaCl-H_2O$, $NaOH-H_2O$, $KOH-H_2O$ and $K_2SO_4-H_2O$ [12] systems.

To develop these studies, we have constructed an installation for *PVT* measurements (with an error of $\mp 0.2-0.5\%$) at temperatures up to 600°C and pressures up to 300 MPa [12]. The volume of the solution (of any concentration), placed into evacuated platinum ampoule, is determined from the measurements of two experiments (with empty and solution-filled ampoule) at equal temperatures; in each experiment the amount of CO_2 filling the autoclave at a given pressure is determined. The difference in the amounts of CO_2 measured in the experiments (at equal T and P) makes it possible to estimate the volume of aqueous solution, using the known *PVT* data for CO_2.

The results of *PVT* measurements of $CaCl_2$ solutions, carried out on the above installation together with V. A. Ketsko, are given in Table 3 in the form of specific volumes (cm^3/g) at 238, 288, 338 and 388°C over pressure interval of 10-60 MPa.

The study of the temperature dependence of volumetric properties over a wide temperature range is hampered because of the nonequivaiency of the state of solutions at constant pressure. Therefore, we considered the effect of temperature on the behaviour of molar volume of constant concentration solutions in equilibrium with vapour. This makes it possible to automatically maintain the uniformity of the state of a system over a wide range of parameters. Such analysis, carried out for NaCl, NaOH, and KOH [10] solutions shows that for these 1:1

Table 3

mol. % \ P, MPa	0	1.4	3.2	5.1	8.3	9.1	13.8	20.6	32.8
511 K									
60	1.162	1.068	0.976	—	—	0.799	0.702	0.640	—
50	1.171	1.076	0.982	—	—	0.804	0.706	0.643	—
40	1.181	1.085	0.990	—	—	0.810	0.711	0.647	—
30	1.192	1.095	0.999	—	—	0.818	0.718	0.653	—
20	1.205	1.195	1.009	—	—	0.824	0.723	0.657	—
10	1.219	1.203	1.020	—	—	0.832	0.729	0.663	—
561 K		(1.45)		(5.2)					
60	1.250	1.144	1.046	0.962	0.856	—	0.742	0.664	0.592
50	1.265	1.157	1.056	0.971	0.864	—	0.749	0.669	0.595
40	1.284	1.172	1.069	0.983	0.874	—	0.757	0.676	0.599
30	1.304	1.188	1.083	0.996	0.885	—	0.765	0.684	0.606
20	1.328	1.205	1.098	1.009	0.897	—	0.776	0.692	0.613
10	1.360	1.227	1.119	1.025	0.902	—	0.789	0.703	0.623
611 K									
60	1.377	1.258	1.132	1.031	0.916	—	0.794	0.705	0.625
50	1.409	1.285	1.154	1.044	0.927	—	0.805	0.715	0.630
40	1.449	1.315	1.178	1.059	0.940	—	0.816	0.726	0.636
30	1.507	1.355	1.207	1.077	0.956	—	0.830	0.738	0.644
20	1.593	1.404	1.244	1.104	0.977	—	0.849	0.751	0.656
661 K									
60	1.589	1.460	1.302	1.164	1.021	—	0.853	—	0.668
50	1.667	1.512	1.339	1.191	1.042	—	0.972	—	0.680
40	—	—	1.388	1.227	1.071	—	0.996	—	0.694
30	—	—	—	1.274	1.112	—	1.029	—	0.710

electrolytes, starting from 10-15 mol%, a linear temperature dependence is observed in molar volumes, this being typical also of the electrolyte melts.

TRANSITION REGION OF CONCENTRATION
OF HYDROTHERMAL SOLUTIONS

An analysis of numerical data on thermodynamic properties and conductivity of hydrothermal solutions reveals specific changes in these properties over some concentration region; we call it the "transition" region because transition from water-like properties of solutions to melt-like ones is observed in it.

Most definite information about molecular nature of the transition region may be obtained by studying intermolecular stretching vibration of water by IR or Raman spectroscopy.

Until recently, the stretching mode of water at high temperatures (up to 350-400°C) were studied only in 1:1 electrolyte solutions [13-16]. Now we have obtained the IR spectra of the O-D stretching fundamental of HDO diluted in aqueous $CaCl_2$ solutions (1.5-9 mol%) at temperatures up to 388°C and pressures up to 160 MPa. The spectra were recorded on a UR-20 grating spectrometer (Carl Zeiss, Jena), using the reflection cell with a sapphire window, similar in design to that described in Ref. [17].

An analysis of our and literature data shows that considerable changes in the spectrum of stretching vibration of water molecules on addition of any electrolyte are observed only up to the concentrations of the transition region. At higher concentrations the state of water molecules practically does not change. In contrast to 1:1 electrolyte solutions where the spectral characteristics were observed to be constant at 10-15 mol% [13-16], in $CaCl_2$ solutions the change in the parameters of the O-D stretching mode of HDO is completed at 3-5 mol% (Fig. 2).

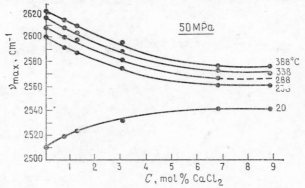

Fig. 2. Concentration dependence of the O-D band maximum in aqueous (HDO/H₂O) $CaCl_2$ solution at 50 MPa for 20, 238, 288, 338 and 388° C

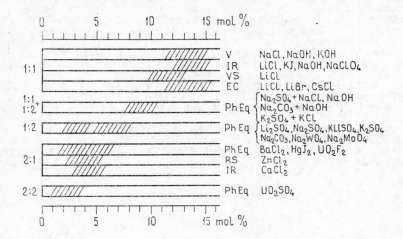

Fig. 3. Concentrations of transition region obtained using different physico-chemical properties of systems: *V* — according to the results of volumetric measurements; *IR* — according to the data of *IR*-spectroscopy; *VS* — according to the data of spectrophotometric studies; *RS* — according to the data of Raman spectroscopy; *PhEq* — according to the data of the phase equilibria studies; *EC* — according to the data of electroconductivity measurements

Thus, the new data confirm the earlier established regularity, i. e. the concentration of the transition region decreases with the increasing charge of electrolyte [10]. These data are in good agreement with the transition concentrations determined from other physico-chemical properties of water-salt systems (Fig. 3).

All the foregoing discussion convincingly confirms the generality of the phenomenon of transition region, revealing that melt-like behaviour of the physico-chemical properties of hydrothermal solutions is related to the destruction of the system of hydrogen bonds between the water molecules prevailing in dilute mixtures, and to the distribution of water among the ions forming the melt-like system of ionic bonds at the concentrations of the transition region.

REFERENCES

1. Martynova O. I. (1964): *Zh. Fiz. khim.* **38**, 1065-1070.
2. Tödheide K. (1966): *Ber. Buns. Ges. Phys. Chem.* **70**, 1022-1030.
3. Valyashko V. M. (1976): In: "Thermodynamics and structure of solutions", No. 4, Ivanovo, 89-106 (in Russian).
4. Puchkov L. V., Fyedorov M. K. et al. (1979): In "Experimental methods of study of hydrothermal equilibria", Nauka, Novosibirsk, 72-81 (in Russian).
5. Gorbaty Yu. E. and Demianets Yu. N. (1982, 1983): *Zh. struct. khim.,* **23**, 73-85; **24**, 66-80.
6. Valyashko V. M. (1981): *Zh. Neorg. khim.* **26**, 3044-3054.

7. Rowlinson J. S. (1980): In "Chemistry and Geochemistry of Solutions at High Temperatures and Pressures", Eds. D. T. Rickard and F. E. Wickman, Perg. Pr., 41-61.
8. Glancy P., Gubbins K. E. (1979): *Farad. Discuss. Chem. Soc.* 66, 116-129.
9. Valyashko V. M. and Kravchuk K. G. (1983): In "Experimental Studies of Endogenous Ore Formation", Nauka, Moscow, 33-50 (in Russian).
10. Valyashko V. M. (1977): *Ber. Buns. Ges. Phys. Chem.* 81, 388-396.
11. Hwang J. U., Lüdemann H. D., and Franck E. U. (1970): *High Temp. High Press.* 20, 651-669.
12. Ketsko V. A. and Valyashko V. M. (1983): In "Experimental Studies of Endogenous Ore Formation", Nauka, Moscow, 50-64.
13. Franck E. U. (1974): In: "Structure of Water and Aqueous Solutions", Ed. W. Luck, 41-61.
14. Franck E. U. (1973): *J. Solut. Chem.* 2, 339-353.
15. Valyashko V. M., Buback M., and Franck E. U. (1980): *Z. Naturforsch.* 35a, 549-555.
16. Valyashko V. M., Buback M., and Franck E. U. (1981): *Z. Naturforsch.* 36a, 1169-1176.
17. Franck E. U. and Roth K. (1967): *Discuss. Farad. Soc.* 43, 108-114.

High Temperature Aqueous Solutions and Energy Considerations—Physical and Chemical Studies

WILLIAM L. MARSHALL

*Chemistry Division, Oak Ridge National Laboratory,
Oak Ridge, Tennessee 37831, USA*

ABSTRACT

Recent studies of high temperature aqueous systems are presented, some of which are directly applicable to the design and operation of steam generators. These studies include (i) the phase behavior of some substances [SiO_2, Na_2SO_4, Na_2HPO_4, K_2HPO_4] that could possibly precipitate from steam when for example the substance is concentrated in metal crevices, (ii) the boiling point elevation of natural solution mixtures, and (iii) the ion-product and (iv) specific electrical conductance of pure water. Determinations and correlations of ionization equilibria are presented for calculations of the concentrations of species present in water and steam. Fundamental approaches for the application of the results allow wide predictabilities. The studies have been used for the design and operation of geothermal power steam generators [to prevent scale formation of amorphous silica solids upon cooling of hydrothermal natural waters] and for direct application to fossil fuel and nuclear power steam generators.

INTRODUCTION

A few years ago, essentially the only properties of water substance thought to be necessary for the efficient operation of a steam generator for power production were the very exact pressure-temperature-volume

145

(*PVT*) properties of water substance from which the thermodynamic functions could be obtained. Over prior years, the continuing *PVT* studies were correlated and standardized through international conferences on *PVT* properties. With the formal organization of the International Association for the Properties of Steam (IAPS) in 1971, it was realized that knowledge was needed of additional properties of water substance and also of water substance containing impurities that were not possible to remove entirely. It was realized also that these impurities, with concentrations even in the "parts per billion" range, possibly could become concentrated because of differences in temperature in particular regions of steam generators, for example, in crevices that might appear from corrosion or by defect. Thus, it appeared that not only the properties of very dilute steam and water solutions should be studied but also those over the entire range of solute concentration, that is, to the concentrations of solutions saturated with solid phases. IAPS expanded its directions of effort to include other properties of water and also those of dilute and concentrated solutions containing solute components appearing both from natural waters and arising from corrosion of steam generator components, for example, dissolved iron oxides.

The present paper reviews some particular studies of high temperature water and steam solutions that have been performed very recently at the Oak Ridge National Laboratory (ORNL) and includes also some recent evaluations of the ion-product and electrical conductance of pure water substance. The emphasis in this paper is on some fundamental studies that are directly applicable to the design and operation of steam generators under several conditions.

PARTICULAR STUDIES

Solubility Behavior of Amorphous Silica and Quartz. Silicon dioxide, consisting of the first and second most abundant elements in the lithosphere, almost invariably is present in all natural waters. Surprisingly, until recently its solubility in aqueous electrolyte solutions had not been studied in much detail at any temperature, although numerous solubility studies in pure water had been made [11]. Because of its abundance in nature and its relevance in waters used for power production, a series of studies has been performed on amorphous silica in several sets of concentrated salt solutions at 25 to 350°C [4, 5, 15, 16, 20-22] and on quartz in concentrated aqueous sodium chloride solutions up to 600°C [6, 7]. These studies have revealed hitherto unsuspected simplicities in the solubility behavior of amorphous silica and quartz in electrolyte solutions. Figure 1 shows plots of the solubility of amorphous silica in several salt solutions [NaCl, NaNO_3, MgCl_2, Na_2SO_4, MgSO_4] at temperatures from 25 to 350°C [4]. Generally, the solubility of amorphous silica decreases with increasing concentration of dissolved salt. However, sodium sulfate at high temperatures produces the opposite effect, which can be explained by the formation of a dissolved silica-sulfate complex [15]. A plot of the *logarithm* of the solubility versus the *molarity* of an

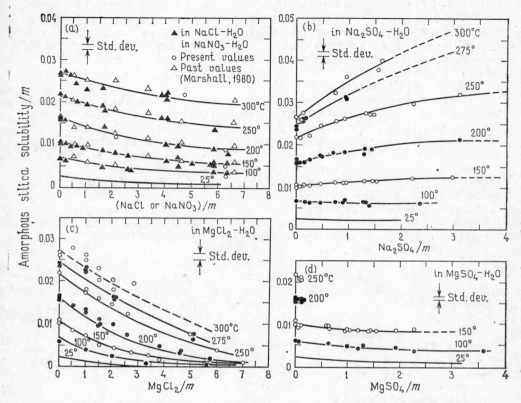

Fig. 1. Solubility of amorphous silica in aqueous salt solutions, 25-300° C

added salt in the solution yields a straight line essentially to the highest attainable concentration (of added salt) [4, 16, 20]. With this observation, a simple additive equation was developed that allows the calculation of the solubility of amorphous silica in natural, concentrated mixed salt solutions at temperatures from 0 to at least 350°C [16].

The solubility of quartz (crystalline SiO_2) has been recently studied in high temperature water and aqueous sodium chloride solutions by Fournier and Potter [7] and Fournier [6]. These studies have been correlated on the basis of the "effective" (free) water present in the solutions. In a related study, the solubilities of amorphous silica in aqueous solutions containing separate salts predominant in natural waters have been correlated as a function of effective water at temperatures up to 350°C [5].

Some of the above studies have been used in the design of steam generators that operate directly from high temperature concentrated geothermal brines, for example, in the Imperial Valley, California, USA

[Union Oil Company of California, USA]. These high temperature brines at 200-300°C generally are in saturation equilibrium with quartz. When they are cooled in a steam generator, silica becomes less soluble and precipitates first as an amorphous silica. Knowledge of the solubility of amorphous silica as a function of temperature and composition of the natural brines allows the proper design of the steam generator to prevent or limit this scale formation by silica.

Phase Behavior of Sodium Sulfate in High Temperature Concentrates. Although water used for steam generation is processed to remove natural inorganic impurities down even to the "parts per billion" range, there may still be present some naturally abundant constituents such as sulfate, chloride, hydroxide, sodium, and hydrogen ions. If by corrosion or defects of the metal surfaces of the steam generator tubes, crevices appear or are initially present, then a concentration of the steam generator fluid within the crevices may occur. This concentration behavior could result because of the gradient in temperature across the steam generator tube walls, thus producing "boiling" concentration, and possible precipitation of solids within the crevices, thereby possibly causing additional corrosion.

For the above reason, a study was made of the solubility behavior of a probable precipitant, sodium sulfate, from solution mixtures of sodium chloride and sulfuric acid at temperatures of 250-400°C [13]. These mixtures comprise the more predominant species that would be expected to be present upon concentration of the (very pure) water in a steam generator.

Synthetic solution mixtures of sodium sulfate, sodium chloride, and sulfuric acid were prepared, and these mixtures were sealed in small fused silica tubes. Each of the several tubes was placed separately in a semi-micro phase study apparatus [25] for visual observation of the solubility point of sodium sulfate in the 250-400°C range. These points are easily obtained because of the rapid approach to solubility equilibrium by sodium sulfate. Since this solid at the high temperatures is less soluble than at 25°C, the tubes were easily filled at 25°C with unsaturated solutions of known composition.

Figure 2 shows the solubility behavior of sodium sulfate at several constant temperatures as a function of the concentration of added sodium chloride. These particular experimental measurements are those of Schroeder, Gabriel, and Partridge [32]. At temperatures below 315°C, the mass action effect incurred by the increasing concentration of sodium ion *decreases* the solubility of Na_2SO_4. At temperatures higher than 315°C, the ionic strength effect as described by Debye-Hückel theory predominates, and the solubility *increases*.

By applying extended Debye-Hückel theory and previously obtained solubility product constants of Na_2SO_4 [23] and ionization constants for HSO_4^- [24] and H_2O [3, 33], the many experimentally obtained solubilities were accurately described as solubility products by simple straight line functions at a given temperature. Examples of these for three temperatures, 250, 300, and 350°C, are shown in Fig. 3 where the slope of each

148

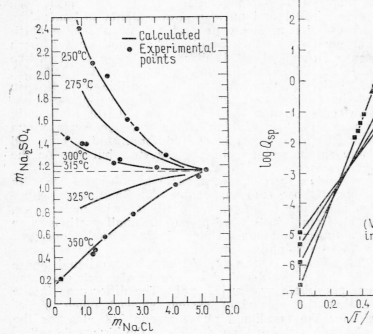

Fig. 2. The solubility of Na$_2$SO$_4$ in aqueous NaCl solutions at 250-350° C

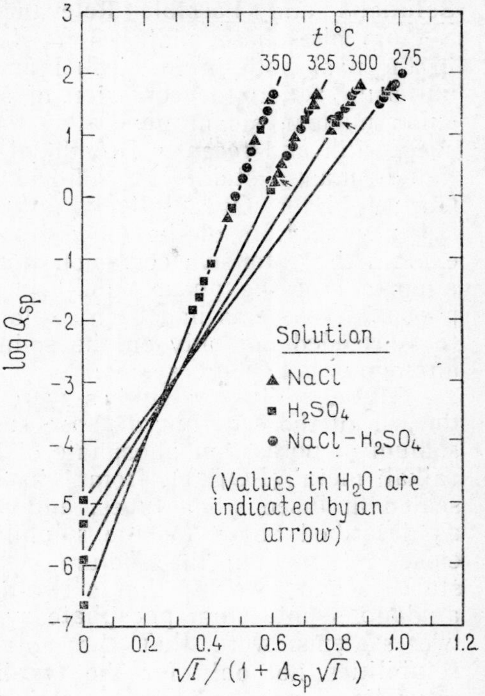

Fig. 3. Plots of the logarithm of the solubility quotient (Q_{sp}) of Na$_2$SO$_4$ in aqueous solutions of H$_2$SO$_4$, NaCl, and H$_2$SO$_4$-NaCl *vs* $\sqrt{I}/(1+A \quad \sqrt{I}$

straight line is the Debye-Hückel theoretical slope obtained from properties of pure water. At each temperature the separate symbols are for solubility products of sodium sulfate in water solutions containing NaCl, H$_2$SO$_4$, or wide ranges of mixtures of NaCl and H$_2$SO$_4$. Extrapolation to zero ionic strength gives the solubility product *constant*. The computer program successfully developed to accurately describe these solubility products can now be used to predict the solubility behavior of sodium sulfate generally in natural water concentrates from an analysis for the mole ratios, Na/SO$_4$ and Na/Cl, and the acidity of the very pure steam generator waters [13]. This knowledge, together with the vapor pressures over these concentrates to be discussed later, can be used to design and to operate steam generators in order to possibly prevent the formation of a solid (Na$_2$SO$_4$) that might enhance corrosion in crevices.

Studies on Two-Liquid Phase Regions in Aqueous Alkali Phosphate Solutions and Possible Relevance to Steam Generator Applications. Sodium phosphate additives to steam generator waters have been used in the past both in fossil fuel and nuclear power plants to control by buffering action the corrosion of steam generator components. In particular nuclear steam generators, this additive has not always been successful, and recently Broadbent, et al., discovered the presence of two liquid phases stable at 265-350°C in the system Na_2HPO_4-H_2O [2]. Garnsey later (speculatively) proposed that the possible formation in steam generators of the concentrated, and highly corrosive, phase might explain the extensive corrosion that sometimes occurred [10]. Since only a portion of the phase equilibria of high temperature aqueous sodium phosphate systems had been explored [2, 28], additional studies appeared to be needed not only on the sodium phosphate system but also on the analogous potassium system.

These additional studies were made at our laboratory [17, 18] with the use of the semi-micro phase study apparatus [25]. Solutions of either sodium or potassium phosphate of known compositions, with mole ratios, (alkali metal/(PO_4^{3-})), from 1 to 3, were prepared. Small portions were sealed in fused silica tubes, and phase changes were visually observed as described above for the solubility of sodium sulfate. These studies closely defined in the sodium system the regions of two-liquid phase stability and revealed that a two-liquid phase region also existed in the analogous potassium phosphate system [17, 18]. The differences in the two systems were that the potassium system showed much higher solubilities not only for the two-liquid phase region but also for the saturation by the solid K_3PO_4 rather than Na_3PO_4 at the high mole ratios, alkali/PO_4. Two liquid phase regions existed for Na_2HPO_4 and NaH_2PO_4 solutions, and mixtures of the two, and also for the corresponding potassium phosphates. The phase behaviors for these two systems are shown in Figs. 4 and 5. Thus the minimum temperature of liquid-liquid immiscibility is 360°C for the potassium containing system compared with about 280°C for the sodium containing system. We had earlier suggested that this much higher temperature of immiscibility for potassium phosphate, as a potential additive to steam generator waters, might have an advantage over the sodium system in eliminating the possibility of occurrence of immiscibility in the general temperature range (250-350°C) for operation of a steam generator [17, 18]. However, others believe that the potassium systems show greater corrosivity than the analogous sodium systems.

Ion Product of Water at High Temperatures. Knowledge of the ion-product of water (K_w), generally defined as the molality of hydrogen ion times the molality of hydroxide ion [although it can also be stated in terms of molarity], is needed for a complete description for any set of hydrolytic reaction equilibria at high temperatures. There have been several past studies of the ion-product of water first beginning with the values obtained by the pioneering work of A. A. Noyes and associates, published in 1907 and 1910 [26, 27]. Much more recently, there are the

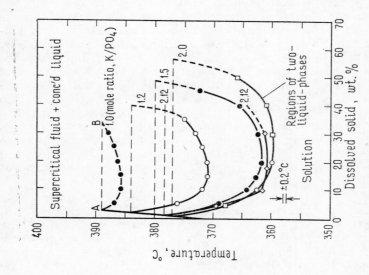

Fig. 5. Two-liquid-phase boundaries for aqueous solution mixtures of potassium phosphate salts for mole ratios, K/PO₄, from 1.00 to 2.12, 360-400°C

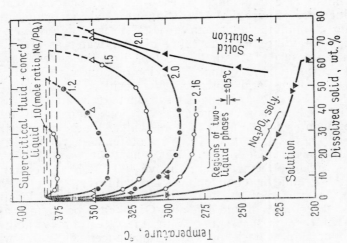

Fig. 4. Two-liquid-phase and solution-solid boundaries for aqueous solution mixtures of sodium phosphate salts for mole ratios, Na/PO₄, from 1.00 to 3.00 at 200 to 300°C

experimental measurements and correlations of Sweeton, Mesmer, and Baes to 300°C along the liquid-vapor curve [33]. Several additional studies in the high temperature region have been referenced and correlated by Sweeton, et al. [33]. The International Association for the Properties of Steam (IAPS) recently published a *Release* on the ion-product of water, and it is conveniently available in a publication by Marshall and Franck [19]. Pitzer believes that the relationship presented does not apply at pressures lower than about 200 bars at 374-800°C, and he has given his conclusions in a recent paper [30]. With the acceptance of the correlations as given by IAPS, but with the question whether to use the Pitzer or the IAPS approach at the lowest pressures, the ion product of water as a function of temperature and (high) pressure appears to be well-defined. Application of this ion-product with conductance measurements described below allow the estimation of the electrical conductance of water substance at temperatures from 0 to 1000°C and at pressures up to 10.000 bars [14].

When the ion product is applied with ionization constants for high temperature aqueous electrolyte equilibria, a more quantitative description of the concentration of species in solution can be made. This type of knowledge may aid in defining the character of corrosion in steam generators, and in possibly reducing corrosivity.

Electrical Conductance of Pure Water. Because of its extremely low conductance, it is very difficult to obtain sufficiently pure water for accurate conductance measurements. These values are needed at high temperatures in order to make proper subtractions of water conductivity from a total measured conductance, thus providing an accurate measurement of impurity conductance. There has been one very careful study of the conductance of pure water by Bignold, et al., at temperatures up to 300°C along the liquid-vapor curve [1]. There is, however, an indirect method for obtaining the coductance of pure water as a function of temperature and pressure. This method is based on the several, rather simple, correlations of the conductance behavior of aqueous electrolytes, observed at this laboratory and summarized recently by Frantz and Marshall [9], and the behavior of the ion-product of water discussed above. The limiting equivalent conductances of ions appear to follow straight line relationships with the density of water. In the supercritical temperature region, they also appear to extrapolate, at zero density, to a common value of conductance close to 1900 ohm^{-1} cm^{-3} equiv.$^{-1}$, which is essentially independent of temperature (above 400°C). By a reasonable assumption that these limiting conductances are additive and that these straight line relationships apply at least to moderately low densities, estimates of the conductance of the H$^+$ and OH$^-$ can be made as a function of temperature and pressure. With these estimates and the calculated ion-products of water, specific conductances of pure water have been calculated [14]. Figure 6 shows examples of these calculations as isobars at temperatures from 0 to 1000°C. Agreement of the calculated values with the above mentioned experimentally determined values of

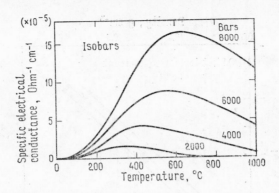

Fig. 6. Calculated specific electrical conductance of water, 0-1000° C, 1-8000 bars

Bignold, et al., [1] are within 3 to 5%, which would appear to be excellent agreement.

Ionization Equilibria; Recent Observations. For a complete description of the nature of water and steam containing impurity electrolytes and their concentrates, the extent of ionization (or association) of the electrolytes should be known. Probably the best way to determine this ionization behavior is through the measurement of electrical conductance. There have been many studies of this nature at this laboratory, following the pioneering work of Franck [8]. From these many studies, some simple correlating relationships have been observed. One correlation for limiting equivalent conductances was mentioned above. Another observed relationship shows that the logarithm of the ionization constants (for 1-1 salt electrolytes at infinite dilution) is a straight line function of the logarithm of the density of water. This same behavior has been observed recently for the ionization of 2-1 electrolytes, $MgCl_2$ and $CaCl_2$, where for the first ionization equilibria, $MgCl_2 \rightleftharpoons MgCl^+ + Cl^- (K_1)$, these salts are found to behave like 1-1 salts [9a]. Plots of $\log K_1$ for these two salts versus log density give straight lines, providing ionization constants and a slope close to those of the previously studied 1-1 salts. Figure 7 shows plots of this type for the first (K_1) and second (K_2) ionization constants of $CaCl_2$. The ionization behavior of hydrochloric acid recently has also been studied extensively [9b], and is compared to the much earlier study by Franck [8c]. The constants and slope for HCl are somewhat different than for the salts, and reflect the strongly different character of the H^+ ion. The overall generalization of behavior for salts (and acids) at temperatures from 250 to 800°C together with knowledge of K_w allows application to the description of species present in high temperature water and steam.

At pressures decreasing below about 200 bars, and above the critical temperature (374°C), Pitzer has taken an approach to indicate that the ionization constants of sodium chloride (his example for study), as for

153

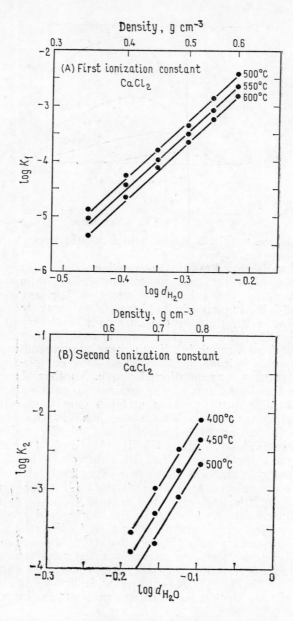

Fig. 7. The logarithm of the first (part A) and second (part B) ionization constants (molal units) of CaCl$_2$ in aqueous solutions plotted against log density (g/cm³) of water; 400-600°C, up to 4000 bars pressure

K_w [31], can diverge from the straight line extrapolated values by several orders of magnitude [30]. By direct conductance measurements, it is probably impossible to confirm experimentally either the calculated or the extrapolated ionization constants below 200 bars in the supercritical temperature region. The experimental conductances become much too low for sufficiently accurate measurements.

Vapor Pressures and Temperature Differences upon Concentrating Na₂SO₄ Impurity in Steam Cycles. The solubility limits of Na_2SO_4 (as an impurity) in concentrates of steam in steam generator tube crevices was discussed above. Also combined with the study of Na_2SO_4 solubility were evaluations of the vapor pressures as a function of temperature and the relative solution concentrations of Na_2SO_4, $NaCl$, and H_2SO_4, which are major contaminants of the very pure waters used for steam generation. To make these evaluations, the osmotic coefficients of Na_2SO_4 obtained by Holmes and Mesmer to 225°C [12] and those for $NaCl$ to 350°C [29] were applied in a simple method for predicting osmotic coefficients of an aqueous electrolyte mixture [13]. The activity of water is directly related to the osmotic coefficient. The vapor pressures are calculated from the activities through an equation of state for steam.

In making the above calculations the concentrations of all species assumed to be formed in the system Na_2SO_4-$NaCl$-H_2SO_4-H_2O, at the given conditions of temperature and formal concentrations, were calculated from the several available ionization constants. Vapor pressures were then calculated for a variety of conditions, and these values were directly related to boiling point elevations. The benefits of this study are equations, applied through a computer, that provide the extent of impurity concentration in a crevice for a given difference in tube wall temperature in a steam generator [13]. The solution within the crevice should concentrate until its vapor pressure is the same as the bulk steam. The maximum temperature differences may be calculated before precipitation of solid Na_2SO_4 occurs within a tube crevice. The resulting computer programs can thus allow the proper design of the steam generator to prevent the precipitation of Na_2SO_4 solid.

CONCLUSIONS

In this paper, particular studies of high temperature aqueous systems directed toward applications for energy production have been presented. In pursuing these studies, some simple correlations have been observed that allow the predictions of steam cycle behavior for the design and operation of both fossil fuel and nuclear power steam generators and also of steam generators using aqueous geothermal sources. Of equal satisfaction is the attainment of fundamental knowledge on the behavior of high temperature aqueous solutions.

ACKNOWLEDGEMENTS

I wish to thank my many colleagues who have been indispensable for the success of the ORNL experimental studies, interpretations, and applications presented in this work. This paper was sponsored by the

Division of Chemical Sciences, Office of Basic Energy Sciences, U. S. Department of Energy, under contract DE-AC0-840R21400 with the Martin Marietta Energy Systems, Inc. Much of the work reviewed was performed at the Oak Ridge National Laboratory and was sponsored by the U. S. Department of Energy Office of Basic Energy Sciences Divisions of Chemical Sciences and of Engineering, Mathematical, and Geosciences and, for some of the Na_2SO_4 studies, by the Electric Power Research Institute, Palo Alto, California (Research Project 623-5).

REFERENCES

1. Bignold, G. J., Brewer, A. D., and Hearn, B. (1971): *Trans. Farad. Soc.* **67**, 2419.
2. Broadbent, D., Lewis, G. G., and Wetton, E. A. M. (1977): *Chem. Soc. Dalton Trans.* 1977, 464.
3. Busey, R. H., and Mesmer, R. E. (1978): *J. Chem Eng. Data* **23**, 175.
4. Chen, C-T.A., and Marshall, W. L. (1982): *Geochim. Cosmochim, Acta* **46**, 279.
5. Fournier, R. O., and Marshall, W. L. (1983): *Geochim. Cosmochim. Acta* **47**, 587.
6. Fournier, R. O. (1983): *Geochim, Cosmochim. Acta* **47**, 579.
7. Fournier, R. O., and Potter, R. W. (1982): *Geochim. Cosmochim. Acta* **46**, 1969.
8. Franck, E. U. (1956): *Z. Physik. Chem. N. F.* **8**, (a) 92, (b) 197, (c) 192.
9. Frantz, J. D., and Marshall, W. L. (a) (1982): *Am. J. Sci.* **282**, 1666; (b) (1984): *Am. J. Sci.* **284**, 651.
10. Garnsey, R. (1979): *J. British Nucl. Energy. Soc.* **18**, 117.
11. Holland, H. D., and Malinin, S. (1975): "The Solubility and Occurrence of Non-Ore Minerals", Chapter 9, in *Geochemistry of Hydrothermal Ore Deposits,* 2nd. Ed. (H. L. Barnes, editor), Wiley Interscience, Publ., New York, 1975.
12. Holmes, H. F., and Mesmer, R. E. (1984): To be submitted for publication.
13. Lietzke, M. H., and Marshall, W. L. (1983): "Sodium Sulfate Solubilities in High-Temperature (250-374°C) Salt and Acid Solutions", Electric Power Research Institute Report *EPRI-NP-3047* (Project 623-5).
14. Marshall, W. L. (1984): Chemistry Division Annual Progress Report, Oak Ridge National Laboratory (Oak Ridge, Tennessee, USA) Report ORNL-6037, p. 30 (1984): Presentation before the International Association for the Properties of Steam Working Groups, Tokyo, Japan, September, 1983.
15. Marshall, W. L., and Chen, C.-T. A. (1982): *Geochim. Cosmochim. Acta* **46**, 367.
16. Marshall, W. L., and Chen, C.-T. A. (1982): *Geochim. Cosmochim. Acta* **46**, 289.
17. Marshall, W. L. (1982): *J. Chem. Eng. Data* **27**, 175.
18. Marshall, W. L., Hall, C. E., and Mesmer, R. E. (1981): *J. Inorg. Nucl. Chem.* **43**, 449.
19. Marshall, W. L., and Franck, E. U. (1981): *J. Phys. Chem. Ref. Data* **10**, 295.
20. Marshall, W. L. (1980): *Geochim. Cosmochim. Acta* **44**, 925.
21. Marshall, W. L. and Warakomski, J. M. (1980): *Geochim. Cosmochim. Acta* **44**, 915.
22. Marshall, W. L. (1980): *Geochim. Cosmochim. Acta* **44**, 907.
23. Marshall, W. L. (1975): *J. Inorg. Nucl. Chem.* **37**, 2155.
24. Marshall, W. L., and Jones, E. V. (1966): *J. Phys. Chem.* **70**, 4028.
25. Marshall, W. L., Wright, H. W., and Secoy, C. H. (1954): *J. Chem. Educ.* **31**, 34.
26. Noyes, A. A., Kato, Y., and Sosman, R. B. (1910): *J. Am. Chem. Soc.* **32**, 159.
27. Noyes, A. A., et al. (1907): *The Electrical Conductivity of Aqueous Solutions,* Carnegie Institution of Washington, Washington, D. C., USA, Publ. No. 63.
28. Panson, A. J., Economy, G., Liu, C.-T., Bulischeck, T. S., and Lindsay, Jr, W. T. (1975): *J. Electrochem. Soc.* **122**, 915.
29. Pitzer, K. S., Peiper, J. C., and Busey, R. H. (1984): *J. Phys. Chem. Ref. Data* **13**, 1.
30. Pitzer, K. S. (1983): *J. Phys. Chem.* **87**, 1120.
31. Pitzer, K. S. (1982): *J. Phys. Chem.* **86**, 4704.
32. Schroeder, W. C., Gabriel, A., and Partridge, E. P. (1935): *J. Am. Chem. Soc.* **57**, 1539.
33. Sweeton, F. H., Mesmer, R. E., and Baes, Jr, C. F. (1974): *J. Solution Chem.* **3**, 191.

Electrical Conductances of Dilute Aqueous Sodium Chloride Solutions at Elevated Temperatures and Pressures

A. A. SILKOV and Yu. N. UDODOV

*A. P. Vinogradov Institute of Geochemistry, Siberian Branch
of the USSR Academy of Sciences, Irkutsk, USSR*

Experimental measurements of electrical conductances of electrolyte solutions at elevated temperatures and pressures have long been of interest to scientists concerned both with theoretical and applied problems of geochemistry, chemistry and power energy.

A multitude of experimental studies on electrical conductance of aqueous electrolyte solutions has been made over a wide range of temperatures and pressures. Sodium chloride solutions have been studied in more detail. But an analysis of the literature data on electrical conductance of NaCl solutions has shown lack of data at temperatures from 50 to 200° C and high pressures.

On the other hand, there is no evidence of efforts to compile data on electrical conductances of dilute aqueous NaCl solutions at elevated temperatures and pressures as well as skeleton tables on the base of these data.

The present work was undertaken to provide some data on electrical conductances of dilute aqueous NaCl solutions at temperatures from 50 to 200° C and pressures up to 300.0 MPa and to use them together with the literature ones for statistical treatment.

Electrical conductance of NaCl was determined using the high-pressure equipment described in Ref. [1]. Sodium chloride solutions with concentrations ranging from 1.347×10^{-3} to 7.56×10^{-3} mol/kg H_2O were

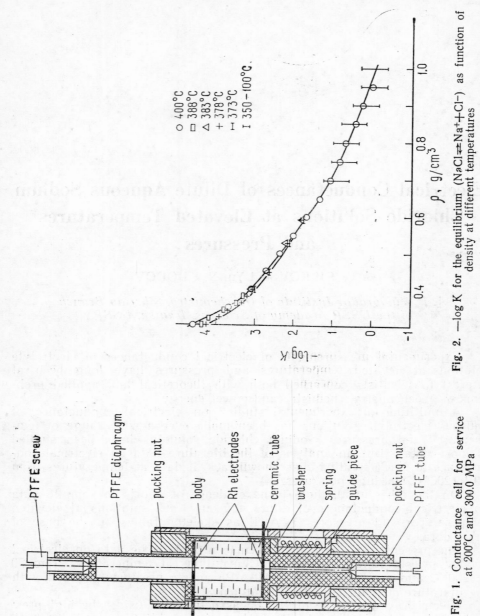

Fig. 2. —log K for the equilibrium (NaCl⇌Na+ +Cl−) as function of density at different temperatures

○ 400°C
□ 388°C
△ 383°C
+ 378°C
I 373°C
I 350 —100°C

Fig. 1. Conductance cell for service at 200°C and 300.0 MPa

PTFE screw

PTFE diaphragm

packing nut

body

Rh electrodes

ceramic tube

washer

spring

guide piece

packing nut

PTFE tube

gravimetrically prepared from ultra pure NaCl and distilled water; their specific conductance was about 1.4×10^{-6} ohm^{-1} cm^{-1} at 25°. The accuracy of making up solutions was 0.01%.

Two types of cells were used: the cell of Lown and Lord [2] with platinated platinum electrodes and the microcell with smooth rhodium electrodes (Fig. 1). The cells' constants were within 0.2248-2.650 cm^{-1}. The cells were calibrated with 10^{-3}-5×10^{-2} mol/dm^3 KCl solutions using the standard procedure [3].

Electrical conductances of NaCl solutions were measured of E-8-2 bridge with an associated audio-frequency alternator Г3-35 and a Ф-582 null-indicator. The average instrumental errors of measured electrical conductances of NaCl were known to be within 0.2%. To eliminate systematic errors due to the effect of polarization, electrical conductances were measured at frequencies of 0.5-10 KHz and extrapolated to an infinite frequency. The uncertainty in values of electrical conductances caused by the effect of polarization, amounted to 0.1-1.6% and increased with increasing temperature and NaCl concentration.

Two series of experiments were conducted. In the first series the background electrical conductances and electrical conductances of NaCl solutions were measured between 50 and 125°C (after every 25°C) by raising pressure up to 300.0 MPa (in 50.0 MPa increments) and back to the initial pressure at fixed temperatures. A run at one temperature continued for 4 hours, because the changes of pressure caused heating or cooling effects usually take 1 hour. The second series of experiments were conducted between 150 and 200°C with speeding up heating because of hysteresis in background conductivity. The background correction was made by taking measurements at fixed time. The thermostat maintained the temperature to ±0.05°, pressure was measured to within 1%.

The experimental data were approximated and the limiting molar conductances of NaCl were computer calculated by the following equations: Onsager [4], Shedlovsky [5], Robinson-Stokes [4], and Chen Mon-Shan [6] using a nonlinear least-squares method. Also the density and dielectric constant of water were calculated from [7] and the viscosity of water from international skeleton tables [8]. The change in concentration of NaCl solutions with density of water was taken into account by the equation: $c = m\rho_{H_2O}$, where c — concentration in mol/dm^3; m — concentration in mol/kg H_2O; ρ_{H_2O} — density of water.

All the above mentioned equations give close values of λ_0 and their errors. Variations of distance parameter $\overset{\circ}{a}$ for Robinson-Stokes equation were 1-5 Å and had slight influence on the accuracy of λ_0 calculation. Therefore, the final fit of experimental data was obtained by Robinson-Stokes equation with a fixed distance parameter $\overset{\circ}{a} = 3.5$ Å (Table 1).

All experimental works on electrical conductances of NaCl solutions at elevated temperatures and pressures have been analysed (Table 2). Some of the literature data with a very low accuracy or with a lack of values as a function of concentration at various temperatures, pressures and densities were not included in the statistical data processing.

Table 1. Limiting molar conductances ($cm^2\ ohm^{-1}\ mol^{-1}$) of NaCl as per our data

T° C \ P, MPa	Sat	50.0	100.0	150.0	200.0	250.0	300.0
50	197.9	196.6	194.6	192.3	189.0	185.7	182.1
75	276.7	271.3	267.1	262.4	257.2	251.5	245.5
100	360.8	351.7	343.7	336.2	328.1	320.5	312.8
125	447.8	434.5	422.7	411.5	400.6	390.2	379.9
150	533.5	515.4	499.2	485.2	471.1	459.0	445.3
175	619.1	595.2	574.3	556.9	539.6	523.2	507.8
200	703.5	672.2	646.4	625.9	605.6	586.4	568.6

Table 2. Literature data used in statistical treatment

Reference	Temp. range, °C	Pres. range, MPa	Molality range mol/kg H_2O	Estimation precision, %
[9]	60	0.1-100	8.6×10^{-2}	0.5-1.0
[10]	100-306	Sat.	5×10^{-4}-10^{-1}	0.24-0.50
[11]	25	0.1-400	2×10^{-2}-10^{-1}	0.10-0.12
[12]	75	0.1-300	10^{-2}	0.15-0.22
[13]	378-393	22-25	10^{-4}-10^{-2}	2-3
[14]	50	Sat.	10^{-2}-10^{-1}	0.05
[15]	300-383	Sat.-40	2×10^{-4}-3×10^{-3}	0.5-1.0
[16]	100-400	0.1-400	10^{-3}-10^{-1}	0.7-15
[17]	25-200	Sat.	10^{-3}-2×10^{-2}	0.17-0.24
[18]	200-300	Sat.	10^{-4}-5×10^{-4}	2.5
[19]	25-75	0.1-100	10^{-3}-10^{-1}	0.27-0.31
[20]	25	0.1-200	2×10^{-4}-2×10^{-2}	0.04-0.1
Authors' data	50-200	0.1-300	10^{-3}-8×10^{-3}	0.2-1.5

Besides, the data up to 400°C were analysed because above this temperature data have been obtained only by Quist and Marshall [16] and no information on densities of dilute NaCl solutions necessary for calculating molar concentration is available.

In most works the data on electrical conductances of dilute NaCl solutions are tabulated in molar conductivity calculated by a simplified equation. We recelculated the available data on molar conductivities of NaCl solutions, taking into account large systematic errors given by this equation at concentrations above 10^{-2} mol/dm³ H_2O.

Up to 200°C the molar concentration of NaCl solutions was computed by the equation [21]:

$$c = m\rho_{H_2O}/(1+k\cdot m\rho_{H_2O}) \quad (1)$$

where k — coefficient, equal to 0.021 for NaCl solution. Above 200°C the molar concentration of NaCl solutions was calculated by a precise equation using data on densities of dilute NaCl solutions at elevated temperatures and pressures [22-24].

Special attention was given to analysing the accuracy of all previous measurements. Wherever necessary, we allowed for the systematic errors, caused by the background conductivity, effect of polarization,

changes in the cell constant with temperature and pressure, etc. Accidental errors were calculated by the equation:

$$\sigma_\lambda = \sqrt{\sum_{i=1}^{n} \left(\frac{\partial \lambda}{\partial X} \right) \sigma_X^2} \qquad (2)$$

where X — parameter, influencing the molar electrical conductance, temperature, pressure, density, concentration, etc.; σ_x — mean square error of this parameter.

It is known that the accurate description of the physico-chemical properties of a system is a common problem in creating the system's model over a wide range of temperature, pressure, and concentration. As to electrical conductances, this is done in two stages:

(1) approximating experimental data by the electrical conductance equations and the equation of mass action law which have a physical meaning, with calculation of parameters λ_0, K and $\overset{\circ}{a}$ at different temperatures, pressures, and densities;

(2) finding empirical dependences of parameters of equations on parameters of state.

The choice of an electrical conductance equation is based on the principle of adequate description of experimental data with consideration of possible ionic equilibria, the work boundaries of this equation depending on the concentration, and the accuracy of experimental data. This is the reason why experimental data have been made on a three-parameter equation, which is a combination of the Robinson-Stokes electrical conductance equation and the equation of mass action law:

$$\lambda = \gamma [\lambda_0 - (\alpha \lambda_0 + \beta) \sqrt{\gamma c} / (1 + \overset{\circ}{a} B \sqrt{\gamma c}) \qquad (3)$$
$$K = \gamma^2 c f_\pm^2 / (1 - \gamma) \qquad (4)$$

where λ_0 — limiting molar conductance; K — NaCl ionization constant; $\overset{\circ}{a}$ — distance parameter; $f_\pm$ — mean molar activity coefficient; γ — degree of NaCl ionization.

The mean molar activity coefficient of NaCl was calculated from the second Debye-Hückel equation [4]. Moreover, the distance parameter for Eq. (3) as well as for Debye-Hückel one used to be the same.

The unknown parameters λ_0, K, and $\overset{\circ}{a}$ of Eqs. (3) and (4) were found by the principle of maximum likelihood [25]. This principle amounts to finding out a set of parameters λ_0, K and $\overset{\circ}{a}$ such that the dependence of the measured property of the system on initial concentrations was best described. As test of goodness of fit to the experimental data we chose the weighted sum of squared deviations in the space of unknown parameters:

$$F = \sum_{i=1}^{n} w_i (\lambda_{i,e} - \lambda_{i,t})^2 \qquad (5)$$

where w_i — statistical weights, equal to $1/\sigma_\lambda$; $\lambda_{i.\,e}$ — experimental values of electrical molar conductance; $\lambda_{i.\,t}$ — theoretical values of electrical molar conductance.

Equations (3) and (4) were solved by the method of successive approximations. The minimum of objective function F was found by the deformable polyhedron method [26]. The algorithm for the calculation of λ_0, K and $\overset{\circ}{a}$ was realized in FORTRAN.

In the free fit (λ_0, K, $\overset{\circ}{a}$) the distance parameter should have a steady dependence on temperature and pressure, and a slight influence on the variation of λ_0 and K. Therefore, we have recalculated λ_0 and K with a regulated distance parameter $\overset{\circ}{a}$ as a function of temperature and pressure.

The temperature and pressure dependence on the distance parameter should be described by the following empirical equations: $\overset{\circ}{a} = 24 D_0/D_p$ when $T = 298.15$ K, $\overset{\circ}{a} = 4.32 + 0.0068(T-323.15)D_0/D_p$ when $T \geqslant 323.15$ K. Here D_0 and D_p are correspondingly dielectric constants of water at saturation pressure and at pressure above this one. Certainly, these equations have no physical meaning, but they are the consequence of experimental data fitting on theoretical equations. We assume that more accurate experimental data on electrical conductances of NaCl solutions would help to solve this problem. The limiting molar conductances of NaCl obtained by fitting both the experimental and the authors' data on Eqs. (3) and (4) are presented in Table 3.

Because of lack of space we have not included the values of limiting molar conductances of NaCl calculated at fractional temperatures, pressures, and densities. Also, we have not taken into account the data of Benson et al. [13] at 378-398°C and densities below 0.30 g/cm³, because

Table 3. Limiting molar conductances (cm² ohm⁻¹ mol⁻¹) of NaCl at integer
temperatures and pressures

T, °C \ P, MPa	Sat., v.p.	50.0	100.0	150.0	200.0	250.0	300.0
25	126.40	127.98	128.35	127.65	126.28	123.96	121.50
50	197.41	195.92	194.49	191.89	188.59	185.17	181.38
75	276.05	271.82	267.22	262.20	257.15	251.20	245.29
100	360.5	351.8	343.5	335.7	327.6	319.9	312.2
125	447.5	434.6	423.4	410.4	399.8	389.5	379.2
150	534.2	519.5	502.0	487.6	473.3	459.0	445.2
175	617.7	595.4	575.1	557.5	539.7	523.4	507.1
200	704.9	673.7	647.3	624.7	603.3	584.5	567.5
250	857	807	759	736	709	688	669
300	1013	943	890	849	815	790	766
350	1200	1051	987	940	900	870	839
400	—	1194	1076	1015	971	935	908

the fitting of these data on Eqs. (3) and (4) gives incorrect values of λ_0, K, and $\overset{\circ}{a}$. This is probably due to very strong association of NaCl at low densities. As a result, the electrical conductance Eq. (3) in limit reduces to $\lambda = \gamma \cdot \lambda_0$ unsolvable for λ_0 and K, and has no physical sense for the distance parameter $\overset{\circ}{a}$. Though we have not solved the problem of the tolerance calculation of λ_0 and K, but we can assert that the presented values of λ_0 (Table 3) have been calculated with an uncertainty comparable to the experimental uncertainty.

The dependence of $-\log K$ of sodium chloride upon density at different temperatures is shown on Fig. 2.

As one can see from this figure, at temperatures up to 400°C and densities of 0.5 g/cm³ the NaCl ionization constant is temperature independent. We cannot determine the NaCl ionization constant at temperatures below 400°C and densities above 0.5 g/cm³ precisely because the available data on electrical conductances are not accurate enough and there is very limited range of concentrations.

Besides, at temperatures above 350°C and densities below 0.3 g/cm³ we have made fit experimental data on the equation (3) taking into account the simultaneous existence in the equilibrium of NaCl molecules and $[Na^+Cl^-Na^+]^+$, $[Cl^-Na^+Cl^-]^-$ triples. But the consideration of triples did not give an increasing accuracy of fitting experimental data, and this evidenced the absence of the assumed $[Na^+Cl^-Na^+]^+$, $[Cl^-Na^+Cl^-]^-$ triples or the inaccuracy of the experimental data.

This paper gives results of the first stage of statistical treatment and fitting experimental data on the theoretical conductance equation and equation of mass action law with the calculation of fitting parameters λ_0, K, $\overset{\circ}{a}$ for NaCl. In order to fit experimental data on electrical conductances of NaCl solutions in a wide range of temperatures, pressures, and densities, it is necessary to have equations describing the parameters of state.

REFERENCES

1. Udodov Yu. N. and Silkov A. A. (1979): In: Methods of Experimental investigations of hydrothermal equilibrium, Nauka, SO, Novosibirsk, 36-41 (in Russian).
2. Lown D. A. and Lord W. (1967): *Journal of Science Instruments,* 44, 1037.
3. Zwollnik L. J. and Fuoss R. (1959): *Journal Am. Chem. Soc.,* 81, 1557.
4. Robinson P. and Stocks P. (1963): Electrolyte solutions, IL, Moscow.
5. Shedlovsky T. (1932): *Journal Am. Chem. Soc.,* 54, 1405.
6. Chen Mon-Shan, Onsager L. (1977): *Journal of Phys. Chem.,* 81, 2017-2021.
7. Hegelson H. and Kirkham H. (1974): *American Journal of Science,* 274, No. 10, 1089-1261.
8. Alexandrov A. A. (1977): *Teploenergetika,* 4, 87-91.
9. Lusanna S. (1897): *Nuovo Cimento,* S(4), 441-459.
10. Noyes A. A. (1907): *Carnegie Inst. Washington Publ.,* 63.
11. Adams L. H. and Hall R. G. (1931): *Journal Phys. Chem.,* Ithaca, 35, 2145-2163.
12. Zisman W. A. (1932): *Phys. Rev.,* 39, 151-160.
13. Fogo J. K., Benson S. W. and Copeland Ch. S. (1954): *Journal Chem. Phys.,* 22, 212-216.
14. Chembers J. F. (1958): *Journal Phys. Chem.,* 62, 1136-1138.

15. Pirson D., Copeland C. S. and Benson S. W. (1963): *Journal Am. Chem. Soc.,* **85,** 1044-1046.
16. Quist A. S. and Marshall W. L. (1968): *Journal Phys. Chem.* 72, 684-703.
17. Smolyakov B. S. (1968): VINITI, No. 776-69. **Dep.**
18. Sirota H. M. and Shvyryaev Yu. V. (1969): *Teploenergetika,* 3, 82-84.
19. Lukashev Yu. M. and Shcherbakov V. N. (1979): NIITEHIM, Cherkassy, No. 3067/79 **Dep.**
20. Fisher F. H. and Fox A. P. (1981): *Journal Sol. Chem.,* 10, 871-879.
21. Udodov Yu. N. and Silkov A. A. (1977): VINITI, No. 3345-77. Dep.
22. Hilbert R., Tödheide K., and Franck E. U. (1980): *Proceedings of the 9th International Conference on the Properties of Steam,* Münich, 1979, Oxford e.a. 616-623.
23. Benson S. W., Copeland C. S., and Person D. (1953): *Journal Chem. Phys.,* 21, 2208-2212.
24. Khaibulin I. Kh. and Borisov N. M. (1963): *Teploenergetika,* 10, No. 3, 12.
25. Kramar G. (1975): Mathematical methods of statistics, Mir, M.
26. Himmelblau D. (1975): Applied non-linear programming, Mir, M., 534.

The Thermodynamic Properties of Sodium Hydroxide-Water System

V. I. ANDROSOV and V. V. VOSPENNIKOV

*Novomoskovsk branch of the Moscow D. I. Mendeleev
Chemico-Technological Institute, USSR*

The sodium hydroxide-water system is one of the most corrosion-dangerous system, and therefore it was used for studying the thermodynamic properties as applied to steam turbine engineering. The following are the special features of sodium hydroxide: low melting point — 594 ± 1 K ($321\pm1°C$), high solubility in water and ability to dissolve in steam. Sodium hydroxide solutions can be present in the whole range of expansion parameters of steam turbine: inlet steam in high-pressure section up to zone of saturation in low-pressure section. Sodium hydroxide exhibits salting out effect that considerably promotes the formation of deposits in the cycles of thermal power plants. It is impossible to prevent sodium hydroxide presence because high-temperature hydrolysis of sodium salts is a way of its formation. To explain and predict the behavior of materials and power-block constructions, it is necessary to know the thermodynamic properties of the real heat-transfer medium, in particular the properties of sodium hydroxide-water system.

Review [1] generalizes the literature data on thermodynamic properties of sodium hydroxide-water system, contains equations for calculating the main properties, and tabulates the obtained data. This paper is the continuation of Ref. [1] and extends the range of investigated parameters. In Eqs. (3.1-3.4) of Ref. [1] use has been made of the properties of water as solvent, that are comparable with those of the solution at equal temperatures or pressures. But the zone of existence of liquid solutions of electrolytes considerably extends beyond the critical point of water.

165

So we propose to compare the properties of solution and water at equal reduced parameters — temperature (τ) and pressure (π). The main difficulty of using this method is the absence of literature data on critical parameters of sodium hydroxide solutions. This difficulty can be overcome by calculating critical parameters using the proposed method. The critical temperature is determined from the partial molar enthalpy of vaporization temperature dependence, which equals zero at the critical point. An analysis of literature data shows that the vaporization enthalpy curves both for water and solutions are almost similar in the whole range of temparature (from room up to critical for water). It allows us to suppose that the water vaporization enthalpy function will be identical in shape for both solvent — water and aqueous solution. The temperature dependence of vaporization enthalpy is presented in the form:

$$\ln[(T_c/T)+\sqrt{(T_c/T)^2-1}] = \sum_{i=1}^{6} a_i \cdot h^i, \tag{1}$$

where $T_c=647.27$ K is the critical temperature, T is the current temperature, h is the specific heat of water vaporization, a_i is the coefficient. Coefficients a_i of equation (1) for water were determined by the least squares method. The error of approximating the data of Ref. [2] on vaporization enthalpy of water by equation (1) is less than 0.1% between 283-543 K; in the nearcritical zone it reaches 1%. As no experimental data on vapour pressures of electrolytes aqueous solutions are available up to critical temperatures, from which partial enthalpy of water vaporization can be determined, we decided to find the error of computing critical temperature depending on the extent to which the massive $\{h_i, T_i\}$ is away from the critical point. The calculation results for water show that the error determining T_c depends on the extent to which the massive is away from T_c. Thus, at maximal T, corresponding to $T/T_c=$ $=\tau=0.95$ the relative error equals $\delta=0.02\%$; at $\tau=0.9$, $\delta=0.07\%$; at $\tau=0.8$, $\delta=0.6\%$. The results obtained were examined for the sodium chloride-water system using Haas's data [3] on partial molar heats of water vaporization. The computed values of critical temperatures for the NaCl-H_2O system are in good agreement with those of Ref. [4]; the disagreement equals about 1 K. Using the procedure described and the data of Ref. [1], we determined critical temperatures for the NaOH-H_2O system, which are presented in Table 1.

Table 1

Concentration of NaOH wt.%	5	10	15	20
Critical temperature, K	684	725	770	819

These are described by the following equation:

$$T_c=647.27+b_1X+b_2X^2, \text{ K} \tag{2}$$

where X — is the concentration of solution, wt. %,

$$b_1 = 6.94;$$

$$b_2 = 0.0833.$$

Equation (2) with an error of about ± 1 K is applicable for NaOH solutions up to 20%. The calculated critical temperatures for the NaCl-H_2O systems allowed us to compare the vapour pressure of solutions and water at equal reduced temperatures. This revealed that vapour pressure over solution was higher than that over pure water in the whole range of reduced temperatures. The more the solution concentration the greater the increase. Critical pressures of the NaOH-H_2O system were obtained by extrapolating the $(P_x - P_o)/P_x = f(\tau)$ function to critical temperature, where P_x and P_o are the vapour pressures over solution and water respectively at equal reduced temperatures. The possibility of such an extrapolation was checked up on the NaCl-H_2O system and the results were found to be in good agreement with Khaibullin's data [5]. The concentration dependence of the critical pressures of the NaOH-H_2O system is:

$$P_c = 22 \cdot 115 + a_1 \frac{X}{100 - X} + a_2 \left(\frac{X}{100 - X} \right)^2 +$$

$$+ a_3 \left(\frac{X}{100 - X} \right)^3, \text{ MPa} \qquad (3)$$

where X is the concentration of solution, in wt. %;

$$a_1 = \quad 0.9576;$$

$$a_2 = -0.47272;$$

$$a_3 = \quad 0.18265$$

Calculated values of critical pressures make it possible to compare reduced pressures of water and solutions at equal reduced temperatures and also their difference as function of concentration. These dependences are shown in Figs. 1 and 2; their analysis permits us to draw the following conclusions. The $\pi_x = f(\pi_0)$ dependence, where π_x and π_0 are the reduced pressures of the solution and water, respectively, is a straight line, the deviation being maximum in the zone of low π, that is, in the low-temperature zone, for $\pi = 0.4$-0.8 the deviation equals 1%. The $\Delta\pi = \pi_x - \pi_0 = f(m)$ dependence is also a straight line, especially in the zone of low τ. But in the zone near $\tau = 0.8$ this function is satisfactorily described by a polynomial of order two; and when described analytically its free term equals zero. The results of investigation may be used to extend concentration and temperature limits for getting reliable values of thermodynamic properties from the restricted experimental data.

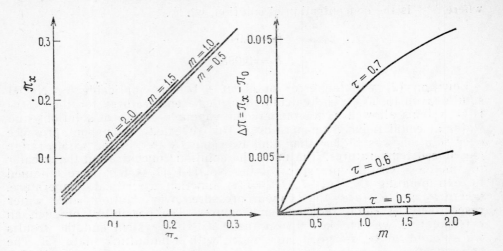

Fig. 1. Comparison of the reduced vapour pressures of water and NaOH — H₂O system

Fig. 2. Concentration dependence of reduced pressure difference of NaOH—H₂O system and water

The data on aqueous sodium hydroxide solutions vapour pressures and partial molar volumes of water in solutions [1] allowed the water activity to be calculated by Eq. (4):

$$\ln a_1 = \ln \frac{P_s}{P_o} + \frac{1}{RT} \int_{P_s}^{P_o} \left(\frac{RT}{P} - V_g \right) dP +$$

$$+ \frac{\bar{V}_1}{RT} (P_o - P_s) \qquad (4)$$

where P_s — the vapour pressure of the solution; P_o — the vapour pressure of pure water; V_g — the molar volume of pure water vapour in equilibrium with the solution at temperature T and vapour pressure P_s; $\bar{V}_1$ — the partial molar volume of water in the solution.

Calculated values are represented in Table 2.

The molar volume of pure water vapour was calculated with the use of Eq. (3.2) [6] for specific volume of steam, the error of determining the integral in Eq. (4) equals 0.05%. The reliability of the obtained data on water activity was estimated by calculating sodium hydroxide activity coefficients, using Eq. (5):

$$\ln \gamma_\pm = - \left[(1 - \varphi) + \int_0^m \frac{1 - \varphi}{m} dm \right], \qquad (5)$$

where m is molality;

$$\varphi = \frac{1000 \cdot \ln a_1}{v \cdot m \cdot M_1}$$ is the osmotic coefficient.

Here v is the number of ions into which electrolyte dissociated, $M_1 = 18$ is the molecular weight of water. The values of $\gamma_{\pm}$ calculated by Eq. (5) for low temperatures were found to be in good agreement with the experimental data of Åkerlöf and Kegeles [7].

Table 2

X, wt.%	Water activity at temperature (K)						
	323	373	423	473	523	573	623
0.2	0.9983	0.9986	0.9988	0.9990	0.9992	0.9994	0.9996
0.4	0.9967	0.9972	0.9976	0.9980	0.9984	0.9987	0.9991
0.6	0.9946	0.9957	0.9964	0.9970	0.9975	0.9981	0.9986
0.8	0.9932	0.9943	0.9952	0.9960	0.9967	0.9974	0.9982
1.0	0.9915	0.9928	0.9940	0.9950	0.9959	0.9968	0.9977
2.0	0.9825	0.9852	0.9876	0.9896	0.9915	0.9933	0.9953
4.0	0.9629	0.9688	0.9736	0.9780	0.9820	0.9860	0.9901
6.0	0.9407	0.9500	0.9578	0.9648	0.9713	0.9777	0.9842
8.0	0.9153	0.9284	0.9396	0.9497	0.9591	0.9682	0.9774
10.	0.8864	0.9040	0.9189	0.9325	0.9453	0.9576	0.9698
20.	0.6948	0.7380	0.7768	0.8130	0.8477	0.8814	0.9141
30.	0.4670	0.5308	0.5909	0.6492	0.7072	0.7650	0.8220
40.	0.2697	0.3373	0.4053	0.4743	0.5455	0.6191	0.6942
50.	0.1334	0.1899	0.2518	0.3182	0.3893	0.4652	0.5451

REFERENCES

1. Martynova O. I., Androsov V. I., and Vospennikov V. V. The thermodynamic properties of sodium hydroxide-water system. Reviews on thermophysical properties of substances/TPhC. Moscow: IVTAN, 1982, No. 4 (36), 4-108 p.
2. Rivkin S. L. and Alexandrov A. A. Thermophysical properties of water and steam. Moscow: Energy, 1980, 424 p.
3. Haas J. L., Jr. Preliminary "steam tables" for boiling NaCl solutions. Physical properties of the coexisting phases and thermochemical properties of the H_2O component. U.S. Geological Survey Open File Report, 1975, No. GD-76-130, 66 p.
4. Marshall W. L. and Jones E. V. Liquid-vapour critical temperatures of aqueous electrolyte solutions. *J. inorg. nucl. chem.*, 1974, v. 36, p. 2313-2318.
5. Khaibulin I. Kh. The tables of thermodynamic properties of gases and liquids. Issue 6. Aqueous and vapourphased solutions. The system water-sodium chloride. Moscow: Publishers of Standards, 1980, 80 p.
6. Alexandrov A. A. The equations for thermodynamic properties of water and steam. Reviews on thermophysical properties of substances/TPhC. Moscow: IVTAN, 1978, No. 3, 92 p.
7. Åkerlöf G. and Kegeles G. Thermodynamics of concentrated aqueous solutions of sodium hydroxide. *J. Amer. Chem. Soc.*, 1940, v. 62, p. 620-640.

Evaluated Thermal Properties of Aqueous Transition Metal Chlorides: Mn, Fe, Co, Ni, Cu and Zn

B. R. STAPLES

Bureau of Mines, U.S. Department of the Interior,
P.O. Box 70, Albany, OR 97321, USA

A critical evaluation of the heat capacities of aqueous transition metal chloride solutions of Mn, Fe, Co, Ni, Cu, and Zn at 298 K is presented. This evaluation was carried out at the U. S. Bureau of Mines to aid in the improvement of the hydrometallurgical processes involving these domestic chlorides. Heat capacities were calculated from direct calorimetric measurements. A least squares program was used to fit data as a function of concentration at each temperature where data were available. A four parameter Pitzer equation describes the apparent molal heat capacity as a function of the molality at each temperature. Values of the relative partial molal enthalpy and heat capacity are tabulated. The scientific literature has been covered through June 1984.

INTRODUCTION

In recent years there have been increased demands for both precise data and predictive methods for thermal property data. In particular, methods are needed for properties that have not been or cannot be measured in certain ranges of concentration or temperature. This paper presents semitheoretical equations that can be used to calculate apparent molal heat capacities as a function of molality for the aqueous chloride solutions of these selected transition metals: Mn, Fe, Co, Ni, Cu, and Zn.

Thermodynamic properties of aqueous solutions are investigated as a part of the Bureau of Mines program to provide a scientific base of

information required to develop innovative technologies for mineral processing, to predict the feasibility of new processes, to minimize corrosion, and to enhance the environmental aspects and energy efficiency of existing processes.

Solubilities and vapor pressures are important properties in many mineral processing techniques. The principal thermodynamic properties of interest for predicting solubilities and vapor pressure are the activity and osmotic coefficient. The values for these properties must be available at any temperature of interest to apply to processes occurring at temperatures ranging from near 0°C (deep sea) to above 100°C (geothermal or industrial). Heat capacities are needed to calculate these properties as a function of temperature.

Only a few correlations of heat capacity data have been reported in the past. Pitzer and Brewer's revision of Lewis and Randall (1961) contains correlations of calcium chloride heat capacity at 298 K. Pitzer (1977) also evaluated data for several different thermal properties of sodium chloride as a function of temperature.

An actual set of temperature dependent equations for calcium chloride activities has only appeared in the literature during this past year (Phutela and Pitzer, 1983), while several other papers are in review [Atkinson et al. (1984), Staples et al. (1984)].

In the following sections I describe data treatment methods and thermodynamic expressions. Details of the critical evaluation procedure are explained and the results of this evaluation are presented for aqueous transition metal chloride solutions at 298 K and over a molality range of 0 to saturation.

METHODS FOR CORRELATION OF HEAT CAPACITY DATA

Although aqueous electrolyte systems at room temperature have received much attention, fewer data are available at other temperatures. Thermodynamic properties, and in particular, activity and osmotic coefficients can be calculated at any desired temperature given the appropriate thermal property data, such as the heat capacity and the enthalpy of dilution as functions of molality.

Silvester and Pitzer (1977) have extended to enthalpy and to heat capacity the equations developed by Pitzer (1973). Pitzer's (1973, 1977) equations provide a reasonably accurate phenomenological representation of aqueous electrolyte properties over a large ionic strength range.

Since the Pitzer equations have become widely accepted, have enjoyed numerous applications to electrolyte data in the literature, and do provide a reasonable and convenient model, I have fit the heat capacity data to the Pitzer equation. In brief, I will develop the basic equations for enthalpy and heat capacity from the Pitzer formulation.

Using our previous equations (Staples and Nuttall, 1978) which describe the excess Gibbs energy for a binary electrolyte

$$G^{ex} = n_1\overline{G}_1{}^{ex} + n_2\overline{G}_2{}^{ex} \qquad (1)$$

where n_1 and n_2 are the number of moles of solvent and solute, respectively, $\bar{G}_1{}^{ex}$ is the partial molal excess Gibbs energy of the solvent, and $\bar{G}_2{}^{ex}$ is the partial molal excess Gibbs energy of the solute. We recall that the total excess Gibbs energy is

$$G^{ex} = n_w v m R T \, (1 - \varphi + \ln \gamma_\pm) \tag{2}$$

where n_w is the number of kilograms of solvent, m is the molality, R is the gas constant, T designates the temperature in K, φ is the osmotic coefficient, and $\gamma_\pm$ is the activity coefficient. It follows that $\bar{G}_1{}^{ex}$ is given by

$$\bar{G}_1{}^{ex} = (v n_2/n_1) R T \, (1 - \varphi) \tag{3}$$

Here v_m and v_x are the numbers of ions of each type in the formula and $v = v_m + v_x$. Similarly, an electrolyte has ionic charges Z_m and Z_x for the positive and negative ions in electronic units.

We now adopt the general equations for φ and $\ln \gamma_\pm$ of a pure electrolyte (Silvester and Pitzer, 1977):

$$\varphi - 1 = -|Z_m Z_x| A_\varphi \frac{I^{1/2}}{1 + b I^{1/2}} + m \left(\frac{2 v_m v_x}{v} \right) (\beta_{mx}^{(0)} + \beta_{mx}^{(1)} e^{-\alpha I^{1/2}})$$

$$+ m^2 \frac{2 (v_m v_x)^{3/2}}{v} C_{mx}^\varphi \cdot \tag{4}$$

$$\ln \gamma_\pm = -|Z_m Z_x| A_\varphi \left[\frac{I^{1/2}}{1 + b I^{1/2}} + \frac{2}{b} \ln (1 + b I^{1/2}) \right] + m \left(\frac{2 v_m v_x}{v} \right)$$

$$\left[2 \beta_{mx}^{(0)} + \frac{2 \beta_{mx}^{(1)}}{\alpha^2 I} \left(1 - \left(1 + \alpha I^{1/2} - \frac{\alpha^2 I}{2} \right) e^{-\alpha I^{1/2}} \right) \right] + \frac{3 m^2}{2} \left(\frac{2 (v_m v_x)^{3/2}}{v} \right) C_{mx}^\varphi \tag{5}$$

where I is the ionic strength $(I = 1/2 \sum_i (m_i Z_i^2)$, and A_φ is the Debye-Hückel coefficient for the osmotic function, A_φ, or the activity function, A_γ, given as

$$A_\varphi = \frac{1}{3} \left(\frac{2 \pi N_0 \rho_w}{1000} \right)^{1/2} \left(\frac{e^2}{D k T} \right)^{3/2} = \frac{A_\gamma}{3} \cdot \tag{6}$$

Here N_0 is Avogadro's number, ρ_w is the density of the solvent, e is the absolute electronic charge, D is the static dielectric constant of pure water, and k is Boltzmann's constant. The Pitzer $b = 1.2$ and $\alpha = 2.0$ for these electrolytes. Since the values of v_m, v_x, and v for MCl_2 (2-1 salts) are 1, 2, and 3, respectively; the factor $(2 v_m v_x/v)$ is 4/3. The relative enthalpy, L, of an electrolyte solution is defined as

$$L = H - H_0{}^\circ \tag{7}$$

where H is the total enthalpy of the solution and $H_0{}^\circ$ is the total enthalpy of the components of the solution in their standard states.

172

The quantities L and G^{ex} are related by

$$L = -T^2 (\partial (G^{ex}/T)/\partial T)_{P,m}, \tag{8}$$

$$L = vmRT^2 [\partial \Phi / \partial T)_{P,m} - (\partial \ln \gamma_{\pm}/\partial T)_{P,m}]. \tag{9}$$

ΦL, the apparent relative molal enthalpy is defined to be

$$\Phi L = \frac{L - n_1 \bar{L}_1^{\circ}}{n_2} = \frac{L}{n_2}. \tag{10}$$

By taking derivatives of equations (4) and (5) as indicated by equations (9) and (10), we obtain

$$\Phi L = v |Z_m Z_x| (A_H/2b) \ln(1 + 1.2I^{1/2}) - 2v_m v_x RT^2 (mB'_{mx} + m^2 C'_{mx}), \tag{11}$$

where

$$B'_{mx} = (\partial B_{mx}/\partial T)_{I,P}, \tag{12}$$

$$B_{mx} = \beta_{mx}^{(0)} + (2\beta_{mx}^{(1)}/\alpha^2 I)[1 - (1 + \alpha I^{1/2}) \exp(-\alpha I^{1/2})] \tag{13}$$

$$C'_{mx} = 1/2 (v_m v_x)^{1/2} (\partial C_M^{'\varphi}/\partial T)_{I,P} \tag{14}$$

A_H the Debye-Hückel coefficient for enthalpy is defined (Bradley and Pitzer, 1978) and (Anathanaswamy and Atkinson, 1984)

$$A_H = -6A_{\varphi}RT^2 [T^{-1} + (\partial \ln D/\partial T)_P + \alpha_w/3], \tag{15}$$

where $\alpha_w = (\partial \ln V/\partial T)_p$ is the thermal expansion coefficient of water. The total relative heat capacity, J, is defined to be

$$J = C_p - C^{\circ}_p = C_p - (n_1 C^{\circ}_{p1} + n_2 \bar{C}^{\circ}_{p2}) \tag{16}$$

where C°_{p1} is the molal heat capacity of pure water and $\bar{C}_{p2}$ is the partial molal heat capacity of the solute at infinite dilution. We can also obtain J from the relation

$$J = (\partial L/\partial T)_{P,m}. \tag{17}$$

The apparent molal heat capacity, Φ_c, is defined as

$$\Phi_c = (C_p - n_1 C^{\circ}_{p1})/n_2. \tag{18}$$

From Eqs. (10) and (17), we conclude that

$$\Phi_c - C^{\circ}_{p2} = (\partial \Phi L/\partial T)_{p,m}. \tag{19}$$

Values of $\bar{C}^{\circ}_{p1}$ can also be calculated from heat of solution data by

$$(\partial \Delta \bar{H}^{\circ}_s/\partial T)_P = \bar{C}^{\circ}_{p2} - C^{\circ}_{p2}(s), \tag{20}$$

173

if $C^{\circ}{}_{p}{}^{'}(s)$, the heat capacity of the pure solidified salt, is known. The derivative with respect to temperature of Eq. (11) gives

$$\Phi_c = \bar{C}^{\circ}{}_{p2} + v\,|Z_m Z_x|\,(A_J/2b)\ln(1+1.2I^{1/2})$$
$$-2v_m v_x RT^2(mB''{}_{mx}+m^2C''{}_{mx}), \tag{21}$$

$$B''_{mx} = \left(\frac{\partial^2 B_{mx}}{\partial T^2}\right)_{P,m} + \frac{2}{T}\left(\frac{\partial B_{mx}}{\partial T}\right)_{P,m}, \tag{22}$$

$$C''_{mx} = 1/2\,(v_m\,v_x)^{1/2}\left[\left(\frac{\partial^2 C^{\varphi}_{mx}}{\partial T^2}\right)_{P,m} + \frac{2}{T}\left(\frac{\partial C^{\varphi}_{mx}}{\partial T}\right)_{P,m}\right], \tag{23}$$

$$A_J = (\partial A_H/\partial T)_P = 28.873 \text{ at } 298.15 \text{ K} \tag{24}$$

Heat capacities of electrolytes are experimentally determined by direct calorimetric measurements on solutions. The total heat capacities are then converted to Φ_C by the relationship

$$\Phi_C = \frac{1000\,C_p - C^{\circ}_p}{m} + M_2 C_p, \tag{25}$$

in which m is the molality, M_2 is the molecular weight of the salt, C_p is the heat capacity of the solution, and $C_p{}^{\circ}$ is the heat capacity of water, then Φ_C is fitted to Eq. (21) by treating $\Phi_C{}^{\circ}(\bar{C}^{\circ}{}_{p2})$ as an adjustable parameter.

The usual standard states apply when the pure solvent is at the same temperature and pressure as the solution. They apply for the solute in the limiting state, for which $\gamma_{\pm}$ approaches unity as the concentration becomes infinitely dilute, at every temperature and pressure.

The specific terms of the Pitzer formulation for Φ_C (Ed. 21) can be derived from the definitions of terms given in other references (Pitzer, 1973). The equation used here to fit the Φ_C data is of slightly different form than Pitzer's (Eq. 21), because of a grouping of terms which we have used to obtain an excellent description of the actual experimental data. The modifications are described here.

A definition of terms is provided:

$$B^J = \left(\frac{\partial^2 B_{mx}}{\partial T^2}\right) + \frac{2}{T}\left(\frac{\partial B_{mx}}{\partial T}\right), \tag{26}$$

$$B_{mx} = \beta^0 + \frac{2\beta^1}{\alpha^2 I}\,f(I), \tag{27}$$

$$B'_{mx} = \left(\frac{\partial B_{mx}}{\partial T}\right) = \left(\frac{\partial \beta^0}{\partial T}\right) + \frac{2}{\alpha^2 I}\left(\frac{\partial \beta^1}{\partial T}\right)f(I), \tag{28}$$

$$B''_{mx} = \left(\frac{\partial^2 B_{mx}}{\partial T^2}\right) = \left(\frac{\partial^2 \beta^0}{\partial T^2}\right) + \frac{2}{\alpha^2 I}\left(\frac{\partial^2 \beta^1}{\partial T^2}\right)f(I). \tag{29}$$

Therefore

$$B''_{mx} = \left(\frac{\partial^2 \beta^0}{\partial T^2}\right) + \frac{2}{\alpha^2 I}\left(\frac{\partial^2 \beta^1}{\partial I^2}\right)f(I) + \frac{2}{T}\left[\left(\frac{\partial \beta^0}{\partial T}\right) + \frac{2}{\alpha^2 I}\left(\frac{\partial \beta^1}{\partial T}\right)f(I)\right], \tag{30}$$

and

$$B''_{mx} = B''_0 + \frac{2}{\alpha^2 I} B'_1 f(I) + \frac{2}{T} B'_0 + \frac{4}{T\alpha^2 I} B'_1 f(I). \tag{31}$$

Replace B''_{mx} in the second term above, then combine terms into m and $f(I)$, noting that $I = km$, then $\frac{m}{I} = 1/k$, where $k = 0.5\,(\nu_m Z^2_m + \nu_x Z^2_x)$.

Using a shorthand notation of "J" functions defined as

$$FJ1 = \frac{\nu Z_m Z_x}{2b} A_J \quad \text{and} \quad FJ2 = 2\nu_m \nu_x RT^2, \tag{32}$$

then the second term of Eq. (21) becomes

$$-FJ2 \left[B''_0 m + \frac{2}{T} B'_0 m + \frac{2m}{\alpha^2 I} \left(B''_1 + \frac{2B'_1}{T} \right) f(I) \right], \tag{33}$$

so

$$\Phi_c = \Phi°_c + FJ1 \ln(1 + bI^{1/2})$$

$$-FJ2 \left(B''_0 + \frac{2}{T} B'_0 \right) m - \frac{2(FJ2)}{\alpha^2 k} \left(B''_1 + \frac{2B'_1}{T} \right) f(I) - FI2\,(\nu_m Z_m) C''_\varphi m^2 \tag{34}$$

here $1/k$ replaces the ratio (m/I). Thus, we fit

$$y = \Phi_c - FJ1 \ln(1 + bI^{1/2}) = \Phi°_c - FJ2\,(B_0)\, m$$

$$- \frac{2FJ^2}{\alpha^2 k} B_1 f(I) - FJ2\,(\nu_m Z_m) C m^2 \tag{35}$$

with these parameters to be determined: $\Phi°_c$, B_0, B_1, and C; where

$$\Phi°_c = \bar{C}°_{p2}, \tag{36}$$

$$B_0 = B''_0 + \frac{2}{T} B'_0, \tag{37}$$

$$B_1 = B''_1 + \frac{2}{T} B'_1, \tag{38}$$

and

$$C = C''_\varphi. \tag{39}$$

A similar breakdown of the C^J term is unnecessary and of little use because the fits obtained are quite good without using additional terms. In particular, for cases where limited heat capacity data are available, additional terms cannot be statistically justified.

The correlations of the heat capacity data are discussed in the next section. Calculations of the relative partial molal enthalpy and relative partial molal heat capacities have been illustrated in a previous publication (Staples, et al., 1984). A means of calculating $\bar{L}_2$ and $\bar{J}_2$ values is necessary to evaluate the thermal functions and to arrive at the activity coefficient at any desired temperature. The $\bar{L}_2$ and $\bar{J}_2$ values are derived

from the variation of enthalpy and heat capacity (specifically, the apparent relative partial molal enthalpy, Φ_L, and the apparent molal heat capacity, Φ_C) with molality and temperature (see Staples et al., 1948).

For completeness, the Φ_L data have been fit to the Pitzer Eq. (11).

DISCUSSION OF RESULTS

The starting point for obtaining the literature data was the NBS Special Publication 537, Goldberg, Smith-Magowan, (1979), "Sources of Thermal Property Data". Many of the reprints of articles were obtained from the NBS files. A computer search of a commercial database containing Chemical Abstracts and other sources was made to update sources of data since 1979.

A general weighting procedure was adopted and applied to all correlations. A weight of zero was automatically assigned to any data point greater than three times the standard deviation of fit. In addition, a more qualitative weighting was applied to a few of the lowest concentrations. These were assigned a low weight where larger experimental uncertainties exist for these dilute solutions. In the few other cases where the quality of data of a very few experimentalists was obviously low, appropriate weights were assigned, usually 0.5 or 0.1. These assignments are discussed in the sections dealing with individual salts. Individual weights, if different from 1.0, are specified in the tables. Tables 1-6 containing data on molality, specific heats, apparent molal heat capacities and the assigned weights are given for each salt. Table 7 summarizes the Φ_C° $(\overline{C}^\circ_{p2})$ values, gives the parameters for Eq. (36), and the standard deviation of each fit, for all six salts.

$MnCl_2$

The primary sources of data for the Φ_C correlation for the $MnCl_2$ were Spitzer, et al. (1979), Kapustinsky (1942a, b), and Marignac (1876). An approximate correction of $+0.002$ was applied to Marignac's data to estimate the 25°C values from those reported at 35.5°C. The data of Kapustinsky were corrected by the factor $0.99888/0.9978 = 1.00108$ to adjust their value for pure water to the modern value of 0.99888. Additional data of Blümcke (1884), Kaganovich and Mishchenko (1952), and Voskresenkaya and Ponomareva (1946) were not considered. The data of Blümcke were measured only from 0 to 100°C. Kaganovich and Mishchenko report only enthalpies of dilution at 50°C for mixtures of $MnCl_2 \cdot 4H_2O$ and $MgCl_2$.

$FeCl_2$

Four sources of data of $FeCl_2$ were found. The measurements of Karapet'yants et al. (1977) comprised the main source of data for the correlation. A recent measurement at 4.43 $mol \cdot kg^{-1}$ by Bernarducci et al. (1984) was in excellent agreement. The data reported by Vasil'ev et al. (1973) appeared to be estimated values reported to three figures in C_p; they were assigned zero weight. Data of Perreu (1941) were of poor

quality and not considered. Blümcke (1884) reported measurements from 0 to 100°C only.

$CoCl_2$

Four sources of heat capacity data for $CoCl_2$ were considered. The data of Galinker and Belova (1963) were given at high temperatures for three molalities. Since their average temperature was about 90°C, no correction to 25°C was attempted. There were too few data points (3) to obtain a fit at each temperature.

Kapustinsky et al. (1953) presented duplicate measurements at seven concentrations from $m = 0.3$ to 2.24 mol·kg^{-1} at 25°C. A correction of 1.00108 was applied to their specific heats as discussed earlier. Mishchenko and Podgornaya (1961) reported heat capacity measurements at 2, 18, 25, 50, and 75°C for a dozen molalities ranging from 0.1 to 4.2 mol·kg^{-1} at 25°C up to molalities above 7 mol·kg^{-1} at the higher temperatures.

Spitzer et al. (1978) report Φ_c values at low concentrations (0.05 to 0.25 m). These authors have done very high precision measurements which reveals itself in all their data for a variety of electrolytes reported in their paper.

Specific heats presented by Vasil'ev et al. (1973) were corrected by a factor of 1.00098. Their data for 10 points (0.09 to 3.64 mol·kg^{-1}) fit well with other data.

$NiCl_2$

A half dozen measurements of C_p were reported in the literature.[*] All were used in the fitting procedure. The specific heats from Kapustinsky et al. (1953) and Vasil'ev et al. (1973) were corrected by the 1.00098 factor noted previously. Marignac's (1876) data on $NiCl_2$ were reported at 39.5°C and no reasonable correction to 25°C could be made to come close to the data of other authors, so a weight of zero was assigned. The high precision data of Spitzer et al. (1978) and Perron et al. (1981) were weighted high.

$CuCl_2$

Marignac (1876) reported specific heats at 35°C for six molalities (0.28 to 5.56 mol·kg^{-1}) but no reasonable correction to 25°C could be deduced. His data were fitted at 35°C. The 25°C data were solely from Vasil'ev, et al. (1974) at nine molalities ranging from 0.09 to 4.87 mol·kg^{-1}.

$ZnCl_2$

Only three sources of data were found for $ZnCl_2$ solutions. The measurements of Blümcke (1884) were reported at 0 to 100°C only and were not considered. The fourteen duplicate measurements of Karapet'yants, et al. (1967) were the sole source of data for the fit at 25°C. Marignac (1876) reported data at 35°C and a fit at that temperature was obtained, but no direct correction to 25°C was made.

---

[*] *Note.* In NBS Spec. Publ. 537 (Goldberg and Magowan, 1977) a reference listed under $NiCl_2$, Karapet'yants et al. (1977), does not contain C_p data for $NiCl_2$ (only for $FeCl_2$), but reports density data for $NiCl_2$. A recent reference, Perron et al. (1981), should be added to the references in the NBS Spec. Publ.

Table 1. Heat Capacity of MnCl₂ at 298.15 K.

m(mol·kg⁻¹)	C_p (J·gm⁻¹·K⁻¹)	Φ_c(J·mol⁻¹·K⁻¹)	Weight
Marignac (1876)			
0.27750	3.9899	−180.569	0.5
0.55560	3.8384	−130.534	0.0
1.11110	3.5731	−95.898	0.0
Spitzer et al. (1979)			
0.03954	4.1495	−232.009	
0.05274	4.1398	−227.719	
0.05862	4.1356	−225.247	
0.07819	4.1215	−220.716	
0.07916	4.1207	−221.798	
0.10560	4.1019	−216.688	
0.11739	4.0937	−213.971	
0.13207	4.0835	−211.510	
0.15665	4.0666	−207.721	
Kapustinsky (1942)			
0.05053	4.1424	−208.223	0.0
0.05053	4.1391	−274.890	0.0
0.05053	4.1454	−149.808	0.0
0.09981	4.1077	−200.713	
0.09981	4.1102	−175.206	0.0
0.09981	4.1089	−187.938	0.0
0.19961	4.0478	−149.504	0.0
0.19961	4.0490	−143.059	0.0
0.19961	4.0419	−179.608	0.0
0.39842	3.9188	−160.746	
0.39842	3.9221	−151.912	
0.39842	3.9167	−166.2?1	
0.59884	3.8069	−142.723	
0.59884	3.8074	−141.220	
0.59884	3.8074	−141.972	
0.79846	3.7001	−134.479	
0.79846	3.7010	−133.319	
0.79846	3.7010	−133.319	

Table 1 (continued)

$n.$(mol·kg⁻¹)	C_p(J·gm⁻¹·K⁻¹)	Φ_c(J mol⁻¹·K⁻¹)	Weight
0.99806	3.5950	−133.028	
0.99806	3.6004	−126.884	
0.99806	3.6013	−125.940	
2.10300	3.1431	−97.208	
2.10300	3.1372	−100.736	
2.10300	3.1368	−100.987	
3.85000	2.5617	−97.784	
3.85000	2.5671	−95.685	
3.85000	2.5617	−97.784	

Table 2. Heat Capacity of FeCl₂ at 298.15 K.

m(mol·kg⁻¹)	C_p (J·gm⁻¹·K⁻¹)	Φ_c (J·mol⁻¹·K⁻¹)	Weight
Karapet'yants et al. (1977)			
0.03180	4.1501	−391.993	0.0
0.24500	3.9928	−255.175	
0.43900	3.8685	−217.572	
0.83200	3.6422	−183.934	
0.95700	3.5782	−174.615	
1.46100	3.3510	−142.221	
1.97200	3.1547	−119.685	
2.68800	2.9317	−92.522	
3.91000	2.6472	−56.295	
Vasil'ev et al. (1973)			
0.10000	4.1045	−227.732	0.0
0.20000	4.0417	−175.492	0.0
0.50000	3.8409	−189.935	0.0
0.80000	3.6777	−160.795	0.0
1.00000	3.5690	−157.976	0.0
1.60000	3.3137	−120.959	0.0
Bernarducci et al. (1979)			
4.43000	2.5726	−36.597	

Table 3. Heat Capacity of CoCl₂ at 298.15 K.

m(mol·kg⁻¹)	C_p(J·gm⁻¹·K⁻¹)	Φ_c(J·mol⁻¹·K⁻¹)	Weight
Kapustinsky et al. (1953)			
0.33360	3.9426	—219.618	
0.33360	3.9438	—215.577	
0.62900	3.7539	—188.939	
0.62900	3.7543	—188.220	
0.83850	3.6413	—168.796	
0.83850	3.6401	—170.456	
1.16010	3.4865	—144.481	
1.16010	3.4853	—145.726	
1.44970	3.3556	—132.527	
1.44970	3.3568	—131.498	
1.77940	3.2271	—116.109	
1.77940	3.2275	—115.819	
2.23500	3.0911	—85.523	
2.23500	3.0895	—86.489	
Mishchenko and Podgornaya (1961)			
0.09950	4.0953	—312.530	0.0
0.19560	4.0108	—340.802	0.0
0.28410	3.9371	—341.177	0.0
0.47490	3.8321	—233.494	0.0
0.73780	3.7003	—168.744	
1.09100	3.5146	—152.969	
1.21680	3.4644	—137.758	
1.50860	3.3439	—119.629	
2.13160	3.0836	—113.652	
2.88660	2.8255	—102.156	
3.62290	2.6485	—78.668	
3.96590	2.6079	—57.626	
4.22430	2.5502	—50.873	
Spitzer et al. (1978)			
0.05230	4.1382	—248.569	0.0
0.07026	4.1246	—242.507	0.0
0.08728	4.1119	—238.943	

Table 3 (continued)

m(mol·kg⁻¹)	C_p(J·gm⁻¹·K⁻¹)	Φ_c(J·mol⁻¹·K⁻¹)	Weight
0.10440	4.0990	—237.403	
0.11750	4.0895	—233.375	
0.14080	4.0723	—231.043	
0.17420	4.0482	—226.986	
0.23490	4.0054	—220.280	
Vasil'ev et al. (1973)			
0.09000	4.1219	—102.521	0.0
0.16000	4.0662	—178.813	0.0
0.20200	4.0402	—163.861	0.0
0.31000	3.9607	—190.917	0.0
0.52600	3.8187	—189.751	
1.01600	3.5469	—161.915	
1.34800	3.3903	—145.161	
2.14900	3.0837	—109.440	
2.71300	2.9183	—85.912	
3.63700	2.7130	—50.896	

Table 4. Heat Capacity of NiCl₂ at 298.15 K.

m(mol·kg⁻¹)	C(J·gm⁻¹·K⁻¹)	Φ_c(J·mol⁻¹·K⁻¹)	Weight
Kapustinsky et al. (1953)			
0.35200	3.9212	—224.903	
0.35200	3.9208	—226.146	
0.54000	3.8024	—205.117	
0.54000	3.8007	—208.433	
0.75270	3.6748	—193.971	
0.75270	3.6765	—191.531	
1.19220	3.4405	—173.785	
1.19220	3.4384	—175.811	
1.62130	3.2430	—157.176	

Table 4 *(continued)*

m(mol·kg⁻¹)	C_p(J·gm⁻¹·K⁻¹)	Φ_c(J·mol⁻¹·K⁻¹)	Weight
1.62130	3.2443	—156.239	
2.03800	3.0844	—137.459	
2.03800	3.0828	—138.497	
Marignac (1876)			
0.27780	3.9710	—235.037	0.0
0.55560	3.7894	—210.546	0.0
1.11110	3.4936	—164.306	0.0
2.22220	3.0924	—88.321	0.0
3.70370	2.8719	19.215	0.0
5.55560	2.6008	52.943	0.0
Spitzer et al. (1978)			
0.04488	4.1433	—265.906	2.0
0.07489	4.1199	—259.676	2.0
0.07571	4.1193	—258.782	2.0
0.09012	4.1082	—256.460	2.0
0.12640	4.0806	—252.272	2.0
0.14980	4.0631	—248.985	2.0
0.15110	4.0619	—250.214	2.0
0.20150	4.0252	—243.129	2.0
Vasil'ev (1973)			
0.04000	4.1567	—26.882	0.0
0.07000	4.1332	—122.239	0.0
0.08100	4.1223	—169.121	0.0
0.12800	4.0893	—173.355	0.0
0.16200	4.0616	—200.043	0.0
0.20200	4.0268	—232.818	0.1
0.31200	3.9515	—218.156	0.1
0.36000	3.9226	—204.719	0.1
0.50000	3.8216	—220.053	
0.86600	3.6080	—192.046	
1.33900	3.3656	—171.521	
1.60100	3.2583	—152.948	
2.10600	3.0728	—127.144	

Table 4 *(continued)*

n(mol·kg⁻¹)	C_p(J·gm⁻¹·K⁻¹)	Φ_c(J·mol⁻¹·K⁻¹)	Weight
2.73300	2.8755	—104.359	
3.30100	2.7486	—77.164	
3.86000	2.6042	—70.552	
Perron et al. (1981)			
0.09554	4.1034	—262.647	2.0
0.19580	4.0287	—246.903	2.0
0.29750	3.9564	—236.617	2.0
0.48690	3.8298	—221.504	2.0
0.73680	3.6787	—202.705	2.0
1.00290	3.5342	—185.203	2.0
1.23200	3.4218	—171.401	2.0
1.48740	3.3066	—158.204	2.0
1.73150	3.2078	—145.301	2.0
1.97440	3.1174	—133.801	2.0
2.23800	3.0292	—121.300	2.0
2.51100	2.9461	—109.301	2.0
2.64100	2.9113	—102.801	2.0
2.74500	2.8797	—100.200	2.0
2.97800	2.8199	—91.001	2.0
3.22700	2.7602	—82.002	2.0
3.42100	2.7180	—74.901	2.0
3.53200	2.6922	—72.100	2.0
3.89200	2.6192	—61.401	2.0
4.53900	2.5010	—45.600	2.0
4.98900	2.4272	—36.601	2.0
5.38400	2.3658	—30.201	2.0

Table 5. Heat Capacity of CuCl₂ at 298.15 K.

m(mol·kg⁻¹)	C_p(J·gm⁻¹·K⁻¹)	Φ_c(J·mol⁻¹·K⁻¹)	Weight
Marignac (1876)			
0.27780	4.0012	−138.145	0.0
0.55560	3.8493	−109.970	0.0
1.11110	3.6158	−52.480	0.0
2.22220	3.2593	−4.147	0.0
3.70370	2.8552	1.501	0.0
5.55560	2.6112	−237.219	
Vasil'ev et al. (1979)			
0.08100	4.1181	−237.219	
0.40000	3.9159	−166.135	
0.64200	3.7848	−138.638	
0.86300	3.6826	−112.503	
1.85800	3.3144	−48.710	
2.40600	3.1562	−28.379	
3.05900	3.0045	−6.236	
3.91900	2.8077	3.069	
4.87000	2.6389	15.532	

Table 6. Heat Capacity of ZnCl₂ at 298.15 K

m(mol·kg⁻¹)	C_p(J·gm⁻¹·K⁻¹)	Φ_c(J·mol⁻¹·K⁻¹)	Weight
Karapet'yants et al. (1967)			
0.25000	4.0150	−110.163	
0.25000	4.0150	−110.163	
0.36000	3.9530	−89.755	
0.36000	3.9510	−95.852	
0.64030	3.8074	−61.683	
0.64030	3.8083	−60.442	
1.44000	3.4685	−20.876	
1.44000	3.4702	−19.485	
2.56000	3.1497	27.079	
2.56000	3.1514	27.961	
4.00200	2.9079	78.606	
4.00200	2.9104	79.576	
5.68000	2.6694	97.971	
5.68000	2.6719	98.755	
7.70000	2.6150	153.232	
7.70000	2.6175	153.900	
10.24000	2.5970	199.415	
10.24000	2.5970	199.415	
13.36000	2.5489	225.342	
13.36000	2.5476	225.077	
14.44000	2.5543	235.587	
14.44000	2.5543	235.587	
15.22000	2.4974	229.861	
15.22000	2.4987	230.114	
16.01000	2.4493	225.750	0.0
16.01000	2.4476	225.418	0.0
17.12000	2.3958	222.329	
17.12000	2.3932	221.841	
Marignac (1876)			
0.27778	4.0125	−53.804	0.0
0.55560	3.9037	35.920	0.0
1.11110	3.6995	72.355	0.0
2.22040	3.3305	71.605	0.0
3.70060	2.9464	68.379	0.0
5.55560	2.5991	69.787	0.0

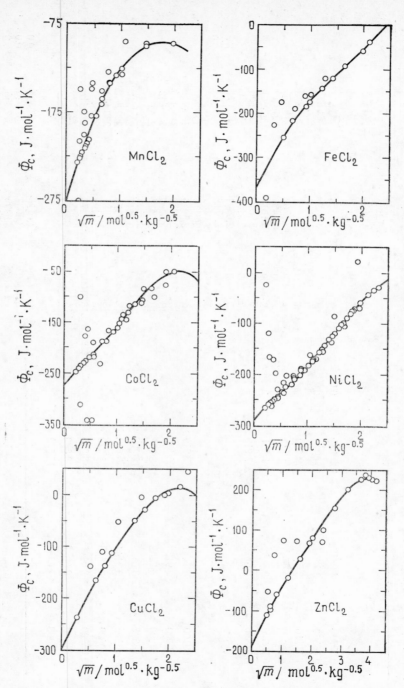

Fig. 1. The apparent molal heat capacity as a function of $m^{1/2}$ at 298 K for six transition metal chlorides

Table 7. Summary of Pitzer parameters for aqueous metal chlorides, at 298.15 K.

Salt	Φ_C^0	B_0	B_1	C
$MnCl_2$	—272.013	—0.117463E-05	0.153173E-03	0.346318E-06
$FeCl_2$	—364.274	—0.763217E-05	—0.214268E-03	0.139662E-08
$CoCl_2$	—274.945	—0.195700E-04	0.373487E-04	0.116233E-05
$NiCl_2$	—290.703	—0.154505E-04	0.416439E-04	0.517962E-06
$CuCl_2$	—295.235	—0.174042E-04	—0.154957E-03	0.101379E-05
$ZnCl_2$	—189.473	—0.100709E-04	—0.899013E-04	0.174255E-06

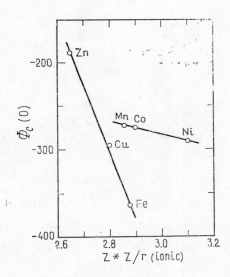

Fig. 2. A correlation of Φ°_C with ionic size for six transition metal ions $(Z = +2)$

SUMMARY

The experimental data are plotted along with the calculated values of Φ_c in Fig. 1 for each salt. The goodness of fit can be observed.

A correlation of coulombic interaction can be made by plotting Φ°_c as a function of Z^2/r, where Z is the charge on the cation, $+2$ in all cases here, and r is the ionic radius based on the recommendations of Staples (1984). Figure 2 shows the observed periodic trend for the six transition metal cations. This kind of correlation of parameters may be useful in predicting values for salts that have no measurement of Φ°_c.

REFERENCES

1. Bernarducci, E. E., Morss, L. R., and Miksztal, A. R. (1979): J. Soln. *Chem.,* **8**, 717.
2. Blümcke, A. (1884): *Ann. Phys.* (Leipzig), **23**, 161.
3. Galinker, I. S., and Belova, N. A. (1963): *J. Gen. Chem.* USSR (Eng. Trans.) **33**, 3047, Zh. Obshch. Khim, **33**, 3119.
4. Kaganovich, Yu. Ya. and Mishchenko, K. P. (1951): *J. Gen. Chem.* USSR (Eng. Trans.) **21**, 29; Zh. Obshch. Khim. **21**, 28.
5. Kaganovich, Yu. Ya. and Mishchenko, K. P. (1952): Dokl. Akad. Nauk. SSSR, **87**, 89.
6. Kapustinsky, A. F. (1942a): *Acta Physicochim.* URSS, **17**, 152.
7. Kapustinsky, A. F. (1942b): *Zh. Obshch. khim.* **12**, 180.
8. Kapustinsky, A. F., Yakushevskii, B. M., and Drakin, S. I. (1953): *Zh. Fiz. khim.* **27**, 588.
9. Karapet'yants, M. Kh., Drakin, S. I., and Lantukhova, L. V. (1967): *Zh. Fiz. khim.* **41**, 2653; Russ. *J. Phys. Chem.* (Eng. Trans.) **41**, 1436.
10. Karapet'yants, M. Kh., Vasil'ev, V. A., and Sanaev, E. S. (1977): Russ. *J. Phys. Chem.* (Eng. Trans.) **51**, 1281; Zh. Fiz. Khim. **51**, 2180.
11. Marignac, C. (1876): *Arch. Sci. Phys. Nat.* **55**, 113.
12. Mishchenko, K. P., and Podgornaya, E. A. (1961): Mishchenko, K. P. and Podgornaya, E. A.: *J. Gen. Chem.* USSR (Eng. Trans.) **31**, 1628; *Zh. Obshch. Khim.* **31**, 1743.
13. Perreu, J., Hebd, C. R. (1941): Seances Acad. Sci. **212**, 701.
14. Phutela, R. C., and Pitzer, K. S. (1983): *J. Soln. Chem.,* **12**, 201.
15. Pitzer, K. S. (1973): *J. Phys. Chem.,* **77**, 268.
16. Pitzer, K. S., and Brewer, L. (1961): Thermodynamics — B. N. Lewis and M. Randall, 2d ed. McGraw-Hill Book Co., New York, NY, 728 pp.
17. Silvester, L. F., and Pitzer, K. S. (1977): *J. Phys. Chem.* **81**, 1822.
18. Smith-Magowan, D., and Goldberg, R. N. (1979): *Natl. Bur. Stand. Spec. Publ.* **537**, US Gov. Printing Office, Washington, DC 20402.
19. Spitzer, J. J., Olofsson, I. V., Singh, P. P., and Hepler, L. B. (1979): *Thermochim. Acta* **28**, 155.
20. Spitzer, J. J., Singh, P. P., McCurdy, K. G., and Hepler, L. G. (1978): *J. Soln. Chem.* **7**, 81.
21. Staples, B. R., and Nuttall, R. L. (1977): *J. Phys. Chem. Ref. Data,* **6**, 385.
22. Staples, B. R. (1984) manuscript in preparation.
23. Vasil'ev, V. A., Sanaev, E. S., and Karapet'yants, M. (1973): Kh. Tr. Mosk. Khim. Tekhnol. Inst. **75**, 13.
24. Vasil'ev, V. A., Karapet'yants, M., Sanaev, E. S., and Novikov, S. N. (1974): Russ. *J. Phys. Chem.* (Eng. Trans.) **48**, 1398.
25. Voskresenskaya, N. K., and Ponomareva, K. S. (1946): *Zh. Fiz. Khim.,* **20**, 433.

Use of the Thermodynamic Perturbation Theory to Determine Non polar Gas Distribution Between Vapour and Water Phases of Heat Power Plants

S. N. LVOV

Leningrad Gorny Institute, Leningrad, USSR

V. G. KRITSKY

*All-Union Scientific Research and Project Institute
of Complex Power Technology, Leningrad, USSR*
and

O. I. MARTYNOVA

Moscow Power Engineering Institute, Moscow, USSR

Water and vapour-dissolved nonpolar gases, in particular, oxygen and hydrogen and their distribution between phases are significant for the process of formation and transfer of corrosion products in thermal and nuclear power plants. In power engineering analytical methods are used to measure gas content in water and vapour; these are based on the known laws governing equilibrium and distribution of gas between liquid and gas phases at the sampling parameters. Therefore, the knowledge of thermodynamic characteristics such as Henry constant, $K^\circ{}_H$, and its dependence on temperature and pressure is necessary for comprehensive thermodynamic analysis of all the interaction processes between constructional elements of heat power circuits of water-cooled plants.

185

At the present time, however, most of experimental data on K°_H refer to temperatures below 350 K [1]; only one work [2] contains reliable data on K°_H at 520 K for Ne, Ar, Kr, Xe, and CH_4. Practically there are no reliable values of K°_H above the solvent saturated vapour pressure. This is, first, due to considerable difficulties associated with carrying out experiments at high temperatures for determining gas concentration in liquid and vapour phases, and, second, due to the nonavailability of precise $PVTX$ relations for appropriate vapour-gas mixtures over a wide range of state parameters.

However, a sufficiently reliable model for the thermodynamic fluid perturbation theory (TFPT) [3], based on Lennard-Jones pair potential (LJ), is available. It allows one to calculate theoretically thermodynamic characteristics of liquids and their solutions over a wide range of temperatures and pressures with a minimum number of empirical parameters.

The expression to define $K_H{}^{\circ}$ is known to have the form [1, 2]

$$K^{\circ}_H = \frac{N_2^{gas} \cdot f^{gas}}{N_2^{liq} \cdot \gamma_2^{liq}} \cdot P = \exp(\Delta\mu_2^{\circ}/RT) \qquad (1)$$

where $N_2{}^{gas}$, $f_2{}^{gas}$ are the mole fractions of gas and its fugacity coefficient in the gas phase; $N_2{}^{liq}$, $\gamma_2{}^{liq}$ are the mole fractions of gas and its activity coefficient in the liquid phase; P is pressure, T is absolute temperature and R is the gas constant. $\Delta\mu_2{}^{\circ}$ in Eq. (1) is the difference between standard values of chemical potential of gas in the liquid and gas phases; based on TFPT, it may be expressed as [3]

$$\Delta\mu_2^{\circ} = \Delta\mu_2^{0,ref} + \frac{4\pi N}{V} \int_{\sigma_{12}}^{\infty} U_{12}(r)\, g_{12}^{ref}(r)\, r^2\, dr, \qquad (2)$$

where $\Delta\mu_2{}^{\circ,ref}$ is the value of $\Delta\mu_2{}^{\circ}$ for the reference system; N is the Avogadro number; V is the solvent mole volume; $g_{12}{}^{ref}(r)$ is the binary function of distribution of the basic system; $U_{12}(r)$ is the pair potential of interaction between nonpolar gas and solvent molecules; r is the distance from dissolved molecule center; $\sigma_{12} = 1/2(\sigma_{11} + \sigma_{22})$; σ_{11} and σ_{22} are the hard sphere diameters of the solvent and solute molecules.

As the pairwise interaction potential we used the potential of the form [4]

$$U_{12}(r) = 4\varepsilon_{12}^{*} \left[\left(\frac{\sigma_{12}^{*}}{r} \right)^{12} - \left(\frac{\sigma_{12}^{*}}{r} \right)^{6} \right] \qquad (3)$$

where
$\varepsilon_{12}{}^{*} = \varepsilon_{12}\xi^2$, $\sigma_{12}{}^{*} = \sigma_{12}\xi^{-1/6}$, $\varepsilon_{12} = (\varepsilon_{11} \cdot \varepsilon_{22})^{1/2}$, $\xi = 1 + 0.25\, \alpha_2{}^{*}\mu_2{}^{*2}\,(\varepsilon_{11}/\varepsilon_{22})^{1/2}$, ε_{11} and ε_{22} are the energy parameters of LJ potential for solvent and solute, respectively; $\alpha_2{}^{*} = \alpha_2/\sigma_{22}{}^3$, $\mu_1{}^{*} = \mu_1/(\varepsilon_{11}\sigma_{11})^{1/2}$ are the reduced polarizability of gas molecules and the dipole moment of a water molecule, respectively; α_2 and μ_1 are the polarizability of gas molecules and the dipole moment of a water molecule, respectively.

186

Using the model of hard sphere solution in the approximation of Percus-Jevick equation as the reference system and integrating numerically the second term of the Eq. (2), we obtained an approximate equation for calculating $\Delta\mu^{\circ}_2$

$$\Delta\mu^{\circ}_2 = \Delta\mu_2^{\circ,ref} - (RTY_{12}/T^*_{12})(21.35 + 3.63Y_{12} - 0.366Y^2_{12}) \qquad (4)$$

where $Y_{12} = \pi N\sigma^*_{12}/6V$, $T^*_{12} = RT/\varepsilon^*_{12}N$, and $\Delta\mu_2^{\circ,ref}$ can be calculated theoretically.

The value ε_{11}, used in Eqs. (3) and (4), is the energy parameter of the effective LJ pair potential for a water molecule, which for a high-temperature fluid is known only at low density, and equals $\varepsilon_{11}/k = 809$ K (k is the Boltzmann's constant) [5]. Assumption is made that ε_{11}, first, is independent of the shape of solute molecule and has the same value for nonpolar gases, and, second, increases exponentially with rise in temperature up to 809 K. Solving the inverse problem on determining ε_{11} at 298 K from the experimental values of $\Delta\mu_2^{\circ}$ available for Ne, Ar, Kr, and Xe [1] yielded $\varepsilon_{11}/k = (238\mp10)$ K and allowed us to obtain an expression for calculating ε_{11} at any temperature of existence of water fluid

$$\varepsilon_{11}/k = 809 \exp(-364.8/T) \qquad (5)$$

In the calculations, use was made of the values of σ_{22} and ε_{22} taken from Ref. [6], as the most reliable ones, and σ_{11} and μ_1 were taken equal to 2.64 Å and 1.80 D, respectively [5].

Table 1. Comparison of experimental $\Delta\mu^{\circ}_2$ [2] with those calculated by Eqs. (4) and (5)

Gas	$\Delta\mu^{\circ}_2$ (kJ/mol) at temperature T (K)				
	298	313	328	400	520
Ne	29.06*	30.65	32.13	38.28	45.71
	29.04	30.65	32.13	38.30	46.10
Ar	26.27	28.12	29.87	36.56	43.85
	26.25	27.80	29.70	36.50	43.90
Kr	24.82	26.78	28.59	35.48	43.61
	24.80	26.50	28.20	35.00	42.90
Xe	23.50	25.49	27.38	34.43	—
	23.46	25.50	27.10	34.20	41.31
CH₄	26.28	28.20	29.99	36.58	43.89
	26.25	28.10	30.00	36.20	44.20

* Numerator denotes experimental data, taken from [2]; denominator denotes values calculated by Eqs. (4) and (5).

Thus, using the Eqs. (4) and (5), $\Delta\mu^{\circ}_2$ and hence K°_H can be calculated theoretically in a wide range of temperatures and pressures.

The calculation accuracy can be estimated by comparing the theoretically calculated values of $\Delta\mu°_2$ with those obtained experimentally. Table 1 shows that the calculated and the experimental values coincide with an accuracy of 0.5 kJ/mol.

Theoretically calculated by the Eqs. (1) and (4) values of $K°_H$ for oxygen and hydrogen at 298-623 K and at the saturated steam pressure are given in Table 2. ε_{22} were taken from Ref. [5] and σ_{22} were calculated by solving the inverse problem and using the values of $\Delta\mu°_2$ at 298 K [1], and were found to be equal to $\sigma_{22} = 3.23$ Å for oxygen and $\sigma_{22} = 2.89$ Å for hydrogen. This was in good agreement with the conventional values [4, 5].

Table 2. Henry constant $K°_H$ for hydrogen and oxygen dissolved in water at 298-623 K

Gas	$K°_H$ (GPa) at temperature T (K)							
	298	323	373	423	473	523	573	623
H_2	7.162	7.950	7.986	6.670	4.831	3.099	1.723	**0.713**
O_2	4.385	5.266	5.917	5.278	3.948	2.553	1.405	0.572

The possibility of calculating theoretically $K°_H$ for nonpolar gases at any temperature and pressure allows one to solve a number of problems of interest to nuclear and power engineering.

Ionizing radiation in the water-cooled reactor core causes water to decompose with the formation of radiolysis products — hydrogen and oxygen — on stoichiometric ratio of 2 : 1 [7]. Further, part of these gases goes into the vapour phase and is carried over in the circuit.

Cohen and Martynova [7, 8] believed that the concentration ratio of radiolytic gases removed with steam should remain practically the same as in the liquid phase. However, the experimental studies do not confirm this assumption [9]; this is usually explained by possible occurrence of processes that proceed with the release of hydrogen or absorption of oxygen which is possible due to either metal corrosion or decomposition of organic impurities present in water.

Based on Eq. (1) and assuming that solubilities of every gas in the mixture are independent of each other, and taking, without considerable error, that $f_{O_2}^{gas}/f_{H_2}^{gas} = \gamma_{O_2}^{liq}/\gamma_H^{liq} = 1$, we can write the expression for the concentration ratio of hydrogen and oxygen in the vapour phase of separating drum, $N^{gas}/N_{O_2}^{gas}$, provided their ratio in the liquid phase is known ($N_{H_2}^{liq}/N_{O_2}^{liq} = 2$)

$$\frac{N_{H_2}^{gas}}{N_{O_2}^{gas}} = \frac{K_H^{H_2}}{K_H^{O_2}} \cdot \frac{N_{H_2}^{liq}}{N_O^{liq}} \tag{6}$$

Having determined $K_H^{H_2}/K_H^{O_2}$ at 560 K from Table 2, we calculate the deviation from stoichiometry in the gas phase as compared to the liquid one

$$\frac{N_{H_2}^{gas}}{N_{O_2}^{gas}} = 1.24 \, \frac{2}{1} = 2.48 \tag{7}$$

Close to this, a value (2.50 ± 0.25) (Table 3) was obtained in analyzing the experimental data [9].

Table 3. Experimental data on gas concentration ratio in saturated steam of separating drums at 560 K [9]

Gas in steam and mole fraction relation	Experimental values		
	maximum	minimum	average
Oxygen (ppb)	5100	8500	6800
Hydrogen (ppb)	660	1400	1030
$\dfrac{\text{Mole } H_2}{\text{Mole } O_2}$	2.35	2.65	2.50

Good agreement between the experimental and calculated values enables us to make an important conclusion that the practically observed deviation from stoichiometric ratio of hydrogen and oxygen concentrations in steam medium of the boiling water reactor depends, first of all, upon nonpolar gas dissolution features at steam generation parameters, namely, the difference in Henry constants which are easily calculated by Eq. (4).

REFERENCES

1. Wilhelm E., Battino R., and Wilcock R. J. Low-pressure solubility of gases in liquid water. *Chem. Rev.,* 1977, v. 77, No. 2, p. 219-262.
2. Grovetto R., Fernandez-Prini R., and Japas H. L. Solubilities of inert gases and methane in H_2O and in D_2O in the temperature range from 300 to 600 K. *J. Chem. Phys.,* 1982, v. 76, No. 2, p. 1077-1086.
3. Boublik T., Nezbeda I., and Hlavaty K. Statistical thermodynamics of simple liquids and their mixtures. Elsevier Scientific, Amsterdam, 1980.
4. Hirschfelder J. O., Curtiss Ch. F., and Bird R. B. Molecular Theory of Gases and Liquids. J. Wiley, New York, 1954.
5. Reid R. C., Prausnitz J. M., and Sherwood T. K. The Properties of Gases and Liquids. McGraw-Hill, New York, 1977.
6. Kaplan I. G. Vvedenye v teoriyu mezhmolekulyarnikh vzaimodeistvii. Nauka, Moscow, 1982.
7. Cohen P. Water coolant technology of power reactors. An AEC Monograph. Gordon and Breach science publishers. New York, London, Paris, 1969.
8. Martynova O. I. Transport and concentration process of steam and water impurities in steam generating systems. Proceedings of the 9th International Conference on the Properties of Steam, Munich, 1979, p. 547-562.
9. Mamet V. A. Povedenye gazovykh i organicheskikh primesey v teplonositele AES c RBMK. *Teploenergetika,* 1982, No. 7, p. 14-17.

Volumetric Properties of Liquid and Vapor K_2CO_3 Solutions at Phase Equilibrium

B. E. NOVIKOV and N. A. KORZHAVINA

Krzhizhanovsky Power Engineering Institute, Moscow, USSR

ABSTRACT

The report presents the results of experimental studies of *p-V-T-x* properties of the H_2O-K_2CO_3 system at vapor-liquid phase equilibrium at temperatures of up to 400°C and liquid phase concentrations of up to 45% by weight. The data obtained are presented in the form of a phase equilibrium diagram. An analytical relationship is proposed for calculating vapor pressure above water solutions of K_2CO_3 in the investigated range of the state parameters.

INTRODUCTION

High-concentration water solutions of K_2CO_3 are used as ionizing additive in the MHD-generator channel. The purpose is to artificially attain the required ionization of the organic fuel-combustion products at temperatures of about 3000 K. To perform physical and engineering calculations when designing and operating the MHD-units, it is necessary to have reliable *p-V-T-x* data of potash-water solutions over a wide range of temperatures and concentrations. However, for the H_2O-K_2CO_3 system rather a limited number of publications is devoted to studies of vapor pressure of water solutions and their densities.

The work by M. I. Ravich, F. E. Borovaya, and E. G. Smirnova [1] contains data on vapor pressure in unsaturated and saturated water solutions of K_2CO_3 over the range from 250 to 450°C, as well as approximate data on pressure of saturated solutions at 500, 550 and 600°C. Besides, G. W. Morey and T. W. Chen [2] present approximate

values of vapor pressure for saturated water solutions of this salt at 374, 400, 500 and 600°C.

The volumetric properties of K_2CO_3-water solutions were investigated only in some works. I. M. Rodnyansky, V. I. Korobkov and I. S. Galinker [3] used unique equipment to measure specific volumes of potash-water solutions in the range of concentrations between 3.34 and 29.3% by weight and of temperatures between 25 and 340°C. However, it is impossible to estimate the measurement accuracy for lack of necessary information in this article. D. M. Ginzburg, N. S. Pikulina, and V. P. Litvin [4] used the pycnometric method for measuring specific volumes of K_2CO_3-water solutions over a wide range of concentrations from 16 to 62% by weight but between 25 and 130°C.

EXPERIMENTAL

The present work was carried out with the use of gamma-ray examination on the experimental unit described in Refs. [5] and [6].

The investigation procedure was as follows. A strictly definite amount of a potassium carbonate-water solution of known initial concentration was pumped into an autoclave from which air was evacuated beforehand. The experiments were carried out on slowly (0.5 deg/min) and continuously heating the solution and registering all the necessary parameters at definite time intervals. Special studies [5] have revealed that spontaneous establishment of the equilibrium state in the liquid-vapor system is extremely difficult, especially in the near-critical region of the state parameters. Therefore, the autoclave was equipped with an electromagnetic stirrer which ensured reliable and rapid achievement of the system phase equilibrium parameters.

Vapor and liquid solution temperature was measured by two chromel-copel thermocouples. Vapor pressure above the solution was recorded by several different range pressure gauges.

To determine densities of vapor and liquid phases, the autoclave was exposed to narrow beams of gamma-rays at places of location of the respective phases. Sn-113 isotopes were used as a source of gamma-rays. Densities of vapor and liquid solutions were calculated by the equation obtained from the exponential law of attenuation of gamma-ray monochromatic beam on passing through the medium being investigated.

The distinguishing feature of binary water-salt systems at the vapor-liquid equilibrium state when heated in a closed volume is that with increase in temperature the composition of each of the co-existing phases continuously changes both owing to evaporation of more volatile solvent (water) and dissolution of less volatile compounds in water vapor. The concentration of matter in the liquid phase was calculated by the equation obtained from considering the material balance of the system when heated within a closed volume.

The errors of measurement of main experimental parameters were as follows: temperature $\pm 0.4°C$, pressure ± 0.04-0.1 MPa depending on the measured pressure, density ± 2 kg/m³.

RESULTS

The above-described procedure was used to investigate p-V-T-x properties of the H_2O-K_2CO_3 system at vapor-liquid phase equilibrium in a wide range of state parameters.

The set of experimental data obtained throughout the investigated range radically differed both in concentration and temperature of separate experimental points since the investigation was carried out with continuous heating of the system. Computer processing of these data enabled us to obtain densities of liquid K_2CO_3 solutions at whole-numbered concentrations and also the corresponding to them values of temperature, pressure and density of their equilibrium vapor solutions.

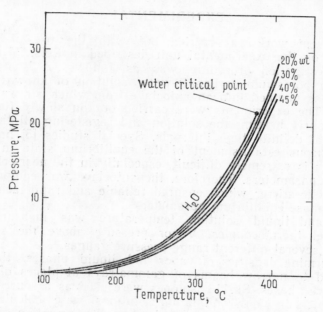

Fig. 1. Vapor pressure above K_2CO_3 water solutions at different concentrations *versus* temperature

The results of investigating vapor pressure of unsaturated water solutions of K_2CO_3 of different concentrations are shown in Fig. 1. An analysis of these data shows that the relationship between pressure and temperature at a constant concentration of liquid phase is approximated by the linear relationship

$$\lg P = A - \frac{B}{T}$$

where A and B are empirical constants; P is pressure, MPa; T is temperature, K.

The values of coefficients A and B for the investigated range of K_2CO_3-water solution concentrations were obtained by computer processing of vapor pressure experimental data by the least-squares method.

Concentration, % by weight	A	B
5.0	4.5308	2065.1
10.0	4.5487	2078.3
15.0	4.5620	2089.6
20.0	4.5713	2100.6
30.0	4.5694	2112.4
40.0	4.5531	2122.7
43.0	4.5450	2125.7
45.0	4.5367	2127.1

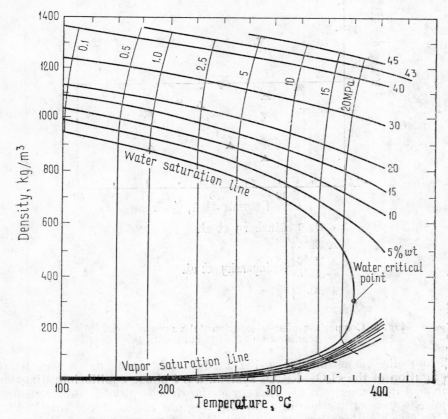

Fig. 2. Phase equilibrium diagram for liquid and vapor K_2CO_3 solutions

Isoconcentrates of potassium carbonate water solutions in $\lg P$ and $1/T$ coordinates are shown by straight lines whose slope slightly increases with increase in concentration. The maximum deviation between the calculated and the experimentally determined vapor pressures is $\pm 0.5\%$.

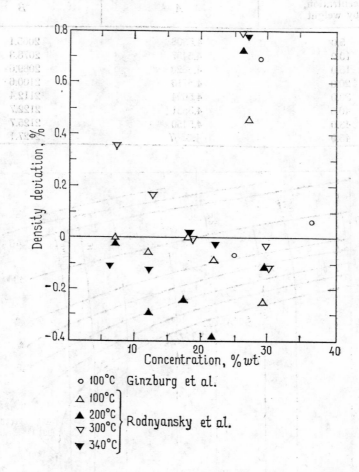

Fig. 3. Comparison of new experimental values of K_2CO_3 water solutions with the literature data

From the experimental p-V-T-x data obtained as a result of investigation of the H_2O—K_2CO_3 system, a ρ-t phase equilibrium diagram shown in Fig. 2 was plotted. Therein is shown a family of isobars. Above the pure water saturation line lies the region of existence of boiling K_2CO_3-water solutions, which is shown by isoconcentrate lines. Below the pure water vapor saturation line are the curves representing the state

of K_2CO_3 vapor solutions in equilibrium with water solutions of constant composition.

The obtained K_2CO_3-water solution densities were compared with the experimental data of Refs. [3] and [4]. The results are shown in Fig. 3. The average deviation of density is $\pm 0.25\%$, maximum deviation for separate experimental points reaches 0.8%.

To obtain a more complete picture of phase equilibriums in the H_2O-K_2CO_3 system, it is necessary to continue investigations aimed at studying the state parameters in the critical region with widening the temperature and concentration ranges.

REFERENCES

1. Ravich, M. I., Borovaya F. E. and Smirnova E. G. (1968): *Zhurnal neorganitcheskoy khimii,* 13(7), 1922-1927.
2. Morey, G. W. and Chen T. W. (1956): *Journal of American Chemical Society,* No. 78, 4245-4249.
3. Rodnyansky I. M., Galinker I. S., Korobkov V. I. (1962): *Zhurnal fizitcheskoy khimii,* 10, 2216-2219.
4. Ginzburg, D. M., Pikulina N. S., and Litvin V. P. (1964): *Zhurnal prikladnoy khimii,* 37, (11), 2353-2357.
5. Khaibullin, I. Kh., Novikov B. E., and Korzhavina N. A. (1981): *Proceedings of the 8th Symposium of Thermophysical Properties,* v. 2, 357-360.
6. Novikov, B. E. (1983): *Teploenergetika,* No. 9, 61-63.

Viscosity of Aqueous Potassium Chloride Solutions Over a Wide Range of State Parameters

R. I. PEPINOV, V. J. YUSUFOVA, N. V. LOBKOVA,
and I. A. PANAKHOV

Azerbaijan I. G. Yesman Research Institute of Energetics,
370143, Baku, USSR

The application of sea and geothermal waters in power engineering calls for the solution of a number of problems, including the determination of thermophysical properties, in particular viscosity, over a wide range of state parameters. As, depending on the location, natural saline waters have different chemical compositions, it is expedient to theoretically calculate their thermophysical characteristics. It is therefore necessary to know the values of thermophysical parameters of salt solutions constituting saline waters over a wide range of state parameters. Potassium chloride is one such component.

Table 1. Parameters and errors of studies conducted by various authors into viscosity of aqueous potassium chloride solutions

Reference	Temperature, °C	Pressure, MPa	Concentration, % mass	Error, %
1	12.5-42.5	0.1	0-0.49	1
2	30-55	0.1	1-5.9	1
3	25-50	0.1	0.1-4	1
4	20-150	—	0.09-3.6	—
5	25-150	0.1-31.0	0.5-4.39	1
6	100-350	20-150	5-25	1.5-2

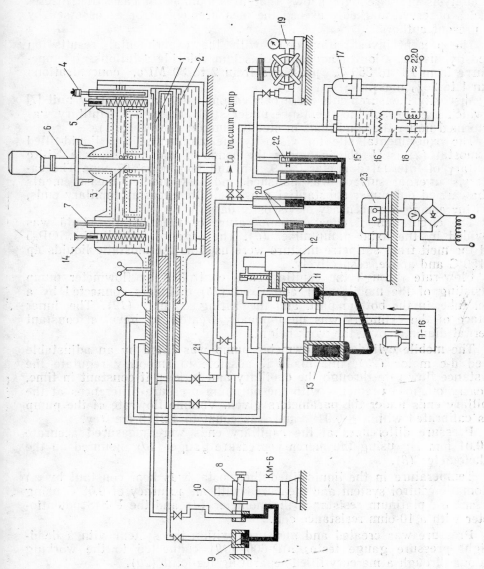

Fig. 1. Diagram of the experimental unit: 1 — capillary tube; 2 — thermostat; 3 — cooler; 4 — resistance thermometer; 5 — controllable heater; 6 — stirrer; 7 — thermocouple; 8-10 — differential pressure gauge; 11-13 — flowmeter; 14 — heater; 15-18 — filling system; 19 — pressure gauge; 20-22 — high pressure vessels; 21 — filters; 23 — d-c motor

to vacuum pump

Parameters and errors of studies conducted by various authors into viscosity of aqueous potassium chloride solutions are listed in Table 1. An analysis of these data shows that in a number of cases the discrepancies observed markedly exceed the measurement error as assessed by the present authors.

The present investigation deals with the experimental results on viscosity investigation of aqueous potassium chloride solutions at temperatures from 25° to 350°C, pressures from 2 to 30 MPa, concentrations from 1 to 20% mass.

Viscosity measurements were carried out using a laboratory unit [7] in which the flow of the substance under study through a capillary was performed under the conditions of steady laminar flow (Fig. 1).

The experimental site for viscosity measurements in a liquid-filled thermostat (2) is U-tube in one of the bends of which there is a Ni-Re capillary of internal diameter 0.349 mm and length 553.07 mm. The capillary's internal surface was polished and this enabled measurements to be carried out under substantial pressure differences at capillary ends, thus increasing the accuracy of viscosity data.

The circulation of constant-temperature or thermostatic liquid was provided by the axial pump-mixer (6). The organosilicon liquid SOP-5 and the melt from saltpetre mixture were used as thermostatic liquids up to 150°C and above, respectively.

Flow-rate through the capillary was controlled by flowmeter pump consisting of the mobile (11) and fixed (13) cylinders connected by a steel tube in the bottom part, and the lifting column (12). The inner surface of the cylinders was mechanically treated to have a constant internal diameter over the work height.

The mobile cylinder of flowmeter pump was lifted by an adjustable speed d-c motor (23); this made it possible to smoothly regulate the substance flow rate through the capillary and to keep it constant in time, to carry out measurements with the optimum pressure differences at the capillary ends under the parameters given. Volume flowrate of the pump was calibrated within $\pm 0.04\%$ and was equal to 1.2113 cm^3/rev.

Pressure differences at the capillary ends were measured accurate to 0.01 mm Hg using the mercury pressure gauge (10) mounted on the cathetometer (8).

Temperature in the liquid-filled thermostat was kept constant by an automatic-control system and measured with an accuracy of 0.03°C using a standard platinum resistance thermometer (4) and the P-348 potentiometer with a 10-ohm resistance coil.

Pressure was created and measured within the system with a dead-weight pressure gauge tester MP-600 (19) connected to the working contour through a mercury-filled separating tee-joint (20).

The outflow time through the capillary tube was recorded by the printing chronograph with an accuracy of 0.01 sec.

Maximum relative error in measuring viscosities of aqueous potassium chloride solutions was $\pm 1\%$ with regard to the reference error.

The viscosities of potassium chloride solutions were measured from isotherms at 2, 5, 10, 20 and 30 MPa.

The solution densities necessary for viscosity calculations were measured to ±0.15% by the hydrostatic method.

Table 2. Viscosity of aqueous potassium chloride solutions, mPa s

t, °C	P, MPa	Concentration, % mass				
		$S=0$	$S=1$	$S=5$	$S=10$	$S=20$
1	2	3	4	5	6	7
25	2	0.8915	0.8914	0.8893	0.8948	0.9298
	10	0.8877	0.8856	0.8870	0.8966	0.9312
	20	0.8852	0.8860	0.8897	0.8988	0.9339
	30	0.8842	0.8930	0.8983	0.9013	0.9360
30	2	0.7995	0.7973	0.7988	—	—
	10	0.7990	0.8007	0.8012	—	—
	20	0.7988	0.8044	0.8034	—	—
	30	0.7990	0.8093	0.8061	—	—
50	2	0.5484	0.5506	0.5625	0.5769	0.6318
	10	0.5501	0.5532	0.5646	0.5789	0.6338
	20	0.5521	0.5553	0.5673	0.5818	0.6362
	30	0.5543	0.5585	0.5700	0.5848	0.6387
75	2	0.3798	0.3832	0.3973	0.4162	0.4593
	10	0.3815	0.3843	0.3990	0.4180	0.4609
	20	0.3836	0.3857	0.4009	0.4201	0.4630
	30	0.3857	0.3874	0.4034	0.4222	0.4651
100	2	0.2829	0.2868	0.3004	0.3198	0.3554
	10	0.2848	0.2884	0.3022	0.3212	0.3569
	20	0.2870	0.2932	0.3045	0.3229	0.3589
	30	0.2894	0.2920	0.3068	0.3245	0.3610
150	2	0.1825	0.1861	0.1971	0.2150	0.2465
	10	0.1846	0.1879	0.1988	0.2165	0.2483
	20	0.1872	0.1902	0.2012	0.2185	0.2505
	30	0.1896	0.1926	0.2036	0.2206	0.2529
200	2	0.1342	0.1384	0.1467	0.1611	0.1937
	10	0.1359	0.1404	0.1483	0.1627	0.1957
	20	0.1384	0.1423	0.1505	0.1645	0.1980
	30	0.1410	0.1447	0.1528	0.1663	0.2003
250	5	0.1060	0.1100	0.1183	0.1295	0.1593
	10	0.1073	0.1114	0.1194	0.1302	0.1602
	20	0.1098	0.1138	0.1216	0.1320	0.1625
	30	0.1125	0.1162	0.1240	0.1341	0.1652
300	10	0.0866	0.0909	0.0993	0.1082	0.1359
	20	0.0899	0.0937	0.1018	0.1101	0.1376
	30	0.0931	0.0968	0.1043	0.1118	0.1396
350	20	0.0690	0.0728	0.0830	0.0903	0.1157
	30	0.0749	0.0786	0.0866	0.0936	0.1194

The results of viscosity measurements for aqueous potassium chloride solutions are given in Table 2. The experimental results obtained are compared with the literature data (Fig. 2).

An analysis of the experimental data has shown that on the 25° and 30°C isotherms for the solutions with 1 and 5% mass concentration an anomalous viscosity variation is observed with increasing pressure and solution concentration.

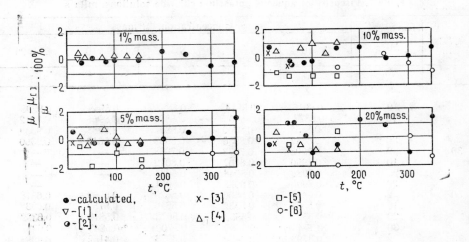

Fig. 2. Comparison of experimental viscosities of aqueous potassium chloride solutions with calculated (Eq. 1) and cited data

Salt dissolution in the solvent (water in our case) usually causes the solution viscosity to increase as compared with the solvent (water) viscosity.

In the case under study, dissolution of potassium chloride in water results in a decrease in the solution viscosity compared to that of pure water. This is explicitly seen on the 25°C isotherm at pressures up to 20 MPa for potassium chloride solutions of 1 and 5% mass concentration, and more vaguely on the 30°C isotherm at pressures less than 5 MPa.

At temperatures above 35°C anomalous variation in viscosity of aqueous potassium chloride solutions is not observed with rise in concentration.

Viscosity of water is known to vary abnormally at temperatures below 35°C with pressure rise; the latter tends to lower its viscosity. This phenomenon is observed also on the 25°C isotherm at pressures up to 10 MPa for aqueous potassium chloride solutions of 1 and 5% mass concentrations. The anomalous variation in viscosity of potassium chloride solutions with rise in concentration and pressure can be explained by a partial destruction of hydrogen bonds in water structure by dissolved salt ions. As a result, the water dipoles become more mobile and the power consumption necessary to form viscous-flow solution as compared with pure water decreases.

The obtained experimental data on viscosity of aqueous potassium chloride solutions were approximated by the Jones-Dole equation:

$$\mu = \mu_0 (1 + AC^{1/2} + BC + DC^2) \tag{1}$$

where μ_0 and μ are the viscosities of water and solution at the same state parameters; A, B, D — equation coefficients; C — concentration of the solution (mol/1000 g H_2O).

As functions of temperature and pressure the coefficients A, B and D were determined from our experimental data by the least-squares technique:

$$A = \sum_{i=0}^{3} \sum_{i=0}^{1} a_{ij} t^i P^j$$

$$B = \sum_{i=0}^{3} \sum_{j=0}^{1} b_{ij} t^i P^j \tag{2}$$

$$D = \sum_{i=0}^{3} \sum_{j=0}^{1} d_{ij} t^i P^j$$

where P is pressure, MPa; t is temperature, °C.

The values of the coefficients of Equation (2) are given in Table 3.

Table 3. Values of the coefficients in Eq. (2)

Coefficients	i \\ j	0	1
a_{ij}	0	$-5.898699 \cdot 10^{-3}$	$1.895525 \cdot 10^{-3}$
	1	$6.393323 \cdot 10^{-5}$	$-4.998739 \cdot 10^{-5}$
	2	$-8.602628 \cdot 10^{-7}$	$3.402237 \cdot 10^{-7}$
	3	$7.779406 \cdot 10^{-9}$	$-7.196612 \cdot 10^{-10}$
b_{ij}	0	$-7.495964 \cdot 10^{-2}$	$-5.884434 \cdot 10^{-4}$
	1	$2.930162 \cdot 10^{-3}$	$2.145393 \cdot 10^{-5}$
	2	$-1.407107 \cdot 10^{-5}$	$-1.749033 \cdot 10^{-7}$
	3	$1.853783 \cdot 10^{-8}$	$3.722229 \cdot 10^{-10}$
d_{ij}	0	$1.917070 \cdot 10^{-2}$	$-1.961889 \cdot 10^{-4}$
	1	$-4.848256 \cdot 10^{-4}$	$2.569577 \cdot 10^{-6}$
	2	$2.876093 \cdot 10^{-6}$	$-5.744710 \cdot 10^{-9}$
	3	$-4.162356 \cdot 10^{-9}$	$-4.327691 \cdot 10^{-12}$

Equation (1) describes our experimental data with maximum error of 1.5%; the majority of points are defined with error not exceeding that of viscosity measurements (Fig. 2).

201

REFERENCES

1. Kaminsky, M. (1957): *Zeitschrift für physikalische chemie,* **12**, 210-1212.
2. Suryanarayana C. V., and Vankasten V. K. (1958): *Bulletin chemical society, Japan,* **31**, 442-448.
3. Stokes R. H. and Mills R. (1965): Viscosity of electrolytes and related properties, Pergamon Press, 105-106.
4. Fabuss B. M. and Korosi A. (1968): Research and development progress report No. 363, Washington, 11-15.
5. Grimes, Kestin J., and Khalifa E. (1979): *Journal of chemical and engineering data,* **24**(2), 121-126.
6. Semenyuk E. M. (1978). Issledovanie vyazkosti vodnykh rastvorov khloridov litiya, natriya, kaliya (Study of viscosity of aqueous lithium, sodium, and potassium chloride solutions), Dissertation for the award of Cand. Sc. degree, Leningrad, 9-10 (in Russian).
7. Pepinov, R. I., Yusufova V. D. and Lobkova M. V. (1977), *Teploenergetika,* **9**, 59-62.

The Thermal Conductivity of Aqueous KCl Solutions at Pressures up to 40 MPa

Y. NAGASAKA, J. SUZUKI, and A. NAGASHIMA

Keio University, Hiyoshi, Yokohama 223, Japan

ABSTRACT

The paper describes a new absolute transient hot-wire apparatus for measurements on concentrated aqueous electrolyte solutions at temperatures extending to 200°C and pressures of 50 MPa. A tantalum wire ($\varnothing$ 25 μm) coated with a thin anodic tantalum pentoxide film (~ 0.18 μm thick) is used as a line heat source. The bridge is arranged to be raised above ground potential in order to protect breakdown of insulation layer. Preliminary measurements on water and KCl solution are reported. The experimental results have an estimated accuracy of ±0.3%.

INTRODUCTION

The thermal conductivity data on aqueous solutions of electrolytes are required in the development and utilization of many industrial energy systems. For example, the data on geothermal brines and sea water, which can be considered as mixtures of aqueous solutions of NaCl, KCl, $CaCl_2$, $MgCl_2$ etc., are needed for geothermal and ocean thermal energy utilization devices and for desalination of sea water. Aqueous LiBr solution is also widely used in absorption refrigerators for air conditioning. However, measurements of the thermal conductivity of these fluids have so far been limited to rather narrow ranges of temperature, pressure and concentration with less satisfactory accuracy.

203

We have developed an apparatus for precise and absolute measurement of the thermal conductivity of electrically conducting liquids by the transient hot-wire method. We have also derived the working equations appropriate to measurements with coated wires [1]. The thermal conductivity of aqueous NaCl solutions has been measured over the temperature range 0-80°C and at pressures up to 40 MPa by employing this apparatus [2]. However, its use was limited at temperatures not exceeding 100°C, since the apparatus used in this first stage of measurements was not originally designed for corrosive fluids such as high-temperature aqueous electrolyte solutions. Thus, in this paper we describe a new, high-precision transient hot-wire apparatus which has been designed specifically to measure the thermal conductivity of concentrated aqueous solutions of electrolytes at temperatures up to 200°C and pressures up to 50 MPa. Finally, some preliminary measurements will be reported for water and aqueous KCl solutions.

EXPERIMENTAL
PRINCIPLE OF MEASUREMENT

In the idealized model, the present transient hot-wire apparatus for the measurement of the thermal conductivity, λ_3, consists of an infinitely long vertical insulated wire immersed in a test liquid. Initially at equilibrium, a step voltage is applied at $t=0$ across the metallic wire. Then, the average temperature rise of the metallic wire, $\Delta\overline{T}_1$, under these conditions can be described as follows.

$$\Delta\overline{T}_1 = \frac{q}{4\pi\lambda_3}\left[\ln t + A + \frac{1}{t}(B\ln t + C)\right] \tag{1}$$

Where q is the heat generation per unit length in the metallic wire and A, B and C denote constants whose explicit expressions are given in Ref. [1]. In the present measurement, the insulation layer is so thin (0.1-0.2 µm) that the deviation from a linear relationship of $\Delta\overline{T}_1$ versus $\ln t$ owing to the $1/t$ term in Eq. (1) turns out to be less than 0.05% of $\Delta\overline{T}_1$. The thermal conductivity, therefore, can be determined by

$$\lambda_3 = \frac{q}{4\pi}\left/\frac{d\Delta\overline{T}_1}{d\ln t}\right. \tag{2}$$

APPARATUS

A cross-sectional view of the hot-wire cell together with pressure vessel is shown in Fig. 1. The present transient hot-wire apparatus is designed to incorporate the following features:

(1) the pressure vessel is made of Hastelloy C276 which has strong corrosion resistance against many electrolyte solutions even at high temperatures,

(2) thin tantalum (Ta) wire, coated with tantalum pentoxide (Ta_2O_5) by the process of anodization, is employed as hot-wire,

(3) resistance of the wire is measured with the aid of four-terminal potentiometric method so as to avoid the error due to lead resistances.

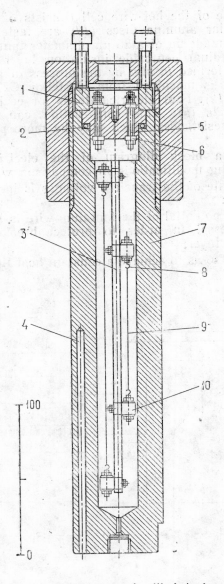

Fig. 1. Pressure vessel with hot-wire assembly: *1* — thrust ring; *2* — PTFE O-ring; *3* — titanium plate; *4* — thermometer well; *5* — terminal; *6* — PTFE seal; *7* — pressure vessel (Hastelloy C276); *8* — platinum hook; *9* — tantalum wire (Ø25 μm); *10* — alumina disc

The main frame of the hot-wire cell consists of titanium plate (3) on which semicircular alumina discs (10) are fastened on both sides. The hot-wire (9) is made up of a 25 μm diameter tantalum wire (99.95% purity) and a tantalum pentoxide insulation layer about 0.18 μm in thickness, which is formed *in situ* by the process of anodization [3]. The pressure vessel (7) is machined from Hastelloy C276 which is sealed with PTFE O-ring (2). Four electrical leads for each wire are brought out through terminals (5). This construction can eliminate possible error due to lead resistances which may be more pronounced at high temperatures.

Fig. 2 shows a block diagram of the electrical system which includes the following modifications in comparison with that of Ref. [1]:

(1) sampling rate of the transient voltages is increased up to about 40 times per second;

(2) the electric potential of the tantalum wire is kept higher relative to the pressure vessel in order to protect breakdown of tantalum pentoxide insulation layer;

(3) R_2 is added so as to achieve a constant heat flux q.

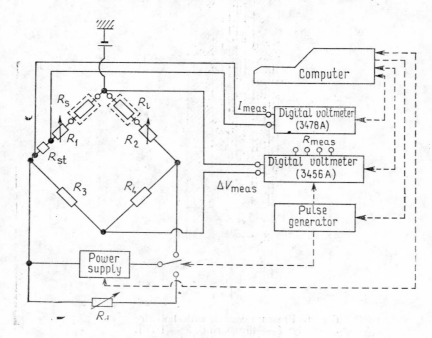

Fig. 2. Block diagram of electrical system

In this diagram, R_l and R_s denote the resistance of long and short wires, respectively. The temperature rise of the metallic wire, whose end effects are compensated using two wires, is converted into voltage change

with the bridge circuit. The voltage change is measured by a digital voltmeter (HP 3456A), which is triggered by a pulse generator with a repetition rate of about 40/s and a resolution of 0.1 μV. Resistance of the wire is also measured by the DVM with the aid of four-terminal technique.

As pointed out by Alloush et al. [3], tantalum pentoxide becomes unstable when it is exposed to negative voltage and can only act as an insulator above ground potential. Therefore, an attempt has been made to always keep electric potential of the wire higher than the ground level. The voltage applied on the bridge was 1 to 5 Volts which has to be small enough comparing to the maximum voltage on the formation of tantalum pentoxide (110 V in this case) and large enough to prevent breakdown of the insulator. This modification may produce the error caused by small leakage of current through insulation layer to ground potential. However, since the leakage current is so small (less than 1 μA) and stable that the error can be neglected.

EXPERIMENTAL RESULTS

Some preliminary measurements have been performed on the thermal conductivity of water and KCl solutions. The solutions were prepared gravimetrically from ion-exchanged distilled water and reagent-grade KCl (purity of 99.5%). In order to show that the apparatus operates in accordance with mathematical description of it, Fig. 3 represents deviations of the measured $\Delta \bar{T}_1$ from the fitted straight line versus $\ln t$ (water; 25°C; 10 MPa). Deviations never exceed 0.03% and the uncertainty of $d\Delta \bar{T}_1/d \ln t$ in this case equals 0.014%. Including all the other factors, the overall accuracy of the present measurements is estimated to be ±0.3%.

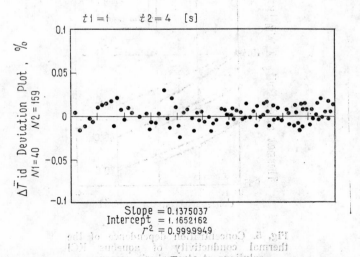

Fig. 3. Deviations of $\Delta \bar{T}_1$ from fitted straight line

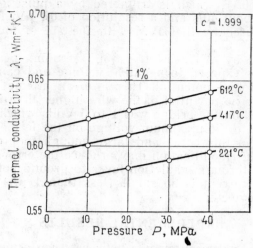

Fig. 4. Pressure dependence of the thermal conductivity of aqueous KCl solution (1.999 molality)

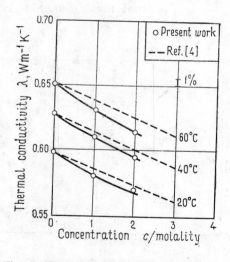

Fig. 5. Concentration dependence of the thermal conductivity of aqueous KCl solutions at atmospheric pressure

Fig. 4 illustrates the pressure dependence of the thermal conductivity of the aqueous KCl solution (1.999 molality). The concentration dependence at atmospheric pressure is shown in Fig. 5. As seen from this figure, the assumption of linear concentration dependence proposed by Riedel [4] is not valid for concentrated solutions.

A further measurement on aqueous KCl solutions at higher temperature is currently under way.

ACKNOWLEDGEMENTS

The authors are indebted to Messers. Y. Wada, T. Matsumura, K. Kawamata and Y. Kita for their assistance in carrying out these measurements.

REFERENCES

1. Nagasaka, Y. and Nagashima, A. (1981): *J. Phys. E.* **14**(12), 1435-1440.
2. Nagasaka, Y., Okada, H., Suzuki, J., and Nagashima, A. (1983): *Ber. Bunsenges. Phys. Chem.,* **87**, 859-866.
3. Alloush, A., Gosney, W. B., and Wakeham, W. A. (1982): Int. *J. Thermophys.* **3**(3), 225-235.
4. Riedel, L. (1951): *Chem. Ing. Techn.* **23**(3), 59-64.

Thermal Conductivity of Aqueous Solutions of Alkali Metals Halides

Yu. L. RASTORGUYEV, B. A. GRIGORYEV, G. A. SAFRONOV,
and Yu. A. GANIYEV

M. D. Millionshchikov Oil Institute, Grozny, USSR

Results of experimental studies on the thermal conductivity coefficient (λ) of aqueous solutions of LiF, NaF and NaBr are summarized in the present paper.

Thermal conductivity was measured by the absolute concentric cylinder method in a cell, whose faces acted as flat layers. The concentric cylinders made of copper were chromium plated and polished. A procedure for introducing corrections to the measured values of thermal conductivity was worked out to fit the construction of the measuring cell. The values of λ were calculated analytically as well as by modelling some units of the set on an electric integrator. To check the validity of introduced corrections, the thermal conductivity of liquids with different thermal conductivity coefficients was measured. Three runs of measurements of λ of water were taken at different layer-thicknesses of the studied substance. Detailed description of the experimental set and the measurement procedure are available in [1-2].

Maximum error in any thermal conductivity measurement (for confidence probability 0.95) did not exceed 1.2-1.5% in the whole range of parameters.

The thermal conductivity coefficients of aqueous solutions of LiF, NaF, and NaBr were measured between 20 and 200°C and at pressures from 0.5 to 100 MPa. The thermal conductivity of LiF solution was measured only along 2 isotherms (25.3°C and 29.6°C). This is because the solubility of LiF decreases with increase in temperature. The

investigated solutions had the following salt concentrations (N_s):
0.00071 mole fraction in H_2O-LiF; 0.0075; 0.0073; 0.01; 0.0139 mole
fraction in H_2O-NaF; 0.0046; 0.0208; 0.0291 mole fraction in H_2O-NaBr.
Measurements were taken along isotherms at 20 MPa intervals,
beginning from minimum pressure.

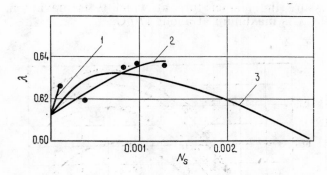

Fig. 1. Dependence of λ of aqueous solutions: *1* — LiF,
2 — NaF and *3* — NaBr on salt concentration N_s,
at $t=30°C$ and $P=0.5$ MPa

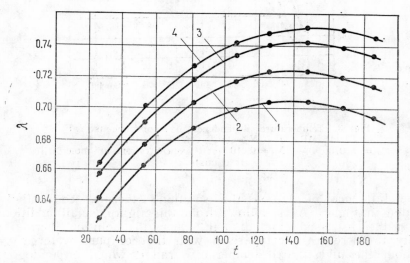

Fig. 2. Temperature dependence of thermal conductivity NaBr solution in
water ($N_s=0.0046$ mole fraction) at different pressures, MPa: *1* — 0.5;
2 — 40.0; *3* — 80.0; *4* — 100.0

Figure 1 shows the dependences of thermal conductivity coefficients
of the solutions investigated at the salt concentration N_s. As is seen, the
variation in λ of solutions depends on the type of salt used and its con-

centration. Thus, in the H₂O-NaF solution λ increases monotonically with
concentration; in the H₂O-NaBr solution, λ of the solution passes through
a maximum at $N_s \approx 0.007$ mole fraction. Figure 2 shows how the thermal
conductivity of NaBr solution ($N_s = 0.0046$ mole fraction) varies with
temperature and pressure. As in the case of water, the λ of solution first
increases with increase in temperature and then becomes maximum. With
increasing pressure the temperature at which λ is maximum increases;
at $P = 100$ MPa, λ is maximum at about 170°C.

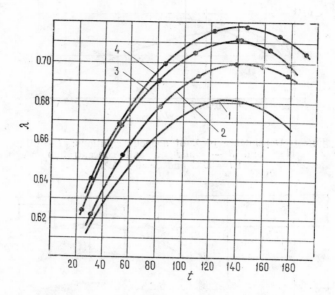

Fig. 3. Temperature dependence of λ of aqueous NaF
at $P = 0.5$ MPa; $1 - N_s = 0$; $2 - N_s = 0.0045$; $3 - N_s$
$= 0.0073$; $4 - N_s = 0.010$ (salt concentration in mole
fractions)

The influence of N_s on the solution's thermal conductivity is
illustrated in Fig. 3. As it follows from this figure, small additions of
salts, in the discussed case, of NaF, cause the solution thermal conductivity to increase in comparison with that of pure water ($N_s = 0$)
throughout the investigated temperature range, which is analogous to
the effect of increasing pressure. Analogy is revealed not only in that
the absolute value of λ varies, but also in that the λ of the solutions
attains maximum values at temperatures exceeding those at which the λ
of water is maximum at the saturation pressure. This is borne out by
the curves of Fig. 4, which shows that the temperature at which the
thermal conductivity coefficient of the solution is maximum, t_m, rises
with increase in the concentration of NaF and NaBr.

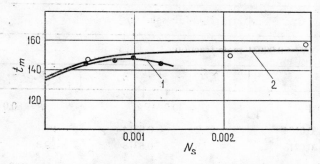

Fig. 4. Effects of additions of salts on temperature at which solution exhibits maximum thermal conductivity at $P = 0.5$ MPa; *1* — NaF; *2* — NaBr

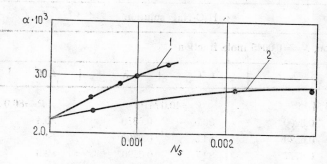

Fig. 5. α versus N_s

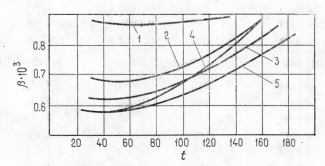

Fig. 6. Dependence of β of NaF on temperature at different N_s, mole fractions: *1* — $N_s = 0$; *2* — $N_s = 0.0045$; *3* — $N_s = 0.0073$; *4* — $N_s = 0.010$; *5* — $N_s = 0.0139$

Table 1. Experimental value of λ, W/m K

H₂O-LiF solution

salt concentration, $N_s = 0.00071$ mole fraction

t, °C	λ	t, °C	λ	t, °C	λ
$P=0.5$ MPa		$P=40.0$ MPa		$P=80.0$ MPa	
25.3	0.6164	25.2	0.6314	25.1	0.6473
29.6	0.6218	29.5	0.6378	29.3	0.6560
$P=20.0$ MPa		$P=60.0$ MPa		$P=100.0$ MPa	
25.3	0.6242	25.2	0.6404	25.1	0.6547
29.5	0.6303	29.4	0.6449	29.3	0.6624

H₂O-NaF solution

salt concentration, $N_s = 0.0045$ mole fraction

t, °C	λ	t, °C	λ	t, °C	λ
$P=0.5$ MPa		$P=40.0$ MPa		$P=80.0$ MPa	
30.6	0.6200	30.5	0.6370	30.4	0.6513
54.5	0.6520	54.5	0.6700	54.3	0.6860
82.8	0.6777	82.7	0.6990	82.6	0.7173
$P=0.5$ MPa		$P=40.0$ MPa		$P=80.0$ MPa	
110.4	0.6934	110.3	0.7142	110.2	0.7312
139.5	0.6996	139.4	0.7255	139.3	0.7423
159.0	0.6980	158.9	0.7245	158.8	0.7442
178.4	0.6922	178.3	0.7202	178.3	0.7409
$P=20.0$ MPa		$P=60.0$ MPa		$P=100.0$ MPa	
30.6	0.6289	30.5	0.6445	30.3	0.6585
54.5	0.6615	54.4	0.6790	54.3	0.6920
32.7	0.6889	82.6	0.7140	82.5	0.7230
110.3	0.7047	110.2	0.7242	110.2	0.7395
139.5	0.7150	139.4	0.7342	139.2	0.7516
159.0	0.7124	158.9	0.7350	158.8	0.7520
178.3	0.7075	178.2	0.7306	178.1	0.7500

Table 1 *(continued)*

salt concentration, $N_s = 0.0073$ mole fraction

t, °C	λ	t, °C	λ	t, °C	λ
$P = 0.5$ MPa		$P = 40.0$ MPa		$P = 80.0$ MPa	
21.6	0.6202	21.5	0.6362	21.4	0.6507
53.3	0.6670	53.3	0.6823	53.2	0.6995
81.5	0.6890	81.4	0.7080	81.3	0.7210
109.0	0.7046	108.9	0.7253	108.9	0.7439
140.1	0.7107	140.1	0.7346	140.0	0.7538
164.6	0.7061	164.5	0.7305	164.4	0.7538
179.2	0.7011	179.1	0.7296	179.0	0.7524
$P = 20.0$ MPa		$P = 60.0$ MPa		$P = 100.0$ MPa	
21.5	0.6277	21.5	0.6432	21.4	0.6580
53.3	0.6470	53.2	0.6906	53.1	0.7072
81.4	0.6983	81.4	0.7164	81.3	0.7345
109.0	0.7162	108.9	0.7352	108.8	0.7528
140.1	0.7240	140.0	0.7450	139.9	0.7665
164.5	0.7165	164.5	0.7430	164.3	0.7631
179.2	0.7170	179.1	0.7412	179.0	0.7606

salt concentration, $N_s = 0.01$ mole fraction

t, °C	λ	t, °C	λ	t, °C	λ
$P = 0.5$ MPa		$P = 40.0$ MPa		$P = 80.0$ MPa	
30.3	0.6356	30.3	0.6531	30.2	0.6676
41.4	0.6560	41.3	0.6704	41.2	0.6859
55.6	0.6747	55.5	0.6908	55.4	0.7080
85.5	0.6940	85.4	0.7120	85.3	0.7302
124.4	0.7148	124.3	0.7361	124.2	0.7561
152.9	0.7167	152.8	0.7376	152.7	0.7627
174.2	0.7122	174.1	0.7375	174.0	0.7625
188.3	0.7064	188.2	0.7360	188.1	0.7610
$P = 20.0$ MPa		$P = 60.0$ MPa		$P = 100.0$ MPa	
30.3	0.6451	30.2	0.6595	30.1	0.6740
41.3	0.6635	41.3	0.6786	41.2	0.6924
55.6	0.6826	55.5	0.6985	55.3	0.7155
85.5	0.7030	85.4	0.7215	85.3	0.7380
124.3	0.7261	124.2	0.7459	124.1	0.7672
152.8	0.7276	152.7	0.7511	152.6	0.7730
174.1	0.7265	174.1	0.7500	173.9	0.7730
188.2	0.7226	188.1	0.7486	188.0	0.7714

Table 1 *(continued)*

salt concentration, $N_s = 0.0139$ mole fraction

t, °C	λ	t, °C	λ	t, °C	λ
$P=0.5$ MPa		$P=40.0$ MPa		$P=80.0$ MPa	
25.5	0.6227	25.4	0.6375	25.3	0.6532
58.2	0.6760	58.1	0.6926	58.0	0.7082

salt concentration, NaF $=0.0139$ mole fraction

t, °C	λ	t, °C	λ	t, °C	λ
$P=0.5$ MPa		$P=40.0$ MPa		$P=80.0$ MPa	
174.1	0.7075	174.0	0.7324	173.9	0.7556
$P=20.0$ MPa		$P=60.0$ MPa		$P=100.0$ MPa	
25.5	0.6300	25.4	0.6453	25.3	0.6601
58.1	0.6847	53.0	0.7013	57.9	0.7160
92.6	0.7110	92.5	0.7305	92.3	0.7485
117.7	0.7208	117.6	0.7375	117.5	0.7570
152.2	0.7263	152.1	0.7469	152.0	0.7716
164.1	0.7271	174.0	0.7427	173.8	0.7655

HH_2O-NaBr solution

salt concentration, $N_s = 0.0046$ mole fraction

t, °C	λ	t, °C	λ	t, °C	λ
$P=0.5$ MPa		$P=40.0$ MPa		$P=80.0$ MPa	
24.6	0.6118	24.5	0.6267	25.4	0.6434
53.1	0.6499	53.00	0.6672	52.9	0.6801
80.5	0.6728	80.4	0.6919	80.3	0.7065
107.5	0.6814	107.4	0.7005	107.3	0.7194
125.7	0.6866	125.6	0.7091	125.5	0.7238
149.3	0.6873	149.2	0.7083	149.1	0.7260
168.8	0.6863	168.7	0.7098	168.6	0.7255
186.5	0.6821	186.4	0.7045	186.3	0.7222
$P=20.0$ MPa		$P=60.0$ MPa		$P=100.0$ MPa	
24.6	0.6191	24.5	0.6347	25.4	0.6492
53.1	0.6598	53.0	0.6735	52.9	0.6868
80.4	0.6832	80.3	0.6998	80.3	0.7134
107.5	0.6909	107.4	0.7097	107.3	0.7265
125.6	0.6960	125.6	0.7157	125.5	0.7318
149.2	0.6975	149.2	0.7167	149.1	0.7346
168.8	0.6967	168.7	0.7162	168.6	0.7345
186.4	0.6953	186.4	0.7131	186.3	0.7316

Table 1 *(continued)*

salt concentration, $N_s = 0.0208$ mole fraction

t, °C	λ	t, °C	λ	t, °C	λ
$P = 0.5$ MPa		$P = 40.0$ MPa		$P = 30.0$ MPa	
26.6	0.6012	26.5	0.6162	26.4	0.6291
57.2	0.6399	57.1	0.6551	57.0	0.6674
88.3	0.6623	88.2	0.6798	88.2	0.6941
126.7	0.6754	126.6	0.6951	126.5	0.7118
167.6	0.6725	167.5	0.6938	167.4	0.7094
$P = 20.0$ MPa		$P = 60.0$ MPa		$P = 100.0$ MPa	
26.6	0.6082	26.5	0.6234	26.4	0.6343
57.1	0.6475	57.1	0.6619	57.0	0.6737
88.3	0.6719	88.2	0.6872	88.1	0.7012
126.7	0.6862	126.6	0.7038	126.5	0.7185
167.5	0.6831	167.4	0.7022	167.3	0.7177

salt concentration, $N_s = 0.0291$ mole fraction

t, °C	λ	t, °C	λ	t, °C	λ
$P = 0.5$ MPa		$P = 40.0$ MPa		$P = 80.0$ MPa	
28.7	0.5946	28.6	0.6085	28.5	0.6208
55.4	0.6295	55.3	0.6458	55.2	0.6571
89.2	0.6517	89.2	0.6678	89.1	0.6824
116.5	0.6618	116.4	0.6808	116.3	0.6943
151.3	0.6657	151.2	0.6850	151.1	0.7001
177.1	0.6648	177.0	0.6852	176.9	0.7003
192.0	0.6601	191.9	0.6847	191.8	0.6972
$P = 20.0$ MPa		$P = 60.0$ MPa		$P = 100.0$ MPa	
28.7	0.6017	28.6	0.6157	28.5	0.6246
55.4	0.6368	55.3	0.6513	55.2	0.6625
89.2	0.6602	89.2	0.6765	89.0	0.6886
116.4	0.6720	116.4	0.6880	116.3	0.7003
151.3	0.6752	151.2	0.6922	151.1	0.7071
177.0	0.6752	176.9	0.6922	176.9	0.7076
192.0	0.6743	191.9	0.6913	191.8	0.7062

The temperature $\alpha = \frac{1}{\lambda} \left(\frac{\partial \lambda}{\partial t} \right)_p$ and pressure $\beta = \frac{1}{\lambda} \left(\frac{\partial \lambda}{\partial P} \right)_t$ coefficients of thermal conductivity were calculated by graphical and analytical treatment of the experimental data.

Figure 5 illustrates $\alpha - N_s$ curves. From these curves it is seen that in the investigated range of N_s the temperature coefficients increase with rise in salt concentration, the α-values of solutions being higher than those of pure water. Besides, comparing α of the solutions containing identical-cation salts (NaF and NaBr) reveals that the higher the temperature coefficient of thermal conductivity, the smaller the anion radius.

Figure 6 shows how the pressure coefficient of investigated solutions of NaF varies with temperature. It is seen that α decreases with increasing temperature: between 40 and 60°C it has a minimum value which with increasing temperature rises almost to the corresponding β-value of pure water. With increase of concentration of the dissolved salt β-minimum shifts towards much lower temperatures; at about 60°C pure water has a minimum value of β.

The results of the experimental investigations on λ of aqueous salt solutions are listed in Table 1.

REFERENCES

1. Y. L. Rastorguyev, Y. A. Ganiyev, and G. A. Safronov (1977), *Inzhenerno-fiz. Zhur.* 33(1), 64-74.
2. Y. L. Rastorguyev, Y. A. Ganiyev, and G. A. Safronov (1977), *Inzhenerno-fiz. Zhur.* 33(2), 275-279.

The Application of the Transient Hot-Wire Method to the Measurement of the Thermal Conductivity of Aqueous Salt Solutions

W. A. WAKEHAM

Department of Chemical Engineering and Chemical Technology,
Imperial College, London SW7 2BY, U.K.

ABSTRACT

The paper describes the results of a preliminary study of the application of the transient hot-wire technique to the measurement of the thermal conductivity of electrically-conducting liquids. The essential feature of the new instrument is the use of an anodic coating formed on the metallic sensing element to provide an electrically-insulating layer contributing little resistance to heat transfer. A prototype of the instrument is described and the results of measurements of the thermal conductivity of water and aqueous solutions of lithium bromide and calcium chloride are presented. At this stage the accuracy of the values reported with the prototype instrument is no better than ±3%, but the viability of the technique is confirmed and the essential features of a high precision instrument are established.

INTRODUCTION

In the last few years the transient hot-wire technique for the measurement of the thermal conductivity of fluids has been refined and successfully applied over an increasingly wide range of thermodynamic states [1-4]. For fluids that are not electrically conducting it is now

possible routinely to attain a precision of $\pm 0.3\%$ in the thermal conductivity in the temperature range 77-500 K and for pressures as high as 700 MPa [1-4]. In the case of electrically conducting liquids it has proved more difficult to achieve a comparable precision because the essential heating element of the instrument should, ideally, be electrically insulated from the fluid while remaining in good thermal contact with it. Two different approaches to the measurement of the thermal conductivity in such liquids have usually been adopted. In the first, employed by Dietz et al. [5] for measurements on water, a bare metallic wire is retained for the heating element but alternating current heating and resistance measurements are used to avoid spurious electrical effects arising from polarization. In the second approach, described by Nagashima and his co-workers [6-8], a thin metallic wire coated with a thin layer of an insulating polyester material [6, 7] or a liquid metal within a glass envelope [8] is used as the heating element. In the latter case, the working equation must be modified to account for the layer of insulation surrounding the heating element in the manner described by Nagasaka and Nagashima [6]. In practice, neither of these two techniques has been able to retain the accuracy characteristic of measurements in insulating liquids. In the particular case of the method involving an insulating layer the metallic wires employed have usually been of greater diameter than the ideal and the insulating layer has contributed a significant correction to the analysis of the data.

In view of the facts that aqueous solutions of ionic salts (particularly alkali halides) have applications in large absorption refrigerators [9] and that molten salts are increasingly used as heat transfer media, it is essential to develop an instrument for the measurement of the thermal conductivity of such electrically-conducting liquids. Furthermore, bearing in mind the earlier remarks with respect to the accuracy of existing instruments as well as their restricted temperature range [6, 7], the development should aim at high accuracy over a wide range of thermodynamic states. The present paper describes the first steps in the development of such an instrument. First, a prototype transient hot-wire thermal conductivity cell has been constructed in which an insulating coating is formed on metallic wire so that it contributes only a small correction to the analysis of the experimental data. Secondly, the prototype cell has been employed in conjunction with existing equipment to perform a preliminary series of measurements of the thermal conductivity of water and aqueous solutions of lithium bromide and calcium chloride at temperatures in the range 297-357 K. At this stage the intention of the measurements is to demonstrate the viability of the technique rather than to provide definitive results.

EXPERIMENTAL

The design of the thermal conductivity cell of the prototype instrument is shown in Fig. 1. The outer wall of the cell is formed by a Pyrex glass tube (1), 160 mm long and 22 mm in internal diameter. At its

upper end the tube is terminated by a pyropholite plug (2) sealed into the tube with adhesive and held secure by the clamping mechanism (3 and 4). The pyropholite plug forms the support for the hot-wire assembly and its adjustment mechanism. The hotwire itself (5) consists of a 25 μm diameter tantalum wire (purity 99.9%). Tantalum has been chosen for the sensing element in place of the more usual platinum [1-4] because it is possible to form a layer of tantalum pentoxide on its surface which is strong, impervious to liquids, electrically insulating, and only 75 mm thick. Furthermore, since the layer is formed electrolytically, it may be deposited *in situ* following assembly of the cell. The details of the process whereby the oxide layer is formed have been given elsewhere [10]. For the present purposes it is sufficient to note that the layer was found to be stable over a large number of temperature cycles of the entire assembly and provided an effective electrical insulation as long as the wire was maintained at a positive potential with respect to any electrolyte in the cell.

At its lower end the tantalum wire is soldered to a 0.6 mm diameter copper lead (6) which passes through an insulating glass tube (7) to the pyropholite plug where it is fixed. At its upper end the tantalum wire is soldered to a helical spring (8) which in turn is connected to a platinum hook (9) mounted in a core (10) carried on the terminal pin (11). The mechanism comprised of the elements (12)-(15) allows vertical movement of the terminal pin (11) without rotation in order to exert a small tension on the wire through the spring (8). The tension serves to maintain the tantalum wire taut at all equilibrium temperatures as well as during the transient heating [11]. All exposed metallic components of the cell are insulated with silicone rubber. The entire cell is mounted in a thermostatically-controlled liquid bath.

For these preliminary studies the cell described above was used in conjunction with an automatic Wheatstone bridge [2] designed for operation with cells containing platinum wires of 7 μm diameter. The auto-

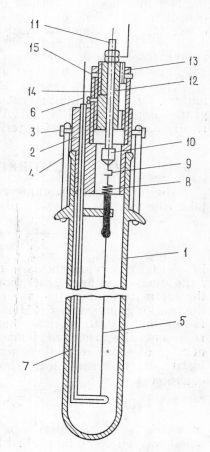

Fig. 1. The design of the thermal conductivity cell for electrically-conducting liquids

matic bridge provides the heating current to the hot wire and records the transient temperature rise of the wire as a function of time over a period of about 1 second after the heating is initiated [2]. The existing bridge was designed to operate with its output near ground potential using two cells of different lengths in order to eliminate automatically the effects of temperature non-uniformities at the ends of the wire, and to reduce electrical noise to a low level [2]. With the present oxide covered wire such an arrangement is not possible since in one of the cells the wire would be at a negative potential with respect to ground and the oxide layer unstable. Consequently, only one cell was employed and an analytic method adopted to correct for the end effects. A second consequence of the use of the existing bridge with the new cells arises from the fact that the resistance of the relatively large tantalum wires is only one quarter of that for the platinum wires. Since it is not desirable to offset this effect merely by using a temperature rise greater than the 4-5 K normally employed [1-4], it is necessary to accept at this stage a poorer resolution in the temperature rise measurements and a corresponding loss in the precision of the thermal conductivity measurements.

The water employed for the measurements was distilled and deionized. The salts lithium bromide and calcium chloride were 'Analar' grade supplied by BDH Ltd. All solutions were manufactured gravimetrically and the estimated uncertainty in the mass fraction is ± 0.001.

WORKING EQUATIONS

The fundamental working equations for the transient hot-wire method takes the form

$$\Delta T_{id} = \frac{q}{4\pi\lambda\,(T_r,\,P)}\,\ln\!\left(\frac{4kt}{a^2C}\right) = \Delta T_w + \sum_i \delta T_i \tag{1}$$

in which ΔT_{id} represents the temperature rise in an ideal instrument, t the time, q the heat generation per unit length of the hot wire, $\lambda\,(T_r,\,P)$ the thermal conductivity of the fluid at a reference temperature T_r and the working pressure P, k is the fluid thermal diffusivity, a the wire radius and C a numerical constant [12]. In addition, the symbol ΔT_w represents the measured temperature rise of the wire and the δT_i a number of small corrections which account for the departures of the real instrument from the ideal. The reference temperature is defined by the equation

$$T_r = T_0 + \Sigma \delta T_i^* \tag{2}$$

in which T_0 is the initial, equilibrium temperature of the fluid and δT_i^* small corrections to be applied to it.

Analytical expressions for all of the corrections δT_i and δT_i^* have been given elsewhere [12]. In particular, we have employed the equations of Nagasaka and Nagashima [6] to account for the presence of the wire

222

coating, the analysis of Haarman [13] to account for end effects in the wire, and the treatment of Castro et al. [14] to deal with the effects of radiation, which were found to be insignificant. No single correction amounted to more than ±0.4% of the wire temperature rise so that their evaluation should introduce no significant error.

INSTRUMENT PERFORMANCE

The working equation (1) shows that the corrected, measured temperature rise of the wire should be a linear function of $\ln t$ within the resolution of the measurements. Figure 2 contains a plot of the deviations

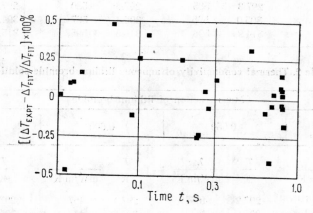

Fig. 2. The deviations of the measured temperature rise of the wire from a least squares fit to the data for an experiment with an aqueous solution of lithium bromide

of the temperature rise measurements from a least squares linear fit to the data for a measurement in an aqueous solution of lithium bromide with a salt mass fraction of 0.55 at a temperature of 23°C. The deviations reveal no systematic character and do not exceed ±0.5% which is consistent with the estimated resolution of each measurement, and representative of all the runs reported here. The magnitude of the random scatter is considerably larger than is usual for measurements in electrically-insulating liquids because of the loss of resolution referred to earlier. Nevertheless, the observed linearity confirms that the instrument operates in accordance with the mathematical model of it. Furthermore, repeated measurements on the same liquid under the same conditions indicate that the reproducibility of the thermal conductivity measurements is one of ±0.7% which is taken to represent their precision. Combining this figure with errors arising from the temperature coefficient of resistance of the wire employed and the effects of the small corrections applied, it is estimated that the uncertainty in the thermal conductivity reported is one of ±3%.

RESULTS

Table 1. Thermal conductivity of aqueous calcium chloride solutions

Mass fraction of calcium chloride							
0.00		0.200		0.250		0.300	
$\dfrac{T}{K}$	$\dfrac{\lambda}{mW/m\ K}$	$\dfrac{T}{K}$	$\dfrac{\lambda}{mW/m\ K}$	$\dfrac{T}{K}$	$\dfrac{\lambda}{mW/m\ K}$	$\dfrac{T}{K}$	$\dfrac{\lambda}{mW/m\ K}$
297.2	608	297.3	585	297.6	580	297.6	572
304.9	614	305.0	596	306.4	587	304.9	577
315.0	619	314.9	598	316.5	595	315.3	585

Table 2. Thermal conductivity of aqueous lithium bromide solutions

Mass fraction of lithium bromide							
0.500		0.550		0.600		0.650	
$\dfrac{T}{K}$	$\dfrac{\lambda}{mW/m\ K}$	$\dfrac{T}{K}$	$\dfrac{\lambda}{mW/m\ K}$	$\dfrac{T}{K}$	$\dfrac{\lambda}{mW/m\ K}$	$\dfrac{T}{K}$	$\dfrac{\lambda}{mW/m\ K}$
297.2	474	297.2	465	297.1	457	297.1	452
305.6	479	305.6	467	305.6	463	305.9	458
334.6	499	314.3	470	315.4	465	315.4	461
356.6	511	334.6	483	335.7	488	337.1	474
		356.6	499	355.5	492		

Tables 1 and 2 list the present thermal conductivity data for water and aqueous solutions of calcium chloride and lithium bromide. For calcium chloride the measurements extend over the temperature range 297-315 K for solutions with mass fractions, below the eutectic composition, which have been used commonly as secondary refrigerants. For lithium bromide the measurements extend to higher temperatures, up to 356 K, and have been carried out for mass fractions typical of those encountered in large absorption refrigerators.

Figure 3 contains a plot of the thermal conductivity of the calcium chloride solutions as a function of temperature and includes the results of earlier measurements on the same solutions by Riedel [15] and Kapustinkski [16]. In addition, the figure includes the IAPS correlation for the thermal conductivity of pure water [17] and the more recent results of Dietz et al. [5]. The comparison of the present results for pure water with earlier work indicates that the instrument yields satisfactory results

within its estimated accuracy of ±3%. For this reason no particular
significance is ascribed to the fact that the temperature dependence of
the thermal conductivity is different from the consensus of earlier results.
For the solutions of calcium chloride there is very good agreement with
the data of Kapustinkski [16] at his single measurement temperature
although the extrapolation of the present results is slightly higher than
the data of Riedel [15].

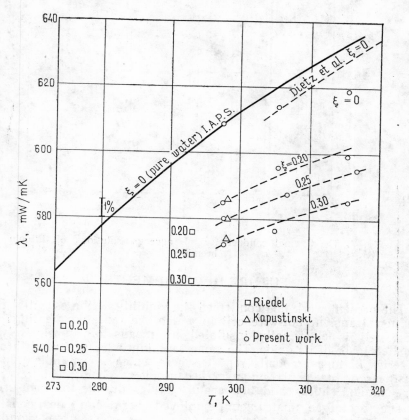

Fig. 3. The thermal conductivity of water and aqueous solutions of
calcium chloride

Figure 4 contains a plot of the temperature dependence of the
thermal conductivity of lithium bromide solutions together with a correlation of the results of the only earlier measurements reported by
Uemura and Hasaba [18]. In this case there are significant differences
between the present data and the earlier results amounting to as much
as 9% at a mass fraction of 0.6. This figure exceeds the mutual
uncertainty of the two sets of data but it would be unwise to make any

statement regarding which set of data is more reliable in view of the relatively large error band at present associated with both sets of results.

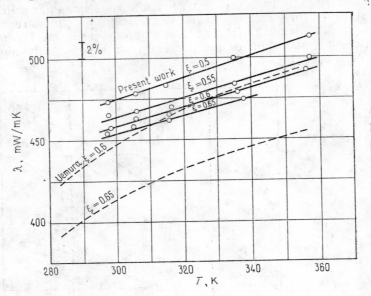

Fig. 4. The thermal conductivity of aqueous solutions of lithium bromide

FUTURE DEVELOPMENTS

The present work has confirmed the viability of the essential features of a new transient hot-wire cell in which the tantalum heating element of the cell is electrically insulated by means of an electrolytically-deposited layer of its oxide. The same tantalum wire was used throughout the series of measurements reported here which is an indication of the robustness of the insulation layer because it was subjected to a number of mechanical shocks and temperature cycles during the measurements.

The prototype instrument employed here can be considerably improved in order to yield an accuracy for thermal conductivity measurements commensurate with that achieved for insulating liquids. To achieve this, it must be possible to operate the instrument with two cells in a differential fashion in order to eliminate effects at the ends of the hot-wires automatically. Furthermore, the resolution of the tempe-rature rise measurements must be improved by about one order of mag-nitude. To satisfy the first condition it is necessary to operate the entire bridge above ground potential because of the decomposition of the insulating layer. This has the inevitable consequence that spurious electrical noise becomes a more severe problem and conflicts with the second condition which requires that the noise level in the instrument be

reduced well below that in current instruments for non-conducting liquids. With sufficient attention to circuit construction, the required noise levels can be met and there is therefore every hope that measurements of the thermal conductivity of electrically-conducting liquids with an accuracy of a few parts in one thousand will be possible.

ACKNOWLEDGEMENTS

The author is grateful to A. Alloush for assistance with the measurements. Financial support for this work was provided in part by the U. K. Science and Engineering Research Council.

REFERENCES

1. Kestin, J., Paul, R., Clifford, A. A., and Wakeham, W. A. (1980): *Physica* **100A**, 349-58.
2. Nieto de Castro, C. A., Calado, J. C. C., Dix, M., and Wakeham, W. A. (1976): *J. Phys. E: Sci. Instrum.* 9, 1073-80.
3. Menashe, J., and Wakeham, W. A. (1981): *Ber. Bunsenges. Phys. Chem.* **85**, 340-6.
4. Roder, H. M. (1981): *J. Res. N.B.S.* 86, 457-493.
5. Dietz, F. J., de Groot, J. J., and Franck, E. U. (1981): *Ber. Bunsenges. Phys. Chem.* 85, 1005-9.
6. Nagasaka, Y., and Nagashima, A. (1981): *J. Phys. E: Sci. Instrum.* **14**, 1435-1440.
7. Nagasaka Y., and Nagashima, A. (1981): *Rev. Sci. Instrum.* 52, 229-32.
8. Hoshi, M., Omotani, T., and Nagashima, A. (1981): *Rev. Sci. Instrum.* 52, 755-8.
9. Gosney, W. B. (1982): *Principles of Refrigeration,* Cambridge University Press, Cambridge.
10. Alloush, A., Gosney, W. B., and Wakeham, W. A. (1982): *Int. J. Thermophys.* **3**, 225-35.
11. Haran, E. N., and Wakeham, W. A. (1982): *J. Phys. E: Sci. Instrum.* **15**, 839-42.
12. Healy, J. J., de Groot, J. J., and Kestin, J. (1976): *Physica* 82C, 392-408.
13. Haarman, J. W. (1982): *PhD Thesis, Technische Hogeschool, Delft.*
14. Nieto de Castro, C. A., Li, S. F. Y., Maitland, G. C., and Wakeham, W. A. (1983): *Int. J. Thermophys.* 4, 311-27
15. Riedel, L. (1950): *Kältetechnik* 4, 99-101.
16. Kapustinski, A. F., and Ruzavin, I. I. (1955): *Zh. fiz. Khim.* 29(12), 2222-9.
17. International Association for the Properties of Steam, Release on Thermal Conductivity of Water Substance (1977): Bradley Associates, Leicester, U. K.
18 Uemura, T., and Hasaba, S. (1963): *Refrigeration (Japan)* 38, 19-24.

Cobalt Corrosion Products Solubility
in Water Solutions, Containing Oxidizing Agents,
at High Parameters

Yu. F. SAMOILOV, T. I. PETROVA, and N. L. KHARITONOVA

Moscow Power Engineering Institute, Moscow, USSR

The main part of unvolatile water contaminants in modern thermal power plants, which are the source of deposit formations, are made up of structural material corrosion products. One of the components, which is inevitably present in structural materials of power plants, is cobalt. Despite its low content, cobalt can adversely affect the operating characteristics of modern power plants.

The modern way to passivate the surface of condensate and feed path of a power plant is to widely use oxidizing agents: oxygen and hydrogen peroxide. However, no data is available on the influence of oxidizing agents on cobalt corrosion products activity in water at high parameters (temperature and pressure), in general, and on solubility, in particular. Thus, the aim of the present work was to determine, by way of experiment and calculations, the solubility of cobalt corrosion products in water solutions, containing oxidizing agents (oxygen, hydrogen peroxide) at high parameters.

In order to determine the solubility of cobalt corrosion products, experimental work had been carried out on a unit which is described in [1]. The test pressure was set at 7.0 MPa, in temperature range from 298 to 548 K. Test solution containing cobalt corrosion products and one of the additives (oxygen or hydrogen peroxide), was pumped (with high

pressure pump) into the heat exchangers where temperature and pressure had been maintained at a preset level. After a constant temperature was achieved, the test solution was stored in the unit for 2 to 9 hours, depending on the temperature. Competence of this method has been proved by M. A. Styrikovitch in [3]. Then samples were taken out to further determine cobalt concentration by photocolorimeter with a nitrozo-r-indicator [2].

The test solution containing cobalt corrosion products was prepared in deionized deoxidized water (conductivity $\leqslant 0.1$ μS/cm, $pH_0 \sim 6.8\text{-}7.1$, oxygen concentration less than 5 ppb) as follows: cobalt chips (containing 99.999% Co) were placed in a plastic container which was then evacuated and filled with deionized water. Concentrated solution of oxygen or hydrogen peroxide was added into the container in such an amount that its concentration in test solution was: 10-1000 ppb.

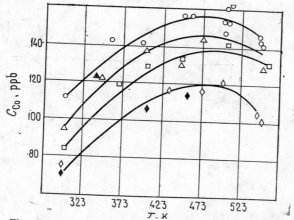

Fig. 1. Temperature effect on the Co corrosion products solubility with O_2 present
$P = 7.0$ MPa: $O - C_{O_2} < 10$ ppb, $\triangle - C_{O_2} = 200$ ppb,
$\square - C_{O_2} = 500$ ppb, $\lozenge - C_{O_2} = 960$ ppb,
$\blacktriangledown - C_{O_2} = 1200$ ppb
$P = 4.0$ MPa: $\blacktriangle - C_{O_2} = 170$ ppb

Figures 1 and 2 show how temperature affects the solubility of cobalt corrosion products in water solutions. An analysis of this dependence reveals that the solubility of cobalt corrosion products in water solutions containing oxygen or hydrogen peroxide is maximal for all tested oxidizers' concentrations. In water solutions containing oxygen, the maximum solubility of cobalt compound corresponds to 498 K. Adding hydrogen peroxide into water shifts the maximum of cobalt corrosion products solubility towards lower temperatures: in solutions that do not contain hydrogen peroxide, the maximum solubility corresponds to 498 K. At a concentration of 350 ppb of hydrogen peroxide the temperature equals 423 K; at 850 ppb it equals 353 K (Fig. 2). Within the whole temperature

range cobalt corrosion products solubility decreases as concentration of oxygen in test solution increases (Fig. 1). Change in concentration of hydrogen peroxide in test solution influences the solubility of cobalt corrosion products: as concentration of hydrogen peroxide is increased

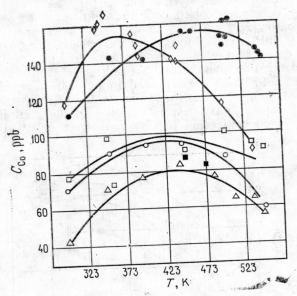

Fig. 2. Temperature effect on the corrosion products solubility with H_2O_2 present
$P = 7.0$ MPa: $\bigcirc$ — $C_{H_2O_2} = 230$ ppb, $\square$ — $C_{H_2O_2} =$ $= 480$ ppb, $\triangle$ — $C_{H_2O_2} = 350$ ppb, $\diamond$ — $C_{H_2O_2} = 850$ ppb
$P = 4.0$ MPa: $\blacksquare$ — $C_{H_2O_2} = 460$ ppb, $\bullet$ — $C_{H_2O_2} = 0$ ppb

from 0 to 350 ppb, the solubility steeply decreases, but a further rise in hydrogen peroxide concentration causes the solubility to increase (Fig. 2).

It is noteworthy that the results of the above experiments on solubility of cobalt corrosion products in water solutions containing oxygen, and the results of other experiments, as per Ref. [4], are in many ways compatible. Thus, in Ref. [4] the tests were conducted at 523 K and 24.5 MPa ($\rho = 0.82$ g/cm³) in solutions, containing oxygen in concentration of about 100 ppb. Fig. 1 shows that despite a great variety of experimental methods used (in Ref. [4], water solution with $pH_0 \sim 7.0$ was let through a layer of cobalt oxides), the results obtained are in good agreement, which proves their accuracy.

We could not compare the solubility of cobalt corrosion products in water solutions containing hydrogen peroxide with other experimental results because of the absence of the latter.

Apart from the experimental work on solubility of cobalt corrosion

products, we attempted to determine also the solubility of the most stable forms of these products. Conditions at which cobalt oxides and hydroxides can remain in water at room temperature can be determined from the pH_0-E_h diagram of Ref. [5], (Fig. 3). The diagram shows that at $pH_0=6$-13 and E_h from —0.4 to +0.6 the most probable thermo-dynamical form of cobalt corrosion products is $Co(OH)_2$ which after introducing oxidizing agents changes first to Co_3O_4 and then to $Co(OH)_3$. Information on mode of existence of cobalt compounds at high parameters is not sufficient, and whatever little there is, it is all rather estimation [6, 7]. Since the aim of the present work was to study the influence of oxidizing agents (in particular, of oxygen) on the solubility of cobalt compounds, it was expedient to determine E_h in dependence of the concentration of oxygen in solution. E_h was calculated by Nernst equation for the following reaction:

$$2H_2O = O_2 + 4H^+ + 4\bar{e}$$

The Henry constant for oxygen was taken from Ref. [8]. Calculations show that at 298 K and $pH_0 \sim 8$, the variation in oxygen concentration from 10 to 1000 ppb is due to the growth of E_h from 0.757 to 0.791 V. On Pourbaix diagram (Fig. 3), E_h-range corresponds to $Co(OH)_3$ area of existence. The diagram shows that $Co(OH)_3$ dissolves and yields Co^{2+} ions. An increase in the concentration of oxygen, i. e. the growth of E_h,

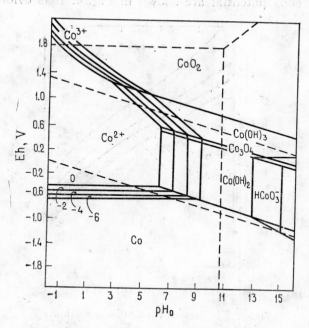

Fig. 3. Potential-pH diagram for Co-H$_2$O system at 298 K
according to Pourbaix [5]

causes the solubility of $Co(OH)_3$ to decrease. The $Co(OH)_3$ solubility was calculated on the assumption that in the process of its dissolution, apart from Co^{3+}, Co^{2+} are formed:

$$Co(OH)_3 + 3H^+ = Co^{3+} + 3H_2O$$

$$Co^{3+} + nH_2O = Co(OH)^{3-n}_n + nH^+$$

$$4Co(OH)_3 + 8H^+ = 4Co^{2+} + O_2 + 10H_2O$$

$$Co^{2+} + mH_2O = Co(OH)^{2-m}_m + mH^+$$

$$1 \leqslant m, \quad n \leqslant 3$$

The applied method of calculations of $Co(OH)_3$ solubility, in general, is analogous to that used for ferrous compounds, described in Ref. [9]. Equilibrium concentrations of Co^{2+} and Co^{3+} ions and of $Co(OH)_3$ molecules were established by a successive approximation method, balance being achieved by H^+ ion concentration in solution. Lack of initial data on thermodynamics of $Co(OH)^{2-}_4$ and $Co(OH)^-_4$ did not allow us to consider the fourth stage of hydrolysis of Co^{2+} and Co^{3+} ions. Lack of accurate initial thermodynamical data was the main difficulty in calculating the solubility of $Co(OH)_3$. Because of limited number of thermodynamical parameters concerning cobalt compounds, we used the most sufficient coordinated system described in Ref. [10].

The results of calculations of $Co(OH)_3$ solubility up to 573 K depending on redox potential are shown in Fig. 4. It is evident from the

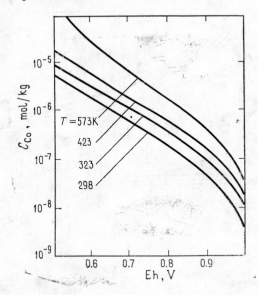

Fig. 4. E_h effect on the $Co(OH)_3$ solubility

figure that at all temperatures the solubility of $Co(OH)_3$ decreases as E_h increases.

It is expedient to compare the obtained data with the results of experiment as well as with the Pourbaix diagram for 298 K. As per experiment, at 298 K an increase in oxygen concentration from $\leqslant 10$ to 1200 ppb causes the solubility to decrease from 110 to 70 ppb (from 1.9×10^{-6} to 1.01×10^{-6} mol/kg); in this case, equilibrium pH_0 of the solution containing cobalt corrosion products was 8.3-8.5. For this equilibrium pH_0 and E_h (from 0.667 to 0.701 V) the solubility of cobalt compounds decreases from 5×10^{-7} to -10^{-7} mol/kg.

An analysis of our present results shows that the solubility of $Co(OH)_3$ decreases with increasing E_h: as E_h grew from 0.6 to 0.7 V, the $Co(OH)_3$ solubility decreased from 2.086×10^{-6} to 7.720×10^{-7} mol/kg, and the equilibrium pH_0 equalled 8.34 and 7.93, respectively. Any discrepancies between the experimental and the calculated solubilities of cobalt compounds in water may be due to some uncertainty in the initial thermodynamical data used in calculations.

ACKNOWLEDGEMENT

The authors are grateful to Professor O. I. Martynova for her encouragement and guidance in conducting this work.

REFERENCES

1. Martynova O. I., Samoilov Y. F., and Petrova T. I. (1979): *Trudy Mosk. Energ. In-ta,* 405, 40-44.
2. Pyatnitsky I. V. (1965). Analytical chemistry of Co, Nauka, Moscow, p. 567.
3. Styrikovich M. A. and Reznikov M. I. (1977). Experimental study of steam generation, Energiya, Moscow, p. 280.
4. Batalina L. N. and Dubrovsky A. I. (1976): *Teploenergetika,* 6, 69-71.
5. Pourbaix M. (1963), Atlas Dequilibres Electrochimiques, p. 644.
6. Asanti P. L. and Kolmeyer E. (1973), About the thermochemical properties of Co compounds, Khimiya, Moscow, p. 324.
7. Ovchinnikova T. M., Ioffe E., and Rotinjan A. L. (1955): DAN, 100, 469-471.
8. Homig H. E. Phisikochemische Grundlagen der Speisewasser Chemie (1963), p. 190.
9. Martynova O. I., Petrova T. I., and Kharitonova N. L. (1983): *Teplofizika vysokikh temperatur* 5, 913-919.
10. Glushko V. P., Thermochemical Constants, VINITI, vol. 6, p. 1.

Solubility of Iron Corrosion Products in the Water-Steam Circuit of a Heat Power Plant at Water Chemistry with H_2O_2 Introduced

Y. F. SAMOILOV and T. I. PETROVA

Moscow Power Engineering Institute, Moscow, USSR

At present, the corrosion processes taking place during operation of thermal power plants cannot be completely avoided. Therefore, when selecting a water chemistry mode for thermal power plants, adequate measures should be taken not only to retard corrosion processes but also to prevent the formation in the water-steam circuit of corrosion products deposits created due to corrosion of construction materials.

In this connection, the problem about corrosion products and additives behaviour in the water-steam circuit and their solubility in water and saturated steam, in particular, arises.

It is known that the solubility of contaminants in the boiling water and their transport into steam depends not only on the system parameters but also on the form of the compounds existence in the boiling water, which, in its turn, is determined by the composition of substance contained therein.

The recent practice in the USSR and abroad is to introduce definite amounts of O_2 or H_2O_2 into the condensate feed-water circuits of power plants. Therefore, a study of solubility of corrosion products of carbon steel (iron) in water and steam in the presence of these oxidizing agents is of practical importance.

This work is dedicated to the study of the effect of hydrogen peroxide (H_2O_2) on the solubility of iron corrosion products in water and steam. The experiments were conducted at 15.5 MPa and between 50 and 344°C. The diagram of the experimental plant used is shown in Fig. 1.

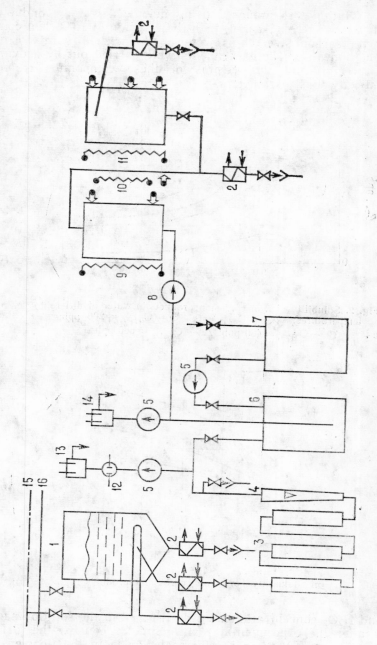

Fig. 1. Experimental plant scheme: *1* — deaerator; *2* — cooling chamber; *3* — ion exchange columns; *4* — flowmeter; *5* — pump; *6, 7* — vessel for preparing solution; *8* — high-pressure pump; *9, 10, 11* — heat exchangers; *12* — conductivity cell; *13, 14* — pH-meter; *15* — steam; *16* — condensate of steam

The test solution was prepared using deionized water (conductivity $\leqslant 0.1 \mu$ S/cm, pH ≈ 7; $C_{o_2} \leqslant 10$ ppb) with the addition of H_2O_2. On coming in contact with carbon steel chips water got enriched with corrosion products of carbon steel. The experiments were carried out by a dynamic method [1], precipitating the surplus of these compounds from the solution at strictly fixed parameters.

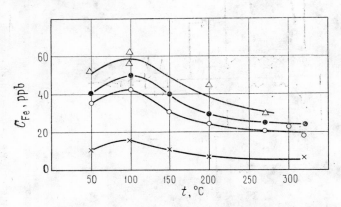

Fig. 2. Solubility of iron corrosion products in water at different temperatures: $\times - C_{H_2O_2} = 0$; $\bigcirc - C_{H_2O_2} = 200$ ppb; $\bullet - C_{H_2O_2} = 400$ ppb; $\triangle - C_{H_2O_2} = 600$ ppb

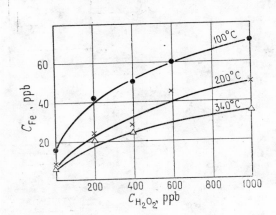

Fig. 3. Influence of concentration of H_2O_2 on solubility of fron corrosion products in water

Iron and H_2O_2 concentration were determined in the solution samples before and after the experimental plant. Simultaneously, the pH and conductivity were measured in special sealed cells in the flow ensuring the absence of contact of water with air. It was estabilished that for all

investigated H_2O_2 concentrations the maximum solubility of iron corrosion products at 15.5 MPa corresponds to about 100°C (Fig. 2), the rise in H_2O_2 concentration in the initial water led to an increase in the solubility of iron corrosion products (Fig. 3).

It should be noted that the solubility of carbon steel corrosion products was investigated not only in water below its boiling point, but also in boiling water and saturated steam. For two-phase system the experimental tests were carried out under 15.5 MPa pressure ($t=344°C$). It was established that for the aforementioned parameters the solubility of iron corrosion products in boiling water and saturated steam was found to be respectively 10 and 4 ppb at H_2O_2 concentration of about 400 ppb. The distribution coefficient represented as a ratio of concentration of iron in saturated steam and boiling water under these conditions is equal to 0.4 (Fig. 4). From this figure it is seen that these

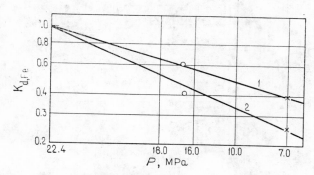

Fig. 4. Distribution coefficient for iron compounds:
$1 - C_{H_2O_2}=0$; $2 - C_{H_2O_2}=400$ ppb

coefficients are in good agreement with the experimental results [2] obtained at 7.0 MPa.

A comparison of distribution coefficients for corrosion products of carbon steel for a system with H_2O_2 shows that these are less than those without H_2O_2.

The presence of complexing agents in water is known to reduce the distribution coefficient [3, 4], with the conversion of molecules into ions; the concentration of molecules decreases. If we presume that only nonpolar compounds go from water to steam, then a decrease in the number of molecules should cause the distribution coefficient to reduce, whereas the molecular distribution coefficient remains constant.

Experimentally it was established [5] that at low temperatures complex compounds are formed in water solutions containing iron corrosion products and hydrogen peroxide.

The decrease in the distribution coefficient may also point to the formation of complex compounds of iron and hydrogen peroxide at high temperatures. Formation of iron compounds and the decomposion pro-

ducts of H_2O_2 may also cause a decrease in the transport of iron corrosion product from water into steam.

The measurements of H_2O_2 concentration show that the increase in water temperature from 50 to 300°C reduces H_2O_2 content in the samples. An abrupt change in H_2O_2 concentration is observed at temperatures above 100°C (Fig. 5); the relative H_2O_2 concentration at 50 and 200°C equals respictively 1.0 and 0.35. Above 250°C the relative concentration of H_2O_2 was less than 0.05. The H_2O_2 content in the boiling water was 8 ppb, and in the saturated steam — 26 ppb. The results shown in Fig. 5 are in conformity with those of Ref. [6].

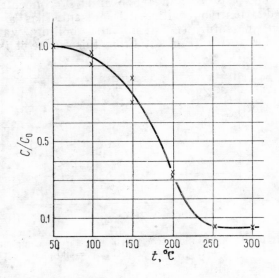

Fig. 5. Influence of temperature on relative concentration of H_2O_2 in water

The coordination number n in the equation $K_d = (\rho_s/\rho_w)^n$ was calculated for iron compounds, using experimental distribution coefficients. For water containing H_2O_2, n was found to be 0.47; for water without H_2O_2, it equalled 0.32. On the basis of the results of the experiment an equation was derived for calculating the iron corrosion product solubility, using the dependence described in [7]:

$$\lg C = n \lg \rho = \frac{\Delta H}{2 \cdot 3\, RT} + B$$

where:
C — contaminant's concentration, mol/kg
n — coordination number
ρ — density, kg/l
ΔH — heat of dissolution, J/(mol K)

R — universal gas constant, J/(mol K)

H — constant of integration, determined by the change in entropy during dissolution.

The heat of dissolution was assumed to be practically independent of temperature between 200 and 400°C.

The solubility of iron corrosion products in water containing H_2O_2 is calculated by the equation: $\lg C_{Fe} = 0.47\,\rho_w + \dfrac{6495}{19.1T} - 7.183$.

This very equation was used for calculating the solubility of iron corrosion product in boiling water at pressures of 4.0-15.0 MPa. The results are shown in Table 1.

It is evident from the table that with decrease in temperature the concentration of iron corrosion products increases.

Table 1. Solubility of iron corrosion products in boiling water

Pressure, MPa	Temperature, K	ρ, kg/l	Concentration of iron	
			mol/kg	ppb
15.5	617	0.594	2.32×10^{-7}	10.0
10.0	584	0.688	2.00×10^{-7}	11.2
7.0	558	0.742	1.78×10^{-7}	13.0
5.0	537	0.778	2.52×10^{-7}	14.1
4.0	523	0.799	2.66×10^{-7}	14.9

REFERENCES

1. Styrikovich M. A. and Reznikov M. I. Experimental Methods to study steam generation processes. 1977, Moscow.
2. Martynova O. I., Samoilov Y. F., Petrova T. I., and Kharlamov V. K. (1982): *Teploenergetika*, 11, 32-34.
3. Margulova T., Martynova O. I., Samoilov Y. F., and Medvedeva R. L. (1974): *Teploenergetika*, 1, 65-69.
4. Martynova O. I., Samoilov Y. F., and Petrova T. I. (1969): *Teploenergetika*, 5, 64-69.
5. Moskvin L. N., Efymov A. A. et al. (1979): *Teploenergetika*, 5, 46-50.
6. Bursik A. (1978): *Energie*, 30(4), 138-141.
7. Martynova O. I. (1964): *Journ. Phys. Chemistry*, 38(5), 1065-1076.

The Solubility of Gases in Water Over a Wide Temperature Range

J. ALVAREZ, R. CROVETTO, and R. FERNANDEZ-PRINI

*Departamento Química de Reactores, Comisión Nacional de
Energía Atómica, 1429-Buenos Aires, Argentina*

SUMMARY

The experimental methods employed to determine the thermodynamics of dissolution of nonpolar gases in water over a wide temperature range are compared. As the temperature approaches the critical point of the solvent, experimental and thermodynamic difficulties become greater. The systems H_2—H_2O and N_2—H_2O have now been studied up to 630 K and the results are discussed. Perturbation theories are shown to be capable of describing the thermodynamics of dissolution of the gases in water.

We have undertaken a research project aiming at the determination of the thermodynamics of dissolution of simple permanent gases in H_2O and in D_2O over almost all the temperature range of coexistence of liquid and vapour. The results of these studies are of physicochemical interest and are also relevant for design and operation of power plants which employ the steam-water cycle and for the primary circuits of nucleo-electric plants employing pressurized light or heavy water as coolant. Since the solubility of gases depends on temperature, pressure and gas phase composition, it is important to determine Henry's constant (k°_H) in order to separate the temperature dependence of solubility from that of the other variables, thus the results can then be applied to different practical situations.

240

In the latter part of this research project we have determined experimentally the solubility of N_2 and H_2 in H_2O over a wide temperature range approaching 630 K. However, this part of the work has been a consequence of observations made in the previous stages where the solubility of inert gases and ethane in light and heavy water [1] and that of ethane in H_2O [2] were determined. These two studies, plus the critical evaluation of all the available data for simple nonpolar gases in these solvents [3], clearly showed that there were severe discrepancies between the available data for the systems N_2—H_2O and H_2—H_2O, which are of great practical importance.

In the temperature range which goes from the melting point of the solvents to a temperature close to their critical point (T_{cl}), the density of the liquid phase and the concentration of gas in this phase change only moderately as the temperature increases; on the other hand, the change of density and concentration in the gas phase mixture is very large. This fact introduces some experimental complications in the determination of Henry's constants for these systems. In view of this, we have considered of great interest to establish the thermodynamics of dissolution of gases in water in a systematic way.

THERMODYNAMIC TREATMENT

The basic thermodynamic equations describing the liquid-vapour equilibrium in the binary systems are:

$$1-y = (1-x) P^{\circ}_1 \Phi^{\circ}_1 \exp[V^{\circ}_1 (P-P^{\circ}_1)/RT]/P\Phi_1 \qquad (1)$$

$$k^{\circ}_H = yP\Phi_2 \exp[-V^{\infty}_2 (P-P^{\circ}_1)/RT]/x = k_H \exp[-V^{\infty}_2 (P-P^{\circ}_1)/RT]/x \qquad (2)$$

In the above equations P is the total pressure, P°_1 the vapour pressure of the solvent, x and y are the gas mole fractions in the liquid and in the gas phase, respectively, and Φ_i denotes the fugacity coefficient and V_i the partial molar volume of component i. The only assumptions in Eqs. (1) and (2) are that the solution in the liquid phase is sufficiently dilute so that the activity coefficients can be taken equal to unity ($x \leqslant 0.01$ up to 630 K), and that the partial molar volumes of solvent (subindex 1) and solute (subindex 2) do not change when the pressure goes from P°_1 to P.

It must be noted that the usual approximations

$$y = 1-(P^{\circ}_1/P) \quad \text{and} \quad \Phi_2 = 1$$

introduce a large error in the value of the equilibrium composition of the gas phase. This, however, is partially compensated in the product $y\Phi_2$ appearing in (2), so that it is worse to correct partially for non-ideality employing, for instance, Lewis and Randall's rule for Φ_2.

Henry's constants are related to the change in solute chemical potential, $\Delta\mu^{\circ}_2$, when it is taken from ideal gas at 0.1 MPa to an infinite dilute solution having $x=1$, i. e.

$$\Delta\mu^{\circ}_2 = RT \ln(k^{\circ}_H/0.1 \text{ MPa}) \qquad (3)$$

From Eq. (3) the other thermodynamic quantities of dissolution may be

241

derived by differentiation. However, the temperature derivatives of the chemical potential which are experimentally obtained, correspond to changes of temperature along the saturation curve (indicated by subindex σ). The derivatives of Eq. (3) over σ are related to the thermodynamic quantities by

$$\Delta H^\circ{}_2 = [\partial(\Delta\mu^\circ{}_2/T)/\partial(1/T)]_\sigma + V^\infty{}_2 T(\partial P^\circ{}_1/\partial T)_\sigma \tag{4}$$

$$\Delta S^\circ{}_2 = [-\partial\Delta\mu^\circ{}_2/\partial T]_\sigma + V^\infty{}_2(\partial P^\circ{}_1/\partial T)_\sigma \tag{5}$$

$$\Delta C^\circ{}_{p2} = \frac{\partial}{\partial T}[\partial(\Delta\mu^\circ{}_2/T)/\partial(1/T)]_\sigma$$
$$- [(\partial\Delta H^\circ{}_2/\partial P)_T - V^\infty{}_2 - T(\partial V^\infty{}_2/\partial T)_P](\partial P^\circ{}_1/\partial T)_\sigma$$
$$+ T(\partial V^\infty{}_2/\partial P)_T(\partial P^\circ{}_1/\partial T)^2{}_\sigma + V^\infty{}_2 T(\partial^2 P^\circ{}_1/\partial T^2)_\sigma \tag{6}$$

As shown by the above expressions, in order to calculate the thermodynamic quantities of dissolution, it is necessary to know the partial molar volume of the solute and its change with pressure, as well as the first and second derivatives of the vapour pressure with temperature. For H_2O and D_2O as solvents, the correction terms in Eqs. (4)-(6) are important because $(\partial P^\circ{}_1/\partial T)_\sigma$ is large for these liquids. Hence, the temperature range in which the thermodynamics of dissolution for these systems can be unambiguously established is limited; at 573 K the corrections amount to 2.8 kJ mol^{-1}, 5 and 40 J(mol K)$^{-1}$ for enthalpy, entropy and heat capacity, respectively.

EXPERIMENTAL METHODS

Different methods have been employed to determine the solubility of gases in aqueous media. We shall briefly describe three of them and comment upon the possibility of employing the data obtained with them for the calculation of the thermodynamic quantities of dissolution.

1. Pressure Change at Constant Volume. The change of pressure upon dissolution of the gaseous solute is recorded in order to establish the amount of dissolved gas. The total volume of the system, that of the equilibration cell and the amount of solvent have to be known. The moles of dissolved gas are calculated from the difference between initial and final pressures. This calculation is strongly dependent on the precise knowledge of the gaseous volume and on the temperature gradient between the equilibration cell and the pressure gauges. This method is not very reliable and cannot be applied for the calculation of thermodynamic quantities of dissolution of gases in water above ambient temperature (up to 20% uncertainty in the solubility at 400 K).

2. Synthetic Method. It is based on the isothermal variation of volume in a moving piston cell which contains the two-phase binary system; the change of the equilibrium pressure in the cell is determined for different values of the volume. This method has been employed in our laboratory with great success for the study of C_2H_6—H_2O up to 473 K [2]. It is especially suitable to take account of the effect of pressure over Henry's constant, but at present it is being adapted for higher temperatures.

3. Analytic Method. It is based on the analytical determination of the amount of gas that is contained in a sample of the liquid phase at (P, T). This determination is carried out in a gas burette system provided with sufficient versatility to keep the experimental error in the measured solubility smaller than 1% over all the temperature range. For this purpose the experimental procedure can be adjusted according to the temperature of the system. The possible choices are: to select the adequate manometer in order to keep a good precision in the pressure measurement, to vary the amount of sample (1-4 g), the temperature of the sampling cell (either boiling nitrogen, 77.15 K or melting chloroform, 209.5 K), its volume and the density of the manometric fluid in the gas burette system.

The cell which was employed is illustrated in Fig. 1. Two temperature probes were employed to guarantee absence of temperature gradients in the cell. These were carefully avoided by variation of the electric power being dissipated in the various sections of the cell. The presence of temperature gradients implies non-equilibrium, furthermore it may give a wrong indication of the value of P°_1 which in turn will

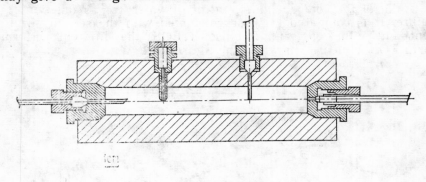

Fig. 1. Cell used in the analytical method

affect very much the value obtained for k°_H as shown by Eqs. **(1)** and **(2)**.

This is the method we have employed to determine the solubility of hydrogen and nitrogen in water between room temperature and 630 K. Before analyzing the results in terms of Henry's constants, it is convenient to discuss the problem of gas phase non-ideality.

EQUATION OF STATE FOR THE GASEOUS MIXTURE

According to Eqs. (1) and (2) in order to describe the thermodynamics of dissolution of gases, it is necessary to know the fugacity coefficients of the two components in the gaseous mixture and consequently an equation of state (EOS) has to be employed.

It is possible to use a second virial coefficient EOS up to about 500 K and still retain an uncertainty smaller than 2% in k°_H which is

our goal. However, in order to cover a wider temperature range, it is necessary to use another EOS. We have found that the Peng and Robinson EOS [4] is adequate for the gas-water gaseous mixtures. The equation is,

$$P = RT/(V-b) - a(T)/[V(V+b) + b(V-b)] \qquad (7)$$

where $a(T)$ and b are parameters of the mixture given by:

$$a = \sum_i \sum_j y_i y_j (1-\delta_{ij})(a_i a_j)^{1/2}$$

$$b = \sum_i y_i b_i$$

a_i and b_i are the two parameters of the EOS for the pure components which are given by Peng and Robinson [4] in terms of critical constants, accentric factor and reduced temperature. The only parameter characterizing the actual mixture is δ_{ij} which we have calculated from the experimental values of the gas — H_2O cross second virial coefficients, which however do not extend to temperatures greater than 500 K [5], [6]. Figure 2 is a plot of the compressibility factor of water (z) as function of temperature. It is compared to those calculated with the second virial coefficient and with Peng and Robinson's EOS, the different performance of both EOS is clearly shown in the figure.

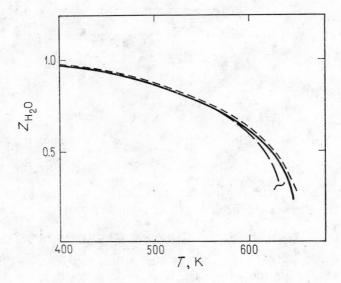

Fig. 2. Compressibility factor of water against temperature: ——————— experimental values; — — — — calculated with second virial coefficients EOS; - - - - - - calculated with Peng and Robinson's EOS

Another variable which must be known to calculate $k°_H$ is the partial molar volume of the solute, Eq. (2), since the solubility is measured at a finite pressure which differs more from $P°_1$ at higher temperatures. We have decided to estimate the effect of pressure on the solubility by means of a theoretical equation, as described below.

RESULTS

The values of $k°_H$ have been calculated according to the procedure described above [8], and $\ln k°_H$ has been fitted to a polynomial in the reciprocal temperature of the form;

$$\ln(k°_H/\text{GPa}) = A_0 + 10^3 \cdot A_1/T + 10^6 \cdot A_2/T^2 + 10^9 \cdot A_3/T^3 \qquad (8)$$

The values of the A_i coefficients for the systems hydrogen- and nitrogen-water, are given in the following Table:

System	A_0	A_1	A_2	A_3	σ_k
H_2-H_2O	—13.307	13.081	—3.580	0.309	0.017
N_2-H_2O	—14.538	14.205	—3.640	0.265	0.014

In order to obtain the coefficients of Eq. (8), we have added to the data obtained in our laboratory three values of $k°_H$ for the temperatures of 278, 298 and 308 K reported by Battino and coworkers [7]. These values are more precise than those in our work since they have been determined with methods applicable to pressures lower than atmospheric. The computer program which we employ to fit the data also calculates the values of the thermodynamic quantities at any specified temperature and, using the matrix of covariance, calculates the uncertainty of each quantity. It may be said in general that if $k°_H(T)$ is known within 2% the uncertainties in the thermodynamic quantities between ambient temperature and 500 K are calculated as 200-700 J mol^{-1} for $\Delta H°_2$, 0.5-2.2 J (mol K)$^{-1}$ for $\Delta S°_2$ and 4-30 J (mol K)$^{-1}$ for $\Delta C_p°_2$.

We have compared our results with all the data available in the literature for the systems H_2—H_2O and N_2—H_2O. These were treated as described previously [3], but using Peng and Robinson's EOS. The data obtained in our laboratory agree very well with the low-temperature data reported in the literature [7] as well as with the studies of Wiebe and coworkers [9] and O'Sullivan and Smith [10]; these measurements had already been noted [3] and are quite precise. The calculation of $\ln k°_H$ with a second virial coefficient EOS at temperatures higher than 500 K overestimates the predicted solubility compared to the value obtained by employing Peng and Robinson's EOS; the difference is larger than experimental uncertainty. Consequently, it is important to use for the high temperature data the Peng and Robinson EOS which allows a consistent description of the whole vapour-liquid equilibrium.

USE OF PERTURBATION THEORY

In terms of the Percus-Yevick EOS for hard spheres [11], the standard chemical potential of dissolution of the gases is given by

$$\Delta\mu^{\circ}_2 = kT[-\ln(1-y) + 3y(R+R^2)/(1-y) + (9/2)(yR/(1-y))^2]$$
$$+ yP^{HS}R^3/\rho_1 + kT \cdot \ln(kT\rho_1) + \Delta\mu'_2 \qquad (9)$$

where ρ_1 is the number density of the solvent, $y = \pi\rho_1 d_1^3/6$ is its packing fraction, $R = (d_2/d_1)$ is the ratio of equivalent hard sphere diameters between solute and solvent and P^{HS} is the hard sphere pressure of the reference fluid (having the same packing fraction of the solvent). The term $\Delta\mu'_2$ in (9) is the contribution of the attractive Lennard-Jones perturbation potential, $u'_{12}(r)$, which is given by,

$$\Delta\mu'_2 = 4\pi\rho_1 \int r^2 u'_{12}(r) g^{\circ}_{12}(r) \, dr \qquad (10)$$

In (10) $g^{\circ}_{12}(r)$ is the radial distribution function corresponding to hard sphere solvent particles surrounding a hard sphere solute particle. Lebowitz [12] has derived an expression for the Laplace transform of $r \cdot g^{\circ}_{12}(r)$ which has been used to calculate $\Delta\mu'_2$.

The rigorous application of this theory requires the calculation of equivalent hard sphere diameters in terms of intermolecular potentials, these diameters are always temperature dependent. When the solvent has strong anisotropic interactions, as is the case of water and other hydrogen-bonded liquids, the application of the rigorous procedure may be hampered by our lack of precise knowledge of the intermolecular potential of the pure solvent. Consequently, for these solvents we have adopted a semiempirical approach to obtain d_i and the energies ε_i corresponding to the minimum in the intermolecular Lennard-Jones potential.

A value of d_1 is fixed based either on a known value from other studies or on the low temperature compressibility of the solvent as given by the Percus-Yevich equation. For water, we have employed 2.70 Å. The values of d_2 and ε_2 are obtained from the literature as detailed in the following Table:

Solute	d_2/Å	(ε_2/k)/K	Ref.
H_2	2.87 (2.75)	29.2	[11, 13]
N_2	3.698 (3.40)	95.05	[11, 13]

The value of ε_{12} is obtained from the geometric mean of ε_1 and ε_2. However, in order to use such combination rule it is necessary to know the intermolecular interaction energy of two water molecules *without the hydrogen bonding contribution*. This quantity was calculated from experimental data on cross second virial coefficients of gaseous binary mixtures containing the solute gas and water vapour [6] and it was found

to be 220 ± 10 K. Employing this set of molecular parameters it was possible to adjust d_2 so that the observed value of k°_H for 298 K could be obtained using Eqs. (9) and (10). These values of d_2 are given in parentheses in the second column of the table. Repeating this procedure over all the experimental temperature range it is possible to obtain the temperature dependence of d_2. The curves of d_2 against temperature are typical of those systems to which perturbation theories have been applied and their slopes are quite reasonable. However d_2 for the systems H_2—H_2O and N_2—H_2O tends to increase abruptly as the temperature goes above 600 K; the origin of this effect is not yet clear. It is considered that it could be due to experimental difficulties in that temperature domain, to the vicinity of the critical point, or to the non-spherical shape of the solute molecules.

The values of v^∞_2 which are required in Eq. (2) to determine k°_H, were calculated with the Percus-Yevick hard sphere equation as derived by Lee [14].

This theoretical procedure of calculation is able to reproduce the changes in the chemical potential of the solute when it is transferred from H_2O to D_2O. This is a relevant feature showing that in spite of the extensive orientational correlation among water molecules by intermolecular hydrogen bonding, a perturbation theory using a hard sphere reference fluid can be applied for the prediction of thermodynamic quantities of dissolution if used semiempirically, i. e. the density of the hard sphere reference fluid is at every temperature equal to that of the real solvent.

EXTENSION OF THE STUDY TO THE CRITICAL POINT OF THE SOLVENT

It is of great interest to extend these studies and the method of data treatment to temperatures closer to the critical point of water. However there are still some theoretical aspects which require answer, e. g. whether the simple perturbation theory can be used close to T_{cl} and which is the value of d_2 as the system approaches T_{cl}.

From the standpoint of thermodynamics it would be very interesting to study a region closer to the critical state of water, because it would cover the gap between the present work and other studies for the same systems in the supercritical range [15], thus extending the thermodynamic description over all the temperature range. On the other hand, a straight forward extension of (2) to $T \to T_{cl}$, always in the limit of $P \to P^\circ_1$, gives at the critical point

$$k^\circ_H = P^\circ_{cl}\Phi^\infty_2(T_{cl}) \tag{1}$$

since $x = y$. Wheeler [16] has shown that Φ_2^∞ is finite at T_{cl}.

This expression is not at present supported experimentally. Moreover, there are reasons to believe that this limit is not rigorously attainable experimentally because, as the critical temperature of the solvent is approached, the solute-solute interactions increase due to higher

247

concentration in the liquid phase and to the longer range of the interactions. Thus, the solute activity coefficient will have to be included in Eq. (2) and extrapolation to $x=0$ will be difficult because at T_{cl} its change with concentration becomes infinite [16]. We consider these points deserve a closer experimental as well as theoretical study.

REFERENCES

1. R. Crovetto, R. Fernández-Prini and M. L. Japas, *J. Chem. Phys.,* **76**, 1077 (1982).
2. R. Crovetto, R. Fernández-Prini and M. L. Japas, *Ber. Bunsenges. Physik. Chem.,* 84, 484 (1984).
3. R. Fernández-Prini and R. Crovetto, AIChE J., in press.
4. D-Y. Peng and D. B. Robinson, "Thermodynamics of Aqueous Systems with Industrial Applications", Ed. S. A. Newman, ACS Symposium Series, 1980.
5. P. Richards, C. J. Wormald and T. K. Yerlett, *J. Chem. Therm.,* **13**, 623 (1981); G. Smith, A. Sellars, T. K. Yerlett and C. J. Wormald, *J. Chem. Therm.,* **15**, 29 (1983).
6. J. Alvarez, R. Crovetto and R. Fernández-Prini, *Z. physik. Chem. (N.F.),* **136**, 135 (1983).
7. E. Wilhelm, R. Battino and R. J. Wilcock, *Chem. Rev.,* **77**, 219 (1977).
8. J. Alvarez, R. Crovetto and R. Fernández-Prini, to be published.
9. R. Wiebe, V. L. Gady and C. Heins, *J. Amer. Chem. Soc.,* **55**, 947 (1933); R. Wiebe and V. L. Gady, *J. Amer. Chem. Soc.,* **56**, 76 (1934).
10. G. O'Sullivan and N. O. Smith, *J. Phys. Chem.,* **74**, 1460 (1970).
11. T. M. Reed and K. E. Gubbins, Applied Statistical Mechanics, McGraw-Hill, Kogakusha, Tokyo, 1973.
12. J. L. Lebowitz, *Phys. Rev.,* **133**, A895 (1964).
13. J. O. Hirschfelder, C. F. Curtiss and R. D. Bird, Molecular Theory of Gases and Liquids, Wiley, New York, 1966.
14. B. Lee, *J. Phys. Chem.,* **87**, 112 (1983).
15. M. L. Japas and E. U. Franck, to be published.
16. J. C. Wheeler, *Ber. Bunsenges, Physik. Chem.,* **76**, 308 (1972).

Solubility of Metallic Ni, Co and Cu in the Presence of Their Oxides, Magnetite and Hematite in High-Temperature Solutions

G. R. KOLONIN, O. L. GASKOVA, and G. P. SHIRONOSOVA

*Institute of Geology and Geophysics, Siberian Branch of
the USSR Academy of Sciences, Novosibirsk, USSR*

Alloyed steels, alloys containing nickel, cobalt, copper and sometimes pure metals, are extensively applied in modern power and heat installations, in autoclaves, and in other production and hydrometallurgical equipment. The inevitable influence of water, water steam and hydrothermal solutions of various compositions requires theoretical evaluation of the corrosion resistance peculiarities of these materials under high-temperature and high-pressure operating conditions. On this basis practical recommendations can be worked out.

Despite the fact that chemical thermodynamics has made a great contribution to the study of the corrosion processes (for example, [1]), the theoretical aspect of the temperature influence remains still open [2]. At the same time, in terms of the data now available, it is possible to find out important peculiarities of interaction of metals and their oxides with high-temperature solutions, including complexing.

The present study concerns nickel, cobalt and copper. The initial thermodynamic constants were borrowed mainly from the handbook by Naumov et al. [3].

Equilibrium calculations were carried out for the temperature interval of 298-573 K. The effect of pressure was ignored, as it is essentially done in all similar studies, because its effect on the saturated

vapour curve is negligible. At the same time, much attention was paid to the effect of oxidation conditions upon the interaction of metals with solutions. To this end, we applied the approach, which was widely used in geochemical researches long ago, when oxidation conditions in a solution are determined by the so-called mineral buffers [4, 5].

The chemical essence of the processes occurring during the interaction of metals with water solutions sharply differs depending on the conditions under which they take place, i. e. whether these processes are reduction or oxidation ones. In the former case, corresponding to the solutions with a rather high content of hydrogen, or to the nearcathod conditions, the process is usually written in the form of a semireaction:

$$Me_{cryst} \rightleftharpoons Me_{aq}^{n+} + n\bar{e} \tag{1}$$

But we prefer to use the following electrical-neutral equivalent of this reaction:

$$\mathbf{Me}_{cryst} + nH^+ \rightleftharpoons \mathbf{Me}_{aq}^{n+} + 0.5nH_2 \tag{1a}$$

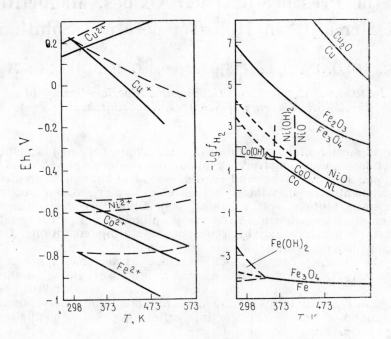

Fig. 1. Effect of temperature on the solution reaction potential of the considered metals at $Me^{n+} = 10^{-5} g^{-ion}/kg$. Solid lines represent the data of [3], dotted lines show the data of [6]

Fig. 2. Effect of temperature on hydrogen equilibrium volatility in metaloxide systems

Fig. 1 shows the temperature effect on this equilibrium. One can see from [3] that the potential, causing the appearance of the important quantities of the metal ion (it was conventionally assumed to correspond to $a_{Me^{n+}} = 10^{-5}$ mol/kg H_2O), appreciably decreases with temperature. At the same time, according to the new data by Zarembo and Puchkov [6], the temperature effect on the electrode potentials of the metals in question is negligible, and even a tendency to an increase of potentials is observed.

The like processes must take place, for example, by the action of the iron-magnetite buffer which, due to the interaction of a metal with water, is responsible for a volatility of hydrogen in the order of 100 MPa that is practically stable irrespective of a temperature (Fig. 2).

In a wide range of temperatures, the volatility is determined by the equilibrium

$$3Re_{cryst} + 4H_2O \rightleftarrows Fe_3O_{4\ cryst} + 4H_2 \tag{2}$$

which in low-temperature conditions should be transformed as follows:

$$Fe_{cryst} + 2H_2O \rightleftarrows Fe(OH)_{2\ cryst} + H_2 \tag{2a}$$

The thermodynamic data for ferrous oxide hydrate were derived from [7] and they were correlated with the other applied values.

As an example of nickel dissolution in the like manner, one can take nickel-bearing steels or nickel bearing alloys containing iron because the latter by virtue of its higher chemical activity must be "potential-determining" in the system. The specific calculation results of the solubility of Ni and Cu in the solutions equilibrated with an iron-containing buffer are presented in Figs. 4 (curves *1-2*) and 5 (curve *1*). The lowest content of the mentioned metals within the whole pH interval is observed precisely in this case. Only in strongly acid solutions, the nickel activity reaches really considerable values. At a temperature up to 573 K, the solubility curves go up approximately by two log. units with a shift of their minimum to lower pH values. General recommendations on the decrease of metal corrosion within the discussed process follow from the very form of reactions (1, 2). They require the redox potential increase in the system.

The second type of the corrosion processes refers to more oxidation or near-anode conditions in a solution, which correspond to oxide stability fields of the metals under consideration. A change from one type of corrosion to another is determined by the equilibrium line position:

$$Me_{cryst} + nH_2O \rightleftarrows MeO_{n\ cryst} + nH_2 \tag{3}$$

It is seen in Fig. 2 that it lies considerably higher for nickel and cobalt than for iron, but it is particularly high for copper. Since $Me(OH)_2$ hydroxides are likely to be the first products of the metallic surface change at low temperatures, according to [2], they are also reflected in Fig. 2.

The nature of the second type of corrosion process can be represented as a dissolution reaction of the corresponding metal oxide:

$$MeO_{cryst} + 2H^+ \rightleftharpoons Me_{aq}^{2+} + H_2O \tag{4}$$

As with the first type of corrosion, the solubility in this case will increase fastly with the growth of the solution acidity (see reaction (1a)). Since, in addition to Me_{aq}^{n+} aqua ions the solution must, in general, contain its other forms (hydroxocomplexes, chlorocomplexes, etc), it is essential to estimate their contribution to the total solubility. This is readily realized, if we apply the Freneous function [8], defined as a ratio of the total concentration of a metal in the solution to its aqua-ion concentration:

$$\Phi_{Me} = \frac{\overline{C}_{Me}}{C_{Me^{n+}}} \tag{5}$$

If we substitute concentrations by activities, we may obtain the following equation:

$$\Phi'_{Me} = 1 + \sum_{p=1}^{P} \beta_{OH, p} \, a^p_{OH} + \sum_{q=1}^{Q} \beta_{Cl, q} \, a^q_{Cl} + \dots \tag{6}$$

It permits calculations of "thermodynamic" complexity Φ' according to the free (excess) activity of hydroxide or chloride-ion, if the stability constants of the stepped complexes on infinite dilution are known ($\beta_{OH, p}$ and $\beta_{Cl, q}$). The p and q indexes designate all possible numbers of the ligands in hydroxo- and chlorocomplexes, respectively. Now combining the expression for the logarithm of equilibrium (1a):

$$\lg K_{1a} = \lg a_{Me^{2+}} + 2pH + \lg f_{H_2} \tag{7}$$

with Eqs. (5) and (6), we obtain the following formula to calculate metal equilibrium activity in solutions:

$$\lg \bar{a}_{Me} = \lg K_{1a} - 2pH - \lg f_{H_2} + \lg \Phi'_{Me} \tag{7a}$$

Returning to reaction (1a) in the systems with a soluble oxide is caused by the necessity to estimate the presence of the metal itself in the immediate proximity to the solution. In other words, equilibrium (1a) in this case is the result of summarising of simultaneously proceeding reactions (3) and (4).

Complex formation constants used in the calculations were derived from Ref. [9] through [18], and the results are presented in Figs. 3-5. Figs. 3a and 4 (curve 3) show the effect of pH and temperature on Ni and Co total activity in the solution equilibrated with MeO/Me pair. The increase of the activity upon alkaline values of pH is connected with the appearance of $Me(OH)_3$ anion complexes in these conditions (in Pourbaix diagrams, the MeO_2 form corresponds to it). Light lines in Fig. 3 show the contribution of various individual forms of Ni and Co to the total activity. The level and the position of the cobalt stability minimum depending on the stability of $Co(OH)_2$ electroneutral complex are shown

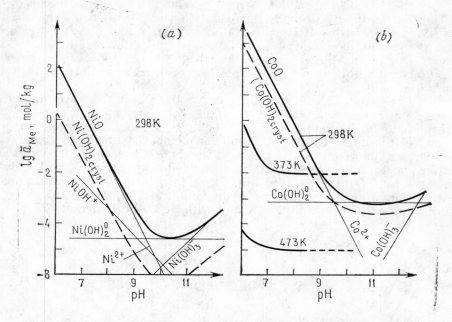

Fig. 3. Solubility of nickel (*a*) and cobalt (*b*) in equilibrium with their oxides (hydroxides) in the conditions of MeO/Me and Me(OH)$_2$/Me buffer pair presence. Light lines show the activity of the individual dissolved forms of nickel and cobalt at 298 K

rather conventionally, since they were obtained by using mutually uncoordinated constants of Co^{2+}_{aq} ion hydrolysis in the first and second steps of Ref. [11, 12]. Both for Co and Ni, the contribution of chloro-complexes to the total activity is quite negligible because of their weakness.

Fig. 4 shows the effect of various redox conditions on the solubility of nickel at three temperatures. Besides the above discussed curves *1, 2*, there are curves *3* and *4* for the metal content in the solution, equilibrated with NiO/Ni and Ni(OH)$_2$/Ni buffer pairs. One can also observe the direction of the solubility shift upon passing to more oxidation conditions determined by Fe$_2$O$_3$/Fe$_3$O$_4$ hematite-magnetite buffer (curves *5*).

The comparison of the level of curves *3* and *5* at different temperatures leads to the conclusion that in high temperature conditions even certain suppression of corrosion takes place. This is probably accounted not only for certain thermodynamic stability of NiO but also for the increase of f_{H_2} with temperature for the considered buffer pairs (see Fig. 2), which finally hinders reaction (1a).

Fig. 5 shows the results of calculations of copper solubility at 298 and 473 K in various acid conditions depending on pH, with allowance

for the effect of chlorocomplex formation. The position of these cerves vs. the ordinate strongly depends on oxidation conditions in a solution: the lowest position, even missing in the diagram for 298 K, is occupied by Fe_3O_4/Fe buffer pair (curve *1* in Fig. 5b), the intermediate one is for hematite-magnetite buffer (curves *2*). The highest position, causing essential solubility in the whole pH interval, corresponds to copper-cuprite buffer (curves *3*).

The reasons of the solubility curve displacement are discussed above. As to the effect of the chloride complexing, which is rather considerable in cuprous systems, it amounts to an increase of the acid branch of the curve by two log. units in comparison with the same one of Cu^+_{aq}-ion upon introduction as little as 0.01 g-ion/kgCl with its further increase by two log. units, provided the chloride ion activity increases by 0.1 g-ion/kg (see Fig. 5a). Fig. 5 also shows the supposed contribution of various cuprous hydrocomplexes to copper solubility in equilibrium with cuprite at 298 K and with Fe_3O_4/Fe buffer at 473 K.

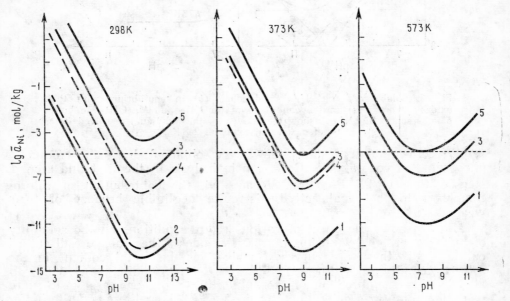

Fig. 4. Effect of various redox conditions, pH and temperature on metallic nickel solubility: *1* — Fe_3O_4/Fe buffer; *2* — $Fe_3O_4/Fe(OH)_2$; *3* — NiO/Ni pair; *4* — $Ni(OH)_2/Ni$ pair; *5* — Fe_2O_3/Fe_3O_4 buffer. The conventional level of the considerable solubility is shown by a point line

It can be deduced from the comparison of Figs. 5a and 5b that, unlike Ni, the increase in temperature leads to an essential rise of copper solubility, i. e. to the loss of its corrosion resistance. The effect is the stronger, the more are reduction conditions in a solution, intensifying from curves *3* to curves *2* and further to curve *1*. The

bifurcated acid branches of the solution curves in Fig. 5b reflect the variants of various authors data for the constants of copper chloro-complexes which are rather restricted and contradictory.

(a)   (b)

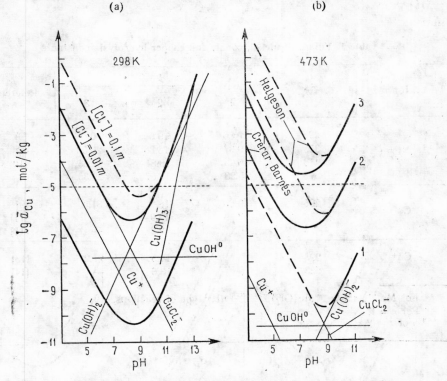

Fig. 5. Effect of various redox conditions and pH on the solubility of metallic copper at 298 K and 473 K: *1* — Fe₃O₄/Fe buffer; *2* — Fe₂O₃/Fe; *3* — Cu/Cu₂O pair

The consideration of the effect of various factors on the corrosion processes of the second type, occurring in the conditions of the oxide stability, i e. in a passivation field allows one to draw a conclusion about the existence of two various groups of metals. Ni and Co represent the situation, when the increase of the solution temperature results in an increase of their oxide stability in equilibrium conditions, and finally hinders the corrosion. The solubility of the other metal oxides, in our case represented by copper, increases with temperature favoring the decrease of their corrosion resistance. As well as in the corrosion of the first type proceeding outside the field of oxide existence, the intensification of the reduction conditions in the solution will favor the decrease of corrosion.

It should be noted in conclusion that chemical peculiarities of metal-solution interaction were considered exclusively in terms of equilibrium

conditions and equilibrium process thermodynamics, ignoring the effect of the solution ion strength. At the same time, rather important but less developed kinetic aspect of the problem is, unfortunately still left outside the report.

Table 1. Values of nickel, cobalt and copper hydrolysis constants *

Ion	T K	$-\lg \beta_n^h$			References
		1	2	3	
Ni²⁺	298	9.86	19	30	[9]
	373	9.37	16.86	26.03	
	473	6.95	13.81	22.68	[10]
	573	5.23	11.67	20.35	
Co²⁺	298	9.65	18.8	31.5	[9]
	373	7.79	13.9	—	[11] **
	473	6.1	12.2	—	[11]
Cu⁺	298	7	14.5	27.2	[13]
	523	5	13.6	—	[14]

* For $Me^{n+} + nH_2O \rightleftarrows Me(OH)_n^{+m-n} + nH^+$ type of reactions.
** For $n=1$.
*** For $n=2$.

Table 2. Values of nickel, cobalt and copper chlorocomplex stability constants

Ion	T K	$-\lg \beta_n^{Cl}$				References
		1	2	3	4	
Ni²⁺	298	—0.24	—	—	—	[15]
	573	0.2 *	—	—	—	
Co²⁺	298	—0.18	6	—	—	[15]
	573	0.09 *	—	—	—	
Cu⁺		2.7	6	5.99	4.7	[17]
		—	4.94	5.8	—	
	298	—	5.14	6.4	—	[16]
	473	2.2	—	—	—	[18]

* Approximate values according to the reaction enthalpy which is assumed to be stable.

REFERENCES

1. Pourbaix, M. J. N. (1966): Atlas of Electrochemical Equilibria in Aqueous Solutions. Pergamon, Oxford.
2. "Corrosion". Handbook edited by Shrayer L. L. (1981), City of London Polytechnic, 632 p.
3. Naumov, G. R., Rizhenko, B. N., Khodakovsky, I. L. (1971). Handbook of thermodynamic values (for geologists). Atomizdat, Moscow, 239 p.
4. Eugster, H. P., Skippen, J. B. (1967) — In: Researches in Geochemistry. John Wiley and Sons, v. 2, p. 492-520.
5. Huebner, J. S. (1971) — In: Research techniques for high pressure and high temperature, Springer-Verlag, p. 123-178.
6. Zarembo, V. I., Puchkov, A. V. (1984) — In: Review on heat physical properties of substances. IVTAN, Moscow, N 2, (46), 102 p.
7. Barin, I., Knacke, O. Thermodynamical Properties of Inorganic Substances (1973), Springer-Verlag, 921 p.
8. Bek, M. Chemistry of Equilibria (1970): Van Nostrand Reinold Company. L., N-Y., Toronto, Melburne.
9. Baes, C. F., Mesmer, R. E.: The Hydrolysis of Cations (1976), John Wiley and Sons, 491 p.
10. Tremaine, P. R., LeBlanc, J. C. (1980): *Journal of Chem.*, v. 56, N. 4, p. 435-438.
11. Giasson, J., Tewary, P. H. (1978): *Can. Journal of Chem. Thermodyn.*, v. 12, N. 6, p. 521-538.
12. Martinova, O. I., Mingulina, E. I., Smirnova, O. Kh., and Kurtova I. S. (1971). Izvestia AN SSSR, "Energetics and transport" series, N. 2, p. 148-153.
13. Kolonin, G. R. (1981) — In: Thermodynamics and structure of hydroxocomplexes in solutions. Abstracts of the Rept. III All-Union conference, Dushanbe. Nauka, Leningrad, 1980, p. 116.
14. Rekharsky, V. I., Pashkov, Yu. N., Kapsamun, V. P. et al. (1982) In: "Geochemistry of ore formation processes". Nauka, Moscow, p. 92-97.
15. Lister, M. W., Rosenblum, P. (1960): *Can. Journal of Chem.* v. 38, p. 1827-1836.
16. Helgeson, H. C. (1969): *Amer. Journal of Sci.,* v. 267, p. 729-804.
17. Ahrland, S., Rawsthorne, J. (1970): *Acta Chem. Scand.,* v. 24, p. 157-172.
18. Crerar, D. A., Barnes, H. L. (1967): *Econom. Geol.,* v. 71, N. 1, p. 772-794.

Temperature and Pressure Dependences of Salt Solubility and Their Relationship with Hydration and Solution Structure

A. K. LYASHCHENKO

Institute of General and Inorganic Chemistry of the USSR Academy of Sciences, Moscow, USSR

B. R. CHURAGULOV

M. V. Lomonosov Moscow State University, Moscow, USSR

At present, experimental data are available on temperature and pressure dependences of salt solubility in water over a wide range of pressures and temperatures. Any unformal systematization or theoretical description of heterogeneous water-salt equilibria should explain the changes in salts solubility and the appearance of four types of polytherms and polybars in equilibrium with the constant solid phase (curves with maxima, minima, increasing or decreasing solubility). For this purpose, use can be made of an approach which establishes relation between thermodynamic parameters characterizing the shape of polytherms and polybars in thermodynamic equations and interparticle interactions and structural changes in solutions. On the basis of known dependences of temperature $\left(t.c.s. = \dfrac{\partial m_s}{\partial T} \right)$ and pressure $\left(p.c.s. = \dfrac{\partial m_s}{\partial p} \right)$ coefficients of solubility on partial enthalpy (ΔH) and partial molar volumetric effect (ΔV) of solubility [1] and assuming division of these values into components corresponding to the interactions in the solid

phase and solution, we analyzed t.c.s. and p.c.s. in water-salt systems and their changes at different temperatures and pressures [2-7].

The values $\Delta\overline{V}$ and $\Delta\overline{H}$ can be represented as follows:

$$\Delta\overline{V}=\Delta V^{r}_{m}+\Delta\overline{V}_{mx}=\overline{V}_{s.s.}-V_{solid} \qquad (1)$$

$$\Delta\overline{H}=\Delta H^{r}_{m}+\Delta\overline{H}_{mx}=\Delta H^{solid}_{ion}+\Delta\overline{H}_{H_2O}+\Delta\overline{H}_{hydr} \qquad (2)$$

where ΔV^{r}_{m} and ΔH^{r}_{m} are the volumetric effect and the enthalpy of salt melting at given P and T, $\Delta\overline{V}_{mx}$ and $\Delta\overline{H}_{mx}$ are partial volumetric effect and the enthalpy of mixing of overcooled electrolyte with the saturated solution (s.s.). V_{solid} and $\Delta H^{ion}_{sol'd}$ refer to the solid substance, the contributions of $\Delta\overline{H}_{hydr}$ and $\Delta\overline{H}_{H_2O}$ correspond to the ion hydration and the breaking of bonds between water molecules in the solution. The components ΔV^{r}_{m} and ΔH^{r}_{m} determine the differences for the particles of the solute due to the difference between solid and liquid states, the contributions of $\Delta\overline{V}_{mx}$ and $\Delta\overline{H}_{mx}$ determine the difference in interactions and structures in solution and in the electrolyte in the liquid state. In the general case, for aqueous solutions of electrolytes the systems can be separated when the first or the second components in $\Delta\overline{H}$ and $\Delta\overline{V}$ predominate. They are quite clearly distinguished by concentrations of saturated solutions. This is due to the differences in structures of aqueous solutions of electrolytes in various concentration regions as well as to a sufficiently clearly defined border between them, which can be identified [4, 8-11].

In the transition concentration region beyond the border of the region of water-like solution structure one can suggest the appearance of similar structural groupings formed by the ions in saturated solutions as well as in crystal hydrates and salts [9-11]. Though energetic and volumetric ratios determining $\Delta\overline{V}$ and $\Delta\overline{H}$ can differ in certain cases, the conformity of ionic and ion-water forms in solution and solid phase suggests that $\Delta\overline{H}_{mx}\to 0$, and $\Delta H\to\Delta H^{r}_{m}$, $\Delta\overline{V}_{mx}\to 0$, $\Delta\overline{V}\to\Delta V^{r}_{m}$. Since the signs of t.c.s. and p.c.s. are determined by those of $\Delta\overline{H}$ and $(-\Delta\overline{V})$ [1], and $\Delta H^{r}_{m}>0$ and $\Delta V^{r}_{m}>0$, in this particular case t.c.s.>0 and

p.c.s.<0. Since $\left(\dfrac{\partial\Delta V^{T}_{m}}{\partial p}\right)_{T}<0$ (which is due to large compressibility of the melt as compared with the corresponding solid salt), then, if the given correlation is observed, $\left(\dfrac{\partial p.c.s.}{\partial p}\right)_{T}>0$, in the same way as in the case of non-polar solid substances solutions in non-polar liquids [12].

Since the derivative $\left(\dfrac{\partial\Delta V^{T}_{m}}{\partial T}\right)_{p}>0$, $\left(\dfrac{\partial p.c.s.}{\partial T}\right)_{p}$ should be less than 0. The experimental data have shown that in more than 20 solid phases p.c.s.<0, $\left(\dfrac{\partial p.c.s.}{\partial p}\right)_{T}>0$, and $\left(\dfrac{\partial p.c.s.}{\partial T}\right)_{p}<0$ [7, 13] according to the conclusion made. As follows from the experimental data, in this concentration region the t.c.s. is also always more than 0.

The specific structure of water in solutions of electrolytes in the first concentration region [4, 8] results in different *t.c.s.* and *p.c.s.* of the salts. The differences are quite distinct and can be identified both by experimental data and in the given scheme. In the equations determining *t.c.s.* and *p.c.s.* through $\Delta\overline{H}$ and $\Delta\overline{V}$ and their components in (1) and (2) such systems represent another extreme case compared with the melt-like heterogeneous equilibria. Due to a sharp difference between interactions and the structures of water solutions and liquid electrolytes, the differences between solid and liquid electrolytes become less important. $\Delta\overline{H}$ and $\Delta\overline{V}$ are determined by $\Delta\overline{H}_{mx}$ and $\Delta\overline{V}_{mx}$ and by their changes either with P and T or when passing from salt to salt. In this case, $\Delta\overline{H}_{hydr}$ is a factor determining the differences of $\Delta\overline{H}$ and, accordingly, *t.c.s.* for various salts. This is due to the fact that variation in hydration energy, when passing from one ion to another with different charge and radius, changes the energy of ion-water interaction to a

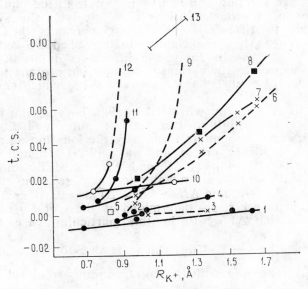

Fig. 1. Change in temperature coefficients of solubility (*t.c.s.*) in a series of salts with ions of different radius: *1* — $MSiF_6$ (M=Li, Rb, Cs, 273-298 K); *2* — $Ln(O_2CH)_3$ (Ln=La, Ce, Pr, Nd, Sm, Eu, Tb, Ho, 298-323 K); *3* — MSO_4 (M=Ca, Ba, 298-323 K); *4* — $M(O_2CH)_2$ (M=Ca, Ba, 298-323 K); *5* — $Sc(O_2CH)_3$ (298-323 K); *6, 7* — MCl (M=Na, K, Rb, Cs, 293-303 K; 353-373 K); *8* — MI (M=Na, K, Cs, 353-373 K); *9* — MF (M=Li, Na, K, 333-353 K); *10* — $M(O_2CH)_2 \cdot 2H_2O$ (M=Mg, Sr, 298-323 K); *11* — $M(O_2CCH_3)_2 \cdot 4H_2O$ (M=Co, Ni, Mn, 298-333 K); *12* — $M(O_2CCH_3)_2 \cdot 2H_2O$ (M=Zn, Mn, 333-353 K; 328-343 K); *13* — MNO_3 (M=Li, Na, K, Rb); $MO_2CH \cdot H_2O$ (M=Li, Cs); MO_2CH (M=Na, K. 298-323 K)

greater extent than that of ion-ion interaction in solid phase because the potential of ion-water interaction in solution decreases faster with increase in the distance [3]. In fact, expected correlation is not observed between $t.c.s.$ of electrolytes in water and lattice energy [2-5]. The temperature coefficient of solubility should grow with increasing lattice energy in series of salts with the same ion. However, in practice $t.c.s.$ falls as is seen from numerous examples. Thus, the higher the degree of hydration of ions and the smaller the variation in structure of the solvent during ion implantation (determined by the contribution of $\Delta \overline{H}_{H_2O}$), the lower the $t.c.s.$ of the salt. This conclusion can be regarded as a general rule, which is evident also from the further analysis of experimental data [2-5]. If $\Delta \overline{H}_{H_2O} \approx const$, for structurally similar ions, the relationship between $t.c.s.$ of salts and hydration characteristics of ions in solutions is clearly seen (Fig. 1, Table 1). If the exothermic contribution of ΔH markedly increases with decrease of ion radii (R), it causes the appearance of $t.c.s. < 0$ (in case of strong hydration of ions) and critical phenomena in saturated solutions of salts. These critical phenomena should be considered as a general characteristic of water-salt systems when ions strongly hydrate in solutions. The critical phenomena can be predicted for the systems of this type which have not been studied experimentally.

A change in $t.c.s.$ with temperature and pressure on the basis of the relationship of $t.c.s.$ with $\Delta \overline{H}$ is due to different variations in the components of $\Delta \overline{H}$ with temperature and pressure. At low temperatures a predominant decrease in $\Delta \overline{H}_{H_2O}$ with increasing temperature and pressure due to distortion of the hydrogen bond network at $\Delta \overline{H}_{on}^{sol'd} \approx const$ and a small variation of $\Delta \overline{H}_{hydr}$ determines the change in $t.c.s.$ sign from plus into minus (column 2, Table 1) and $\left(\frac{\partial t.c.s.}{\partial T} \right)_p < 0$.

Also, $\left(\frac{\partial t.c.s.}{\partial p} \right)_T$ should be less than 0, which is proved experimentally for sulphates of barium, strontium, cadmium, zinc, and copper; calcium carbonate; chlorides of cadmium and zinc; bromides of cadmium and zinc; nitrates of zinc and copper [6, 14, 15, 16, 17]. In the second temperature region at the saturated vapour pressure (when $\Delta \overline{H}_{H_2O} \to 0$), a decrease of $\Delta \overline{H}_{hydr}$ predominates, which should result in inverse correlations (that is, $\frac{\partial t.c.s.}{\partial T} > 0$, column 3, Table 1). If complex formation takes place, the corresponding parameter should be introduced into the equations which makes the dependence more pronounced and widens the temperature interval where the dependence is observed [18]. At elevated temperatures the role of pressure is also inverse. An increase in pressure strengthens the network of bonds between water molecules due to the decrease in free volume. The growth of $\Delta \overline{H}$ in this case should lead to $\left(\frac{\partial t.c.s.}{\partial p} \right)_T > 0$. The inverse effect of pressure on $t.c.s.$ at high temperatures was found from the experimental data (for sulphates

of barium, calcium, sodium, potassium, and lithium; sodium carbonate)
[6].

Table 1. Electrolytes with different types of *t.c.s.* (the compounds are represented
for which the data were obtained at low and elevated temperatures and low pressures
under the conditions of three-phase equilibrium)

Negative *t.c.s.* (strong hydration of ions)	Change of sign of *t.c.s.* from plus to minus with increasing temperature	Change of sign of *t.c.s.* from minus to plus with increasing temperature	Positive *t.c.s.* (weak hydration of ions and complexes)
	LiF, NaF, CaF$_2$, BaF$_2$		KF, RbF, NH$_4$F CsF·H$_2$O
Mg(OH)$_2$, Ca(OH)$_2$		LiON	NaOH, KOH, RbOH, CsOH, LiOH·H$_2$O
Li$_2$SO$_4$Na$_2$SO$_4$ (mon)	K$_2$SO$_4$, Ag$_2$SO$_4$	Na$_2$SO$_4$ (rhomb) Li$_2$SO$_4$·H$_2$O	(NH$_4$)$_2$SO$_4$ Cs$_2$SO$_4$, Tl$_2$SO$_4$, Na$_2$SO$_4$·10H$_2$O
MgSO$_4$, CaSO$_4$, ZnSO$_4$, MnSO$_4$ CdSO$_4$, FeSO$_4$, CoSO$_4$, NiSO$_4$, BeSO$_4$·H$_2$O,	SrSO$_4$, BaSO$_4$ CaSO$_4$·2H$_2$O		MgSO$_4$·6H$_2$O FeSO$_4$·4H$_2$O, FeSO$_4$·7H$_2$O, MnSO$_4$·5H$_2$O
FeSO$_4$·H$_2$O ZnSeO$_4$ MgMoO$_4$·2H$_2$O	CaSeO$_4$·2H$_2$O CaMoO$_4$	Na$_2$SeO$_4$	MnSO$_4$·7H$_2$O Na$_2$MoO$_4$, K$_2$MoO$_4$, Na$_2$MoO$_4$·2H$_2$O Na$_2$CrO$_4$, Na$_2$MoO$_4$·10H$_2$O, K$_2$CrO$_4$, MgMoO$_4$·5H$_2$O, K$_2$Cr$_2$O$_7$
CaWO$_4$		Na$_2$WO$_4$	Na$_2$WO$_4$·2H$_2$O Na$_2$WO$_4$·10H$_2$O
Li$_2$CO$_3$, NaCO$_3$ MgCO$_3$, CaCO$_3$			K$_2$CO$_3$, Na$_2$CO$_3$·10H$_2$O, Na$_2$CO$_3$·7H$_2$O
Na$_3$PO$_4$, Na$_3$PO$_4$·H$_2$O			K$_3$PO$_4$, Na$_3$PO$_4$·8H$_2$O (NH$_4$)$_2$HPO$_4$ LiClO$_4$, KClO$_4$, Cd(ClO$_4$)$_2$

Similarly to *t.c.s.*, in considering the correlations of $\Delta\overline{V}$, *p.c.s.* and structural changes in solution, one can single out principal tendencies of pressure dependence of salts' solubility in the region of water-like structure of solution [7]. In this case, the changes of $\Delta\overline{V}$ are greater than those which correspond to ΔV^{r}_{m}, that is, $\Delta\overline{V}$ is determined by $\Delta\overline{V}_{mx}$. Thus, $\Delta\overline{V}$ is determined by different packings for different types of

structures in solution and the solid phase. Typical differences in $p.c.s.$ are due to the specific features of volume effects of dissolution related to hydration of ions and the presence of water structure in saturated solutions. On the basis of the solution structure model [4, 8] they can be characterized by certain structural processes [7, 19]. In this case

$$\overline{V}_{s.s.} \approx (a + b\gamma) V_{H_2O} + \Delta V_{el} \tag{3}$$

where a and b represent the number of sites and voids occupied by ions or hydrated complexes, γ characterizes the fraction of water molecules filling the voids, ΔV_{el} is the contribution of electrostriction which occurs only at elevated temperatures. At low temperatures (below 80-100°C) $p.c.s.$ can be represented by the ratio containing structural parameters

$$p.c.s. = -\frac{(a + b\,\gamma) V_{H_2O} - V_{solid}}{(\partial \overline{G}_2 / m_2)} \tag{4}$$

If $a > 0$, $b = 0$, the substitution solutions are formed in the openwork water structure (ammonium and cesium chlorides [4, 19]). Accordingly, $\overline{V}_{s.s.} > V_{solid}$ and $p.c.s. < 0$. At normal or elevated pressures, the most typical variant for salts is the situation when a is small and $b > 0$ [4, 8, 19] (a may be even negative due to compression of water during the formation of hydrate complexes). It predetermines that $\overline{V}_{s.s.} < V_{solid}$, $\Delta V < 0$ and $p.c.s. > 0$. It means that in most cases in the region of water-like saturated solutions should be observed $p.c.s.$ of sign opposite to that of $p.c.s.$ in the region of highly concentrated saturated solutions. Since $\overline{V}_{s.s.}$ changes with increase in P and T to a greater degree than V_{solid}, the change of $p.c.s.$ with P and T depends on the change of $\overline{V}_{s.s.}$ (when the change of the saturated solution molality m_s is not great as the latter also influences $p.c.s.$ and $t.c.s.$). The value $r.p.c.s. = 1/m_s$ $p.c.s.$, which considerably weakly depends on m_s, should correlate with ΔV and structural characteristics in a more straight way. Since $\left(\frac{\partial \overline{V}_{s.s.}}{\partial p}\right)_T > 0$ (both on the basis of the structural model [20] and according to experimental data) at $V_{solid} \approx$ const, then $\left(\frac{\partial \Delta V}{\partial p}\right)_T > 0$; and $\left(\frac{\partial p.c.s.}{\partial p}\right)_T < 0$ (and $\frac{\partial r.p.c.s.}{\partial p} < 0$ in all the cases). This is the cause of the appearance of maxima on the polybar of solubility. Table 2 compares the signs of $p.c.s.$ and the $\left(\frac{\partial p.c.s.}{\partial p}\right)_T$ and $\left(\frac{\partial r.p.c.s.}{\partial p}\right)$ derivatives for water-salt systems, for which experimental data are available. These experimental data confirm the trends observed.

At low temperatures $\frac{\overline{V}_{s.s.}}{\partial T}\bigg)_p > 0$ at $V_{solid} \approx$ const. Accordingly, $\left(\frac{\partial p.c.s.}{\partial T}\right)_p$ should be less than 0 (and $\frac{\partial r.p.c.s.}{\partial T} < 0$ in all cases when m_s changes). At elevated temperatures, remote electrostriction changes

the sign of $\left(\dfrac{\partial \overline{V}_{s.s.}}{\partial T}\right)_p$ and therefore the sign of $\left(\dfrac{\partial p.c.s.}{\partial T}\right)_p$ should also change. This inversion is confirmed by the experimental data for barium, calcium, lithium, sodium, and potassium sulphates and for sodium carbonate [6].

Table 2. Signs of *p.c.s.* **and** $\left(\dfrac{\partial p.c.s.}{\partial p}\right)_T$, $\left(\dfrac{\partial r.p.c.s.}{\partial p}\right)_T$ **derivatives of salts in water, experimentally studied up to pressure** P, **MPa.**

Solid phase	T, K	p.c.s.	$\left(\dfrac{\partial p.c.s.}{\partial p}\right)_T$	$\left(\dfrac{\partial r.p.c.s.}{\partial p}\right)_T$	P, MPa
Region of water-like structure of saturated solutions					
CaF$_2$	298	+	+	−	100
CaCO$_3$	274-298	+	+	−	100
SrSO$_4$	275-308	+	+	−	100
CaSO$_4$	373-573	+	+	−	100
BaSO$_4$	298-573	+	+	−	140
LiSO$_4$, NaSO$_4$, K$_2$SO$_4$	473-630	+		−	150
(Li, K)SO$_4$, Na$_2$CO$_3$	630-673	+	(+, −)	−	150
K$_2$SO$_4$, CuSO$_4$·5H$_2$O, NaCL	298	(+, −)*	−	−	1000
KI	298	+	−	−	1000
Tl$_2$SO$_4$	303	+	−	−	150
NH$_4$ClO$_4$	263-293	(+, −)	−	−	100
CsBr	298	−	+	−	100
CdI$_2$	298	−	+	−	1000
CdSO$_4$·H$_2$O; CdSO$_4$·8/3 H$_2$O$_1$	298-348	+	−	−	1000
CdCl$_2$H$_2$O ZnSO$_4$·H$_2$O	323-348	+	−	−	1000
CdBr$_2$	348	(+, −)	−	−	1000
Region of highly concentrated saturated solutions					
NH$_4$NO$_3$	298	−	+	+	1000
Cd(NO$_3$)$_2$; Cd(NO$_3$)$_2$·4H$_2$O	348	−	+	+	700
ZnCl$_2$; ZnCl$_2$·4/3H$_2$O; ZnCl$_2$·2,5H$_2$O; ZnCl$_2$·4H$_2$O	298-348	−	+	+	1000
ZnBr$_2$; ZnBr$_2$·2H$_2$O; ZnBr$_2$·3H$_2$O	298-348	−	+	+	1000
ZnI$_2$; ZnI$_2$·2H$_2$O	298	−	+	+	1000
CdCl$_2$·2,5H$_2$O; CdCl$_2$·4H$_2$O	298-348	−	+	+	1000
CdBr$_2$·4H$_2$O	298-323	−	+	+	1000
ZnSO$_4$·6H$_2$O$_{11}$	298-348	−	+	+	1000
Zn(NO$_3$)$_2$·2H$_2$O; Zn(NO$_3$)$_2$·4H$_2$O; Zn(NO$_3$)·6H$_2$O;	323-348	−	+	+	1000
Cu(NO$_3$)$_2$·3H$_2$O Cu(NO$_3$)$_2$·6H$_2$O	298-348	−	+	+	1000

* (+, −) means a change of sign of *p.c.s.* or of $\left(\dfrac{\partial p.c.s.}{\partial p}\right)_T$ derivative at elevated pressure.

264

Thus, the method proposed can be used to identify the common peculiarities of topology of polytherms and polybars of solubility in water-salt systems as well as the patterns of their relative variations in a number of salts. Also, it makes it possible to suggest a general classification of temperature and pressure dependences of salt solubility in water on a molecular basis.

REFERENCES

1. V. A. Kirillin, A. E. Sheindlin, and E. E. Shpilrein, Thermodynamics of Solutions, Moscow, Energiya Publishing House, 1980, p. 287 (in Russian).
2. A. K. Lyashchenko, *Zh. fiz. khimii*, v. 50, No. 3, p. 701-706 (1976).
3. A. K. Lyashchenko, *Zh. Strukturn. Khimii*, v. 16, No. 5, p. 785-791 (1975).
4. A. K. Lyashchenko, *J. Reser. Inst. Catal. Hokkaio Univer. Japan*, v. 25, No. 3, pp. 129-157 (1977).
5. A. K. Lyashchenko, S. M. Portnova, and E. V. Petrova, *Zh. Neorgan. Khimii*, v. 27, No. 11, p. 2986-2989 (1982).
6. A. K. Lyashchenko and B. R. Churagulov, *Zh. Neorgan. Khimii*, v. 26, No. 5, p. 1190-1197 (1981).
7. A. K. Lyashchenko and B. R. Churagulov, *Zh. Neorgan. Khimii*, v. 28, N. 2, p. 456-465 (1983).
8. A. K. Lyashchenko, *Zh. Strukturn. Khimii*, v. 9, N. 5, p. 781-787 (1968).
9. A. K. Lyashchenko and A. A. Ivanov, *Zh. Strukturn. Khimii*, v. 22, N. 5, p. 69-75 (1981).
10. A. K. Lyashchenko and A. A. Ivanov, *Koordinaz. khimiya*, v. 8, N. 3, p. 291-296 (1982).
11. A. F. Borina, A. K. Lyashchenko, and V. R. Timofeeva, *Koordinaz. Khimiya*, v. 10, N. 2, p. 204-212 (1984).
12. E. P. Doane and H. G. Drickamer, *J. Phys. Chem*, v. 59, N. 5, p. 454-457 (1955).
13. A. K. Lyashchenko and B. R. Churagulov. In: Summaries of Papers Presented at the 6th All-Union Conference on Physico-Chemical Analysis, Kiev, 1983, p. 44.
14. B. R. Churagulov, In: Chemical Thermodynamics, Moscow, Moscow State Univer., 1984, p. 193-212.
15. L. A. Monyakina and B. R. Churagulov, *Zh. Fiz. khimii*, v. 56, N. 10, p. 2436-2439 (1982).
16. A. N. Yakushenko and B. R. Churagulov, *Zh. Fiz. khimii*, v. 58, N. 4, p. 910-913 (1984).
17. A. N. Yakushenko and B. R. Churagulov. In the book: Summaries of the 6th All-Union Mendeleev Discussion. Kharkov, part 2, p. 388 (1983).
18. A. K. Lyashchenko and B. R. Churagulov. In the book: Summaries of the Papers Presented at the 6th All-Union Conference on "The Problems of Solvation and Complexing in Solutions", Ivanovo, p. 208, (1981).
19. A. K. Lyashchenko, *Izv. Akad. Nauk SSSR, Ser., Khim.*, N. 12, p. 2631-2638, (1975).
20. A. K. Lyashchenko and B. R, Churagulov, *Zh. Strukt. khimii.*, v. 21, N. 6, p. 60-68, (1980).

The Chemistry of Polyelectrolyte Interactions in Boiler Water Applications

J. A. KELLY

Technical Director, Boiler Chemicals Research,
Nalco Chemical Company, Naperville, Illinois 60566, USA

INTRODUCTION

Synthetic polymers have been used for many years to control scale and deposition in steam generating systems. They have been used as supplements to carbonate, phosphate and chelate cycle chemistries resulting in a greater degree of boiler cleanliness than the unassisted treatments.

This paper presents the chemistry of polyelectrolyte interactions in boiler water applications. Emphasis is placed on laboratory research and field application results obtained during the development of all-polymer technology [1].

POLYELECTROLYTE INTERACTIONS

Synthetic polymers came into use in the mid-fifties replacing naturally occurring sludge conditioners such as tannins and lignins. This was due in large part to the greater thermal stability of the poly-electrolytes and their high activity in conditioning deposit forming precipitates.

Thermal gravimetric analysis is used to evaluate thermal stability by monitoring weight loss as a function of temperature. Figure 1 illustrates the results for a low molecular weight polyacrylate. Polymers

of this type show some weight loss at approximately 260°C with major effects at approximately 420°C.

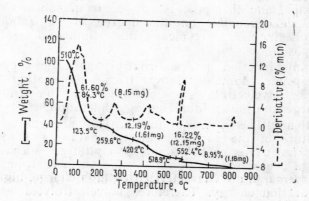

Fig. 1. TGA profile for a typical polyacrylate

An important structural parameter for polyelectrolytes is molecular weight. Figure 2 shows the molecular weight distribution obtained for a polyacrylate.

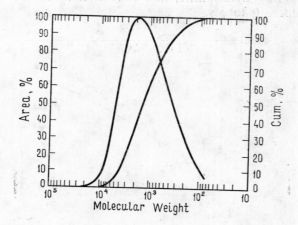

Fig. 2. Polyacrylate CPC profile

Polyelectrolytes incorporating carboxyl groups in the polymer matrix have been used primarily in combination with phosphate. Low molecular weight polyacrylates, polymethacrylates and polymaleic anhydrides are typical anionic polymers used in water treatment. They function by crystal modification and dispersion mechanisms when used in conjunction with phosphate.

Crystal modification takes place when the polyelectrolyte is adsorbed while diffusion and molecular ordering of the crystal lattice occurs during normal crystal growth [2, 3]. This results in alterations in crystal size and shape, and this process can occur in solution or at the heat transfer surface depending on the degree of saturation. Precipitate formed in this manner are typically more fluid and form less adherent or dense deposits.

Dispersants function in scale control applications by keeping solid particles suspended so that removal from an operating system can be accomplished. Polyelectrolytes are adsorbed onto the surface of particles, altering charge and thus minimizing agglomeration.

Adsorption efficiency is a function of particulate surface area, physical nature of the particle, and the nature and solubility of the dispersant. Amorphous or gelatinous precipitates, such as iron oxides, provide large areas for adsorption at the surface. Typically, molecules useful as dispersants are characterized by high charge, high polarizability and low hydration energy. Therefore, polyelectrolytes are favored for primary adsorption.

The adsorption of a polymer molecule on a particle is an equilibrium process. The functional groups become adsorbed which brings the rest of the molecule into close proximity resulting in further adsorption. The deposit mechanism is altered due to changes in particle surface charge, particles remaining hydrophilic, and reduction in the intermolecular and electrostatic binding forces.

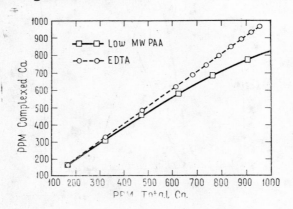

Fig. 3. Calcium binding ability

A new method for controlling scale deposition has been developed which utilizes the ability of polyelectrolytes to maintain calcium and magnesium in soluble form [4]. Polymers can complex hardness ions much in the same manner as chelates. Low molecular weight polyacrylate-based polymers contain carboxylate groups which coordinate with calcium and magnesium. The resulting interaction produces a charged

complex. Although long considered a weak interaction, electrostatic/coulombic effects allow the polyacrylate type polymers to be competitive with chelates.

High charge and polarizability are important characteristics of the polymer used as a complexing agent. The large number of charge sites on the polymer enhances the electrostatic attraction for cations such as calcium. This effect also decreases the probability of release of ions as well. The latter effect is more statistical in that there are a greater number of charged groups available for binding with polymers relative to micromolecular chelates. Figure 3 shows the calcium binding efficiency of a polyacrylate relative to EDTA.

Research into polymer-iron and polymer-hardness interactions has produced a new all-polymer treatment program. It has been shown to produce chelate-clean boilers with corrosion characteristics normally associated with phosphate programs. The following summarizes the laboratory and field results obtained.

LABORATORY RESULTS

The first step in evaluating a new boiler treatment involved a battery of bench testing techniques. Results showed an all-polymer approach at stoichiometry prevented the precipitation of calcium phosphate, calcium carbonate and magnesium silicate at high pH and high total solids.

Figure 4 depicts these results as a function of dosage for the all-polymer program under simulated boiler conditions. As can be seen, 100 percent transport of calcium and magnesium was obtained. The hardness transport peaks first at a threshold level where crystal modifi-

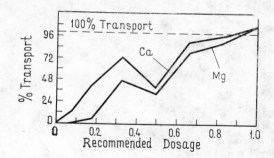

Fig. 4. Hardness transport at 4.1-6.9 MPa with all-polymer treatment program

cation and dispersion mechanisms predominate. Prior to recommended dosage, the polymers begin to complex calcium and magnesium. As shown, the polymers preferentially react with calcium ions. This is in agreement with accepted theory for the complexation mechanism. At recommended stoichiometric ratios, hardness complexation is accomplished using all-polymer.

(b)

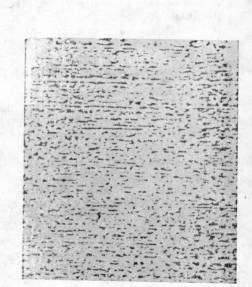

(c)

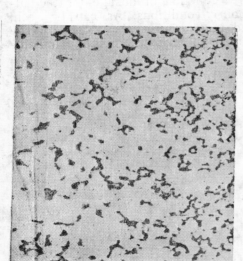

(a)

The all-polymer concept was then tested in laboratory scale boilers [4]. Evaluations included varying pressures from 0.35 to 12.4 MPa and varying heat flux from 346 to 1,103 kW/m². The majority of tests were run following ASME recommendations. However, feedwater and chemical upsets including varying feedwater hardness, silica, iron and polymer were evaluated. Figure 5 shows the heat transfer surfaces obtained with all-polymer, EDTA, phosphate, and no treatment.

One testing sequence evaluated corrosion potential as a function of iron in the boiler with no iron in the feedwater. Table 1 lists these results and shows the greater corrosion potential of chelates relative to the new all-polymer program.

Table 1. Comparison of corrosion of the all-polymer with chelate programs at 4.1-6.9 MPa

Treatment	Dosage *	ppm Fe in the blowdown
All-polymer	1.0	0.4
All-polymer	2.0	0.4
EDTA	0.6	0.6
EDTA	1.0	0.8
EDTA	1.25	2.5
EDTA	2.0	3.1
NTA	0.6	0.4
NTA	1.0	0.6
NTA	1.25	0.8
NTA	2.0	1.5

* Chelate in stoichiometric dosage relative to hardness. All polymer in recommended dosage per ppm hardness.

The high thermal stability of the all-polymer program allowed 100 percent transport of feedwater hardness, silica and iron at pressures up to 12.4 MPa. However, decreasing thermal stability was found at pressures greater than 10.3 MPa. EDTA recovery was found to be 74 percent and NTA, 94 percent at 6.9 MPa. NTA recovery was 100 percent and EDTA was 85 percent at 4.1 MPa. Polymer recovery was 100 percent at both pressures.

Additional tests showed that all-polymer performance was not reduced by excessive oxygen levels in the feedwater. Table 2 lists the results obtained relative to EDTA, which was seriously affected.

Mild on-line cleaning can be achieved with the polymer program. Greater than 100 percent transport was frequently observed in boiler tests. Experiments at 6.9 MPa conducted to elucidate the mechanism showed that the deposit was penetrated. The thickness of the scale was reduced from 0.4 mm to 0.1 mm.

It was also found that 98 percent of the hardness, phosphate and silica removed was soluble. Iron oxides were dispersed. Cleanup rates were high for calcium initially and finally for magnesium and iron.

271

These results confirmed the selectivity of the polymer formulations for calcium followed by magnesium. On-line cleanup is not recommended with the all-polymer program for the same reason as this limitation is imposed on chelate programs.

Table 2. Effect of oxygen on the performance of the all-polymer and EDTA program at 4.1 MPa

Treatment	ppb O_2 in feedwater	% Transport		
		Ca	Mg	SiO_2
All-polymer	20	110	118	107
All-polymer	40-60	101	105	95
All-polymer	100	100	88	105
EDTA	20	93	24	52
EDTA	40-60	89	13	55
EDTA	100	31	2	57

The third evaluation procedure involves a testing regimen in our advanced research boiler system [4]. This unit can simulate most problems found in industry. In addition, on-line monitoring of chordal thermocouples embedded in heat transfer tubes allows testing the effect of simulated feedwater upsets on program performance. The all-polymer program was compared with chelate, chelate-phosphate and chelate-polymer programs. Results showed superior performance for the all-polymer program in scale prevention, corrosion protection and metal passivation. Table 3 lists the transport obtained with the all-polymer program as a function of dosage. Although transport values were under 100 percent at less than recommended dosage, the scaling rate was nil. Hydrogen and blowdown iron studies showed no corrosion potential for the all-polymer program at dosages up to ten-fold recommended.

Table 3. Transport results of the all-polymer treatment program at 4.1 MPa using the advanced research boiler system

Treatment Dosage, recommended x	% Transport		
	Ca	Mg	SiO_2
1.0	101	109	96
2.4	100	123	101
1.7	109	109	100
3.5	102	106	96
0.9	102	93	93
0.4	95	71	93
1.3	102	110	109

Steam purity studies showed no organic carbon contribution to the steam at pressures up to 10.3 MPa. It also showed slight improvement or no effect on silica and sodium concentrations in the same pressure range.

Feedwater studies across the high pressure stage heaters at 232°C showed no effect on corrosion rate. Under conditions generating corrosion in this area, iron levels ranging from 0.1 to 0.3 ppm were the same with and without product.

The results reported in this section briefly summarized the laboratory work. The following sections summarize the field evaluations and follow-up done to verify all-polymer performance.

FIELD RESULTS

The efficacy of any new boiler treatment is established by performance in field applications. Polymers have shown excellent performance in controlling hardness deposits as an overlay on phosphate, carbonate and chelate programs. The use of a polymer as a dispersant for iron oxide has also given excellent results.

The application of an all-polymer program is gaining wide acceptance. In this case, polymers are used in the absence of other hardness control agents. Over 1500 applications of this concept have been made and several systems have been operating on all-polymer programs for over three years.

Six field evaluations were initially conducted at pressures ranging from 0.86 to 10.3 MPa. These evaluations included chemical processing, refining, steel and paper industries. The low pressure test was run where steam was used for heating and cooling. The results for five of these evaluations are summarized in Table 4.

Table 4. Initial case histories — results before and after all-polymer treatment

| MPa | % Transport | | | | ppb Hydrogen in the steam | |
| | Hardness | | Iron | | | |
	before	after	before	after	before	after
10.35	17	109	29	112	2.5	2.5
6.73	64	171	35	178	3.3	3.1
5.52	148	130	884	192	3.5	<1
4.14	61	156	111	158	8-10	3
4.14	122	186	66	183	6-8	1-2

Feedwater contaminants were analyzed on-site by colorimetric procedures including millipore and digestion techniques for iron. Daily grab samples were analyzed in the laboratory using atomic absorption (AA) and colorimetric procedures. Concentration cycles were determined by chloride, sodium, conductivity or product where applicable. Figure 6 illustrates the hardness control aspect of the all-polymer program. Table 5 shows hot side deposit weight analyses after all-polymer treatment.

273

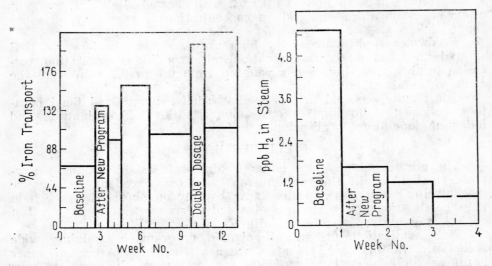

Fig. 6. Hardness transport during a field evaluation

Fig. 7. Hydrogen in the steam during a field evaluation

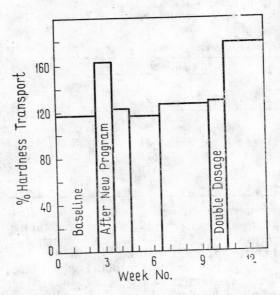

Fig. 8. Iron transport during a field evaluation

274

Table 5. Hot side deposit weight analysis after all-polymer treatment

Pressure, MPa	Hot side deposit weight, g/sq. ft.	Months on all-polymer program
8.8	8.2 (Reduction of 1.8)	13
8.8	5.9	13
8.6	6.2	12
6.7	2.5	11
6.2	1.3	8
6.2	6.7	12
4.1	1.6	11
4.1	8.7	10
4.1	0.5	8
4.1	3.5	8

Table 6. Corrosion results before and after all-polymer treatment

Pressure, MPa	Previous program	Average steam hydrogen value, ppb	
		before	after
4.1	EDTA-NTA	10	1.5
4.1	NTA-polymer	7	1.5
4.5	EDTA-polymer	55	<1
5.5	NTA-polymer	>10	<1
8.6	EDTA-PO$_4$	15	2
9.3	EDTA-polymer	15	1
9.3	PO$_4$-polymer	15	1

Boiler corrosion reactions generate hydrogen in the absence of oxygen. Corrosion rates based on hydrogen in the steam improved in all cases, some more significantly than others. Figure 7 depicts the effect on corrosion rates in a four-week evaluation. The high pressure application (10.3 MPa) in conjunction with phosphate showed no corrosion above background levels. This result was obtained even as hardness transport increased from 17 to 109 percent and dosage changed to four times recommended. Table 6 summarizes corrosion results after treatment at several locations.

Steam purity testing consisted of total organic carbon (TOC), sodium and silica. TOC was determined on grab samples using conventional instrumentation. Silica was analyzed on the same sample colorimetrically and by AA. Sodium was analyzed using a continuous monitor and checked in the laboratory with AA. These studies showed slight improvement or no effect in all the evaluations. Iron control is also an important aspect of the all-polymer program. Figure 8 shows the improved iron transport with the all-polymer program. This iron

transport increase was consistent with a 60 percent reduction in corrosion rates based on hydrogen in the steam.

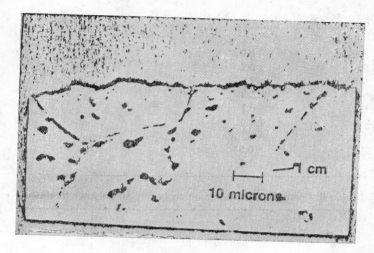

Fig. 9. Tube surface

Superior performance of the all-polymer program has also been obtained when comparisons are made in the passive films generated during both laboratory and field tests. Passivation is characterized by a thin dense protective metal oxide layer on a metal surface. In boilers, the most passive surface is represented by magnetite which is normally black and crystalline. It is distinguishable from the base metal by color. A protective magnetite film is illustrated in Figure 9. This micrograph was obtained from a tube section in a boiler operating on the all-polymer program for 12 months.

SUMMARY

Synthetic polymers have been used as dispersant, crystal modifiers and threshold inhibitors for many years to control deposition in boilers. Polymers have been applied up to this time as supplements to existing phosphate, carbonate or strong chelate hardness control programs. This paper has summarized the results on a new all-polymer approach to hardness and iron control.

REFERENCES

1. Lorenc, W. F., Kelly, J. A., Mandel, F. S., (1984) U. S. Patent No. 4, 457, 847.
2. Reddy, M. M. and Nancollas, G. H., (1971) *Journal Colloid and Interface Science,* 36(2).
3. Berg, E. W. (1963) Physical and Chemical Methods of Separation, New York: McGraw Hill.
4. Kelly, J. A., Nimry, B. N., Flasch, G. W. (1983) Corrosion/83, National Association of Corrosion Engineers, Paper No. 1.

Impure Steam Near the Critical Point

J. M. H. LEVELT SENGERS, C. M. EVERHART,
G. MORRISON, and R. F. CHANG

*Thermophysics Division, National Bureau of Standards,
Gaithersburg, Maryland 20899, USA*

ABSTRACT

The thermodynamic properties of dilute near-critical mixtures are given according to a classical and a nonclassical model. Measurements obtained in dilute solutions of NaCl in near-critical steam are discussed in the light of these models. Questions are raised regarding the validity near the critical point of extended Debye-Hückel formulations recently proposed by Pitzer et al.

INTRODUCTION

That even small impurities can have a large effect on the extensive properties of near-critical fluids was first pointed out at the beginning of this century by Keesom and Verschaffelt at Leiden [1-3]. They used a classical corresponding-states model to show that at constant P, T the density change with impurity concentration is determined by the slope of the critical line and the (diverging) compressibility of the solvent. Properties of dilute near-critical mixtures were studied experimentally in the USSR by Krichevskii and coworkers [3] in the 1960's and theoretically by Anisimov et al. [4], well before a theoretical paper by Wheeler [5] in 1972 drew attention in the West to the peculiarities of dilute mixtures: the divergence of the partial molar volume of the solute, the path-dependence of the partial molar volume of the solvent, and the critical anomalies in the derivatives of the activity coefficients. A careful classical analysis of the behavior of the partial molar volumes and

enthalpies was published by Rozen in the USSR in 1976 [6]. Wheeler and Rozen found independently that the partial molar volume and enthalpy of the solute become proportional to the solvent's compressibility if the limit $x \to 0$ is taken first, x being the mole fraction of the solute. Rozen showed that near the critical point of the solvent the proportionality factor can be calculated from the initial temperature and pressure slopes of the critical line, $\lim_{x \to 0} dT/dx|_{CRL}$ and $\lim_{x \to 0} dP/dx|_{CRL}$, respectively.

Recently, Chang et al. [7] extended Rozen's classical analysis to higher derivatives such as partial molar specific heats, and determined the small-x dependence of these properties along a variety of paths. In parallel, they presented a non-classical analysis based on the Leung-Griffiths model [8] and pointed out the reasons for the inconsistencies Rozen encountered in his approximate nonclassical treatment.

The purpose of this paper is to discuss, in the light of the models of Chang et al. [7], a number of experimental results that have been obtained in dilute aqueous mixtures at high temperatures and pressures. The best studied aqueous mixture is that of sodium chloride in water. For this mixture, there are measurements of the critical line by Sourirajan and Kennedy [9] and by Marshall and Jones [10]; of PVT in the region around the critical line by Khaibullin et al. [11]; of supercritical PVT, compressibilities and partial molar volumes by Benson et al. [12]; of the apparent molar specific heat $C_{p\varphi}$ by Wood et al. [13]; and of the enthalpy of dilution by Busey et al. [14].

The classical and nonclassical models we have developed do not take into account the ionic nature of aqueous salt solutions. A key question is whether the presence of ions modifies the behavior predicted by our models. Although we cannot claim to have a definitive answer to this question we will present arguments that the number of ions may be small; that the inclusion of Debye-Hückel terms in the free energy will leave most of our conclusions unaffected; and that the limiting and extended forms of the Debye-Hückel theory as proposed by Pitzer and his school [15] give incorrect results if extrapolated to the critical point.

CRITICAL BEHAVIOR OF DILUTE MIXTURES

Partial molar, apparent molar and excess properties have evolved naturally in the description of liquid mixtures at atmospheric pressure (Fig. 1). The unusual behavior of these properties near the solvent's critical point is caused by the fact that the solvent is no longer a liquid, but rather a highly compressible near-critical fluid. We draw the equivalent of Fig. 1 for this case in Fig. 2 for a temperature slightly above the critical point and for a nonvolatile solute such as salt in steam. The divergences of $\overline{V}_2$ and $V_{2\varphi}$, the apparent molar volume, are immediately obvious from the way they are constructed, if one remembers that the mixture's critical point will move to the extremum of the coexistence curve and to $V_c(0)$ as $x \to 0$. The enthalpy will show the same unusual behavior of its derivative $\overline{H}_2$ as does the volume. For second derivatives

such as the specific heat C_p the difficulties are compounded because the pure-solvent C_p diverges strongly at the critical point. This will cause the x-derivative, reflected in $\overline{C}_{p2}$, to diverge far more strongly than $\overline{V}_2$. In the work of Chang et al. [7] the nature of these divergences was investigated by assuming, with Rozen [6], that the Helmholtz free energy $a(V, T, x)$, near the solvent's critical point, can be expanded in terms of x, $\delta V = V - V_c$ and $\delta T = T - T_c$. We denote derivatives of $a(V, T, x)$ by

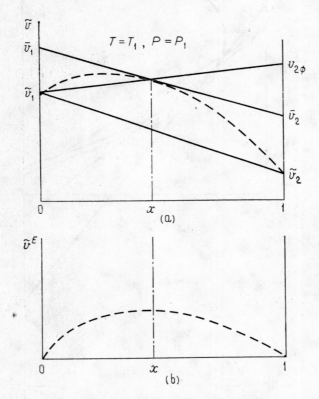

Fig. 1. Graphical display of (a), the partial molar volumes $\overline{V}_1$ and $\overline{V}_2$ and apparent molar volume $V_{2\sim}$, and (b), the excess volume V_E in a binary liquid mixture

repeated subscripts. Thus, $a_V = -P$, $a_{VVT} = \partial^3 a / \partial V^2 \partial T = \partial (V K_T)^{-1} / \partial T$, $a_{Vx} = -(\partial P / \partial x)_{VT}$, with P the pressure, K_T the isothermal compressibility. A superscript c denotes that these derivatives are taken at the pure-solvent's critical point. The coefficients a^c_{VV} and a^c_{VVV} are zero. The partial molar volumes $\overline{V}_1 = V - x (\partial V / \partial x)_{PT}$ and $\overline{V}_2 = V + (1 - x) (\partial V / \partial x)_{PT}$ are calculated by means of the derivative

$$(\partial V / \partial x)_{PT} = V K_{Tx} (\partial P / \partial x)_{VT} = -a_{Vx} / a_{VV} \tag{1}$$

27

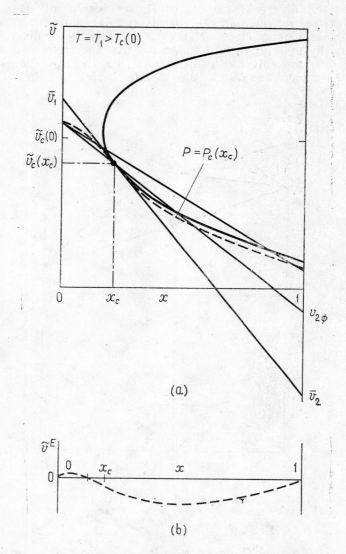

$T = T_1 > T_c(0)$

$\bar{v}_1$

$\tilde{v}_c(0)$

$\tilde{v}_c(x_c)$

$P = P_c(x_c)$

0 x_c x 1

$v_{2\phi}$

(a)

$\bar{v}_2$

$\tilde{v}^E$

0 0 x_c x 1

(b)

Fig. 2. Partial, apparent and excess volumes for the case where the solvent is supercritical, $T > T_{c1}$, and the solute far below its critical point. A two-phase region opens up on addition of the solute. The construction of $\bar{V}_1$, $\bar{V}_2$, and V^E is indicated for the critical point of the mixture. $\bar{V}_2$ and V_2 diverge to $-\infty$ as T approaches T of the solvent. The dashed curve is the critical isotherm-isobar

In the limit $x \to 0$, $(\partial V/\partial x)_{PT}$ becomes proportional to K_T, the compressibility of the solvent. The latter diverges strongly at the critical point. The quantity $(\partial P/\partial x)_{VT} = -a_{Vx}$ is finite in a classical expansion. Its limiting value at the critical point equals [6, 7]

$$-a^c{}_{Vx} = \underset{x \to 0}{\mathrm{Lim}}\,(\partial P/\partial x)^c{}_{VT} = dP/dx\,|^c{}_{CRL} - dP/dT\,|^c{}_{CXC}\,dT/dx\,|^c{}_{CRL} \qquad (2)$$

Here the subscript CRL indicates derivatives taken along the critical line, the subscript CXC the slope of the solvent's vapor pressure curve. Thus, the slopes of the critical line and the pure-solvent properties determine the values of $\overline{V}_1$ and $\overline{V}_2$. The path-dependence of these properties is analyzed by expanding a_{VV} in Eq. (1):

$$(\partial V/\partial x)_{PT} = -a^c{}_{Vx}/[a^c{}_{VVx}x + a^c{}_{VVT}(\delta T) + a^c{}_{VVVV}(\delta V)^2/2] \qquad (3)$$

On the critical line, for instance, δT and δV vary linearly in x, thus $(\partial V/\partial x)_{PT}$ and $\overline{V}_2$ diverge as $1/x$. On any path on which δT and $(\delta V)^2$ approach zero at least as fast as x, $\overline{V}_2$ diverges as $1/x$. The amplitude, however, is path-dependent and $\overline{V}_1$ does not approach V_c since $x(\partial V/\partial x)_{PT}$ remains finite. On the critical isotherm-isobar, however, Chang et al. [7] showed that $\delta V \propto x^{1/3}$. The term in $(\delta V)^2$ dominates in the denominator of Eq. (3); $(\partial V/\partial x)_{PT}$ and $\overline{V}_2$ diverge as $x^{-2/3}$ and $\overline{V}_1$ approaches V_c. If experiments are carried out at constant T, P near but not at T_c, P_c, the denominator in Eq. (3) has the form $x + C_1 + (C_2 - C_3x)^{2/3}$ with $C_1 - C_3$ constants depending on the values of $T - T_c$ and $P - P_c$. In order to obtain the limiting value of $\overline{V}_2$ one might plot $\overline{V}_2^{-1}$ versus x, but some curvature should be expected.

The partial molar specific heat $\overline{C}_{p2} = C_{Px} + (1-x)(\partial C_{Px}/\partial x)_{PT}$ is expected to behave as the similarly defined "partial molar compressibility" $\overline{K}_{T2}$. The latter is easily expressed in terms of derivatives of $a(V, x, T)$

$$V\overline{K}_{T2} = \frac{1}{a_{VV}} + \frac{(1-x)}{(a_{VV})^2}\left[a_{VVx} + a_{VVV}\frac{a_{Vx}}{a_{VV}}\right] \qquad (4)$$

where $1/a_{VV}$ equals VK_{Tx} and both a_{VV} and a_{VVV} approach zero at the solvent's critical point. Near this point, the term multiplied by $(a_{VV})^{-2}$ will dominate. Chang et al. [7] show that $\overline{K}_{T2} \propto 1/x^2$ on the critical line and $\propto 1/x^{5/3}$ on the critical isotherm-isobar. For completion, we also give the expression for $\overline{C}_{P2}/T$ in terms of derivatives of a:

$$\frac{\overline{C}_{P2}}{T} = \frac{C_{Px}}{T} - a_{TTx}(1-x)$$

$$+ \frac{(1-x)}{a_{VV}}\left[a_{VTT}\,a_{Vx} + 2a_{VT}\,a_{VTx}\right]$$

$$+ \frac{(1-x)}{a_{VV}^2}\left[-2a_{VT}\,a_{Vx}\,a_{VVT} - (a_{VT})^2\,a_{VVx}\right]$$

$$+ \frac{(1-x)}{a_{VV}^3}(a_{VT})^2\,a_{Vx}\,a_{VVV} \qquad (5)$$

It is clear that its structure is very similar to that of $\overline{K}_{T2}$ in Eq. (4). On isotherm-isobars where these properties are dominated by the $(a_{VV})^{-2}$

281

terms, which behave as $[x + C_1 + (C_2 - C_3 x)^{2/3}]^{-2}$, the limiting values might be obtained from a plot of $\bar{K}_{T2}^{-1/2}$ resp. $\bar{C}_{P2}^{-1/2}$ vs x, but again some curvature should be expected.

The nonclassical formulation differs from the classical analysis just sketched in several respects. If the limit $x \to 0$ is taken first, $\bar{V}_2$ will diverge as K_T, $\bar{C}_{P2}$ as $(K_T)^2$, but these quantities diverge with the nonclassical exponents γ and 2γ, respectively, with $\gamma = 1.24$. Both a^c_{vvvv} and a^c_{vvT} are zero. The nonclassical Helmholtz free energy cannot be expanded at the critical point. Derivatives such as $(\partial x / \partial \mu_2)_{PT}$, with μ_2 the chemical potential of the solute, and with two "fields", P and T, fixed at differentiation, diverge strongly, γ-like. Derivatives such as K_{Tx}, C_{Px} and $(\partial P / \partial x)^{-1}_{VT}$, in which one "density" is fixed, and which are finite in the classical model, diverge weakly, α-like, ($\alpha \simeq 0.1$) at the critical line. Therefore both a_{vx} and a_{vv} are zero on this line, but their ratio, equal to $(\partial V / \partial x)_{PT}$, remains finite. Finally, quantities such as $(\partial P / \partial T)_{vx}$ remain finite; this constant-x isochoric slope, however, is predicted to approach the slope of the critical line, in contrast to classical behavior where the constant-x isochore intersects the critical line. Rozen, introducing nonclassical critical exponents in the expansion of the Helmholtz free energy, ignored both the weak anomaly of K_{Tx} and the tangency of the constant-x isochore and the critical line. Even though both conditions may elude experimental observation, ignoring them leads to formal contradictions.

The results of our analysis of the nonclassical Leung-Griffiths model leads to the following results. On paths asymptotically tangent to the pure-fluid axis, $\bar{V}_2$ and K_{Tx} diverge more slowly than $1/x$. On the critical isotherm-isobar for instance, $\bar{V}_2$ and K_{Tx} behave as $x^{-\gamma/\beta\delta}$ or $x^{-0.79}$, where $\gamma = 1.24$, $\beta = 0.325$ and $\delta = 4.82$ are the nonclassical critical exponents. $\bar{K}_{T2}$ and $\bar{C}_{P2}$ behave as $x^{-2+1/\delta}$ along this path. On the critical line, $\bar{V}_2$ is finite but diverges as $1/x$ while K_{Tx} and C_{Px} are infinite. On paths asymptotically tangent to the critical line, $\bar{V}_2$ diverges as $1/x$, $\bar{K}_{T2}$ and $\bar{C}_{P2}$ more strongly. On intermediate paths, $\bar{V}_2$ and K_{Tx} behave as $1/x$, $\bar{K}_{T2}$ and $\bar{C}_{P2}$ as $(1/x)^{3-\gamma}$.

In determining the limiting behavior of partial molar properties near the solvent's critical point it is essential that the path of approach be carefully specified and the order in which limits are taken be stated.

CRITICAL BEHAVIOR OF DILUTE SALT SOLUTIONS

Electrolyte chemistry has evolved as a separate branch of physical chemistry because the long-range interactions between the ions play a key role in determining the thermodynamic behavior of solutions. In this paper we will not address the question of the effect of long-range forces on the nature of the critical point. We will take the position that if the forces are short-ranged our nonclassical model will prevail; if they are long-ranged, the classical model should be used. In either case, our predictions are at odds with those of Pitzer et al. [15].

Electrolyte chemistry has, in addition, developed its own lore, the conventional theoretical approach taking the solution to be dilute and the solvent a medium of negligible compressibility. When the traditional treatments are extended to the vicinity of the critical point, trouble looms. The problems are many. If the experimenter is not aware that partial molar properties behave capriciously in all dilute mixtures near the critical point of the solvent, he may be at a loss as to how to extrapolate his data and explain his experimental results. Furthermore, in electrolyte solutions the properties measured are usually the apparent molar properties F_φ, defined as

$$F_\varphi = (M_2 m + 1) F_s - F_w \qquad (6)$$

where F is a thermodynamic property, M_2 the molar mass of the salt, m the molality, F_w the value of F for 1 kg of pure water at the same pressure and temperature as the mixture, and F_s the value of F for the mixture of m moles of salt in 1 kg water. In terms of mole fractions

$$x F_\varphi = \tilde{F} - \tilde{F}_w (1 - x) \qquad (7)$$

where $\sim$ denotes a molar property. It is readily shown that $F_\varphi \to \bar{F}_2$, the partial molar value of F for the solute, in the limit $x \to 0$. Now these apparent molar properties, direct results of measurements, prove particularly resistant to analysis near the solvent's critical point. Whereas the x-dependence of partial molar properties is derived easily from a low-order expansion of the Helmholtz free energy, as we showed for $\bar{V}_2$ in Eq. (3), the apparent ones require a higher-order expansion; only their $x \to 0$ limit, $F_\varphi = \bar{F}_2$, can be readily obtained. Since the partials $\bar{V}_2$, $\bar{H}_2$ diverge strongly at the critical point, and $\bar{K}_{T2}$, $\bar{C}_{P2}$ even more strongly, both partial and apparent molar properties become strong functions of x on isotherm-isobars and the extrapolation of the measured apparent molar properties to $x = 0$ becomes a very uncertain endeavor. The third problem is that the traditional way of extrapolating is guided by the desire to obtain the correct limiting slope from the Debye-Hückel theory [13-15]. It will be shown, however, that on the critical isotherm-isobar the Debye-Hückel form used is at odds with the predictions from our model. The fourth problem that we face is the question of whether there are free ions present in the first place. From their conductivity measurements, Quist and Marshall [16] derived the conventional equilibrium constant for the dissociation of NaCl. With the hypothetical 1 m solution as a standard, they found $\log K$ to be of the order of -4 in supercritical steam at the critical density, and of the order of -1 at twice the critical density. The lowest molalities in the experiments we discuss are of the order of 0.1. At these molalities in supercritical steam, no more than 10% of the NaCl molecules may be ionized. In liquid steam at around 600 K, however, NaCl will be ionized at the lower molalities. In measuring the x-dependence of thermodynamic properties at constant P and T, a "cross-over" may be occurring from ionized to non-ionized states as the density decreases or the molality increases.

Finally, the question of the validity of the Debye-Hückel theory near the critical point of steam must be faced. The square of the Debye

length $1/k$ is proportional to DT/ρ. At the critical point of steam, D is 40 times smaller, ρ 3 times smaller and T a factor of 2 higher than in room-temperature water. The Debye length thus being 2.5 times smaller than in ordinary water, the applicability is proportionally restricted. Apart from this effect, the basic assumption of the Debye-Hückel theory, a charged particle in a homogeneous dielectric medium, becomes questionable when the density is low and its fluctuations large.

It is our contention that the anomalous thermodynamic effects reported by many experimenters in high-temperature dilute aqueous solutions are of the same nature as those found in dilute nonaqueous mixtures when the solvent approaches its critical point. In all these cases, the key to understanding is the initial slopes of the critical line, which determine how much further from or closer to criticality the mixture is compared to the pure solvent.

EXPERIMENTS IN NEAR-CRITICAL NaCl SOLUTIONS

The critical line of NaCl in steam has been measured by several experimenters. In Fig. 3, we show the T_c versus molality slope from the measurements of Sourirajan and Kennedy [9] and of Marshall and Jones [10]. The effect of the solute is drastic: a 1-molal salt solution ($x = 0.018$) has a critical temperature 50°C higher than that of steam. Unfortunately, the uncertainty of at least 30% in the initial slope precludes estimation of the crucial quantity $a^c{}_{vx}$ in Eq. (2).

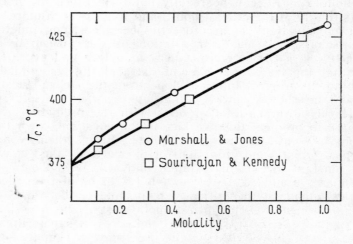

Fig. 3. Initial slope of the critical line in T-m space for NaCl in steam. The discrepancy between these two data sets is substantial

Benson, Copeland et al. [12] measured the equation of state of dilute near-critical salt solutions at 385, 390 and 396°C. In the same apparatus

they measured the equation of state of pure steam, thus reducing the effects of systematic errors. At 1 percent in weight and higher the salt solutions were sub-critical and separated into two phases. At 0.2 weight percent, however, ($x = 6 \times 10^{-4}$), the salt solution was supercritical. The apparent molar volume V_φ was calculated from these data. Fig. 4 shows how V_φ, an approximation to $\overline{V}_2 (x=0)$ at this low molality, displays the deep minimum at densities near the critical density of steam ($\rho_c = 322$ kg/m³) that we expect on the basis of Eq. (1). Highly similar results have been recently obtained in nonaqueous solutions. See, for instance, the data of Eckert et al. [18] on $\overline{V}_2$ for naphthalene in carbon dioxide. Contrary to Eckert's $\overline{V}_2$ data, which correlate with K_T, the Benson data seem to track more closely the "symmetrized compressibility" $\rho^2 K_T$, see Fig. 4. We calculated this quantity from the equation by Haar et al. [17]. From the ratio of $V_{2\varphi}$ (interpreted as $\overline{V}_2 (x=0)$) and VK_T, we obtain an estimate of a^c_{Vx} of approximately 10^3 MPa, with x in mole fraction.

Large anomalies have been reported in the apparent molar specific heat $C_{P\varphi}$ of NaCl solutions by Wood and coworkers [13]. The authors obtained the specific heat of the salt solution relative to that of steam at the same pressure and temperature in a twin flow calorimeter, in the temperature range of 300-600 K, at $m = 0.1$ to 3 and at pressures of 1, 10 and 17.7 MPa. The $C_{P\varphi}$ values at 17.7 MPa and at the high temperatures become violently negative as $m \to 0$. The authors extrapolated their $C_{P\varphi}$ values to $x=0$ by means of a polynomial in $(T-T_c)^{-1}$ and $m^{1/2}$. Any resemblance of Debye-Hückel limiting behavior is, however, visible only at temperatures below 350 K and the extrapolated values are therefore in some doubt. Our model suggests that $\overline{C}_{P2}^{-1/2}$ should be plotted versus m (Sec. 1). A plot of $C_{P\varphi}^{-1/2}$ vs m, however, shows substantial curvature. This is not surprising, because $C_{P\varphi} \neq \overline{C}_{P2}$ for finite m, and because the 600 K, 17.7 MPa point is not close to the steam critical point of 647 K, 22 MPa. We have tried to analyze the $x=0$ limit of the $C_{P\varphi}$ data at this point. At these conditions, even the $m=0.1$ salt solution is still at a density of roughly twice the critical, and thus quite far from criticality. We have estimated the sizes of most of the terms in the classical expressions Eq. (5) in the limit $x \to 0$. In this limit, all derivatives not containing x are calculated from the properties of pure steam [17] at the measured P, T. We use our estimate of 10^3 MPa for a^c_{Vx}. At this point, $\overline{C}_{P2}^\infty$ is negative and 20 times larger in magnitude than $\overline{C}_P$ for pure water. The value of a_{VV} is one order of magnitude smaller than that of water near room temperature. We find that two terms in Eq. (5) dominate; the last one has the correct sign and magnitude, while the term $-2a_{VT}a_{Vx}a_{VV}/(a_{VV})^2$ is about half that of $C_{P\varphi}^\infty/T$ but positive. This confirms our view that the violent divergence of $\overline{C}_{P2}^\infty$ is caused by the terms of order $1/(a_{VV})^2$ diverging at the critical point.

Unfortunately, although Eq. (5) gives guidelines on how to analyze $\overline{C}_{P2}$ at finite x, it does not suggest how to analyze the measured quantity $C_{P\varphi}$, neither does our prediction that $C_{P\varphi} (x=0)$ correlates with $(VK_T)^2$

bear out convincingly. At temperatures below 600 K, the $(a_{vv})^2$ terms are apparently not sufficiently large that all others can be neglected.

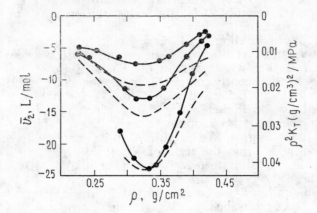

Fig. 4. ($\cdot$) Apparent molar volume $V_{2\varphi}$ of a supercritical solution of NaCl in steam [12], The data do not correlate well with VK_T (Eq. 4) but have the same appearance as the "symmetrized compressibility" [19], $\rho^2 K_T$, dashed curve. $\overline{V}_2$ data obtained in non-electrolytes in near-critical solvents look very similar [18]

The enthalpy-of-dilution measurements of Busey et al. [14] became available to us at too late a stage to be included in the present analysis. These data extend well into the supercritical region and should be easier to analyze than $C_{P\varphi}$ data.

DISCUSSION

Wood et al. [13] offered three different explanations for the large negative values of $C_{P\varphi}$ they observed. They used, respectively, the Debye-Hückel model, the Born model and a classical corresponding-states model. The Debye-Hückel theory, extended by Pitzer et al. to the critical point of steam, has the following features. The Gibbs free energy contains a term of order $m^{3/2}$, which is of higher order than the leading term proportional to m in our model. The limiting behavior of μ_1, $\overline{H}_1$, $\overline{V}_1$ and $\overline{C}_{P1}$ is as $m^{3/2}$; that of μ_2, $\overline{H}_2$, $\overline{V}_2$ and $\overline{C}_{P2}$ as $m^{1/2}$. The coefficients of the leading powers of m are finite for μ_1, μ_2 but diverge as K_T of steam for $\overline{H}_1$, $\overline{H}_2$, $\overline{V}_1$, $\overline{V}_2$ and as K_T^2 for $\overline{C}_{P1}$ and $\overline{C}_{P2}$. A large contribution to the diverging coefficients originates from the derivatives $(\partial \ln D/\partial P)_T$ and $(\partial \ln D/\partial T)_P$ which, because of the smooth dependence of the dielectric constant D on the density, behave like the strongly-diverging compressibility and expansion coefficients. Although the anomalies predicted by the Debye-Hückel theory as extended by Pitzer et al. have similarities with our results, there are serious disagreements. In our model, the correlations with the compressibility of steam hold only if the limit $x \to 0$

is taken first. Pitzer et al., however, would predict infinities in $\bar{H}_2$, $\bar{V}_2$ and $\bar{C}_{P2}$ all along the critical isotherm-isobar, which cannot be correct. On this same curve, Pitzer et al. predict an $x^{1/2}$ dependence for $\ln \gamma \pm$ whereas our model predicts an $x^{1/6}$ dependence, which dominates at small x. We conclude that the Debye-Hückel approach, apart from the obvious physical difficulties concerning application near the steam critical point, requires a careful restating so that limits are taken in the proper order *.

The Born model [13, 15] gives the electrostatic contribution to the free energy as a simple function of the dielectric constant, and therefore of the density; if treated carefully, it should give the same results as our model. The difficulty remains that the number of ions may be small, and that the same near-critical effects occur in non-ionic mixtures.

As to the corresponding-states model, it should, in principle, give results equivalent to our classical model. In our opinion, it has the gist of the correct physics in it, in recognizing that because of the slope of the pseudocritical line the mixture is at a different distance from (pseudo) criticality than the solvent at the same P, T. It also implies, however, that the observed effects are typical of any impurity, ionic or not.

ACKNOWLEDGEMENT

The authors have profited from discussion with D. Smith-Magowan and J. Hubbard.

REFERENCES

1. Keesom, W. H. (1901): *Comm. Leiden.* **75**, *Comm. Leiden.* **79**.
2. Verschaffelt, J. E. (1904): *Comm. Leiden Suppl.* **10**.
3. Krichevskii, I. R. (1967): *Russian Journal of Physical Chemistry.* **41** (10) 1332-1338.
4. Anisimov, M. A., Voronel, A. V., and Gorodetskii, E. E. (1971): *Zh. Eksp. Fiz.* **60**, 1117-1130; *Soviet Physics JETP* **33** (3), 605-612.
5. Wheeler, J. C. (1972): *Berichte der Bunsen-Gesellschaft.* **76** (3/4) 308-318.
6. Rozen, A. M. (1976): *Russian Journal of Physical Chemistry.* **50** (6) 1381-1385.
7. Chang, R. F., Morrison, G., Levelt-Sengers. J. M. H. (1984): *Journal of Physical Chemistry.* **88**, 3389-3391.
8. Leung, S. S., and Griffiths, R. B. (1973): *Phys. Rev.* **A8**, 2670-2683.
9. Sourirajan, S. and Kennedy, G. C. (1962): *American Journal of Science.* **260**, 115-141.
10. Marshall, W. M., and Jones, E. V. (1974): *Journal of Inorg. and Nucl. Chem.* **36**, 2313-2318.
11. Khaibullin, I. Kh., and Borisov, N. M. (1965): *Russian Journal of Physical Chemistry.* **39** (3), 361-364.
12. Benson, S. W., Copeland, C. S., and Pearson, D. (1953): *The Journal of Chemical Physics.* **21** (12) 2208-2212. Copeland, C. S., Silverman, J., and Benson, S. W. (1953): *The Journal of Chemical Physics.* **21** (1) 12-16.
13. Smith-Magowan, D., and Wood, R. H. (1981): *J. Chem. Thermodynamics.* **13** 1047-1073. Wood, R. H., and Quint, J. R. (1982): *J. Chem. Thermodynamics.* **14** 1069-1076. Gates, J. A., and Wood, R. H. (1982): *J. Phys. Chem.* **86** (25) 4948-4951.
14. Busey, R. H., Holmes, H. F., and Mesmer, R. E. (1984): *J. Chem. Thermodynamics.* **16**, 343-372.

* Prof. Pitzer, in a private communication after the Moscow Conference, and also in the lecture published in this Volume, has pointed out that the equations in Ref. [15] were never intended by him to be applied near the critical point of steam.

15. Rogers, P. S. Z., and Pitzer, K. S. (1982): *J. of Physical and Chemical Reference Data.* **11** (1), 15-81. Pitzer, K. S., Peiper, J. C., and Busey, R. H. (1984): *J. Phys. Chem. Ref. Data* **13** (1) 1-39.
16. Quist, A. S. and Marshall, W. L. (1968): *J. Phys. Chem.* **72**, 684-703.
17. Haar, L., Gallagher, J. S., and Kell, G. S. (1984): "NBS/NRC Steam Tables" Hemisphere Publishing Corp. New York (McGraw-Hill International Book Company).
18. Eckert, C. A., Ziger, D. H., Johnston, K. P., and Ellison, D. H. (1983): *Fluid Phase Equilibria* **14**, 167-175.
19. Levelt-Sengers, J. M. H., Kamgar-Parsi, B., Balfour, F. W., and Sengers, J. V. (1983): *J. Phys. Chem. Ref. Data* **12**(1), 1-28.

Experimental Investigation of Density
and Viscosity of Aqueous Boron Solutions

S. L. RIVKIN, E. A. KREMENEVSKAYA,
S. N. ROMASHIN, and O. M. TRAKTUEV

All-Union Heat Engineering Institute (VTI), Moscow, USSR

Recently, at the VTI Physical Laboratory a number of experimental investigations cf density and viscosity of aqueous boron solutions have been carried out in a wide range of parameters of state.

1. DENSITY EXPERIMENTS

The density of aqueous boron solutions has been measured by the VTI-developed method using a constant volume piezometer on the experimental installation previously used for investigating PVT-relations for ordinary water [1-3] and heavy water [4, 5] in liquid and vapor phases as well as within the critical region of parameters of state.

The cited papers present detailed description of the experimental technique and installation, as well as the analysis of errors in measuring the experimental values.

As in previous investigations, high accuracy measurement instruments have been employed in this work: a Class 0.05 dead-weight piston gauge and standard platinum resistance thermometer in combination with a Class 0.002 d.c. compensator, and 0.1-mg-sensitive laboratory balance for determining the mass of the substance.

The density of aqueous boron solutions was experimentally obtained for two concentrations: 20 and 40 g/kg of water within the range of temperatures of 20 to 350°C and pressures up to 30 MPa. To determine

10 Зак. 960

289

the initial amount of the substance in the piezometer, according to the measurement technique, densities of the solutions of the given concentrations at atmospheric pressure and 20°C are required.

Such data were obtained by us during special pycnometer experiments, the results of which are given in Table 1. The error of pycnometer measurements for solution density is 0.005%.

Table 1

Concentration, g/kg of water	Concentration, % (mass)	ρ, g/cm³	v, cm³/g
10	0.992	1.00170	0.99830
20	1.971	1.00522	0.99481
40	3.894	1.01232	0.98783

The densities of aqueous boron solutions of stated concentrations were measured at high pressures on 20, 75, 100, 125, 150, 200, 250, 300, 325 and 350°C and the measurements were conducted 2-3 times on each isotherm. According to the estimations performed, the measurement error for the density of solutions lies between 0.02% at lower temperatures and up to 0.07% at maximum temperature of 350°C.

The results of such measurements are given in Table 2.

Of interest is to compare the values of the density ρ of solution and the density ρ_w of water (i. e., solvent) at equal pressures and temperatures. The preliminary analysis has shown that simplest correlations are obtained if the difference of stated densities are considered vs. pressure, temperature and concentration (c).

The graphical treatment of the measurement results on coordinates $(\rho-\rho_w)/c, p$ and $(\rho-\rho_w)/c, t$ has shown that the value of $(\rho-\rho_w)/c$ at temperatures lower than 280°C does not depend on pressure and concentration and is a function of temperature only, with the temperature dependence $A = (\rho-\rho_w)/c$ passing through the minimum at a temperature close to 215°C (see Fig. 1). The pycnometer results, in the stated diagrams, showed good agreement with the high parameter experimental results.

For temperatures below 280°C the following simple equation is suggested:

$$\rho = \rho_w + Ac \qquad (1)$$

where A is temperature function

$$A = 3\,55 \times 10^{-1} - 5.2 \times 10^{-2}\left(\frac{t}{100}\right) + 2 \times 10^{-3}\left(\frac{t}{100}\right)^2$$

$$+ 4 \times 10^{-3}\left(\frac{t}{100}\right)^3,$$

ρ and ρ_w are given in m³/kg, c is solution concentration in g/kg of water.

290

Table 2. Density of aqueous boron solutions

P, MPa	ρ_{exp}, kg/m³	ρ_{calc}, kg/m³	$\delta\rho$, %	P, MPa	ρ_{exp}, kg/m³	ρ_{calc}, kg/m³	$\delta\rho$, %
			Concentration 40 g/kg of water				
	75.00°C			14.819	885.9	886.2	—0.04
28.865	998.8	1000.0	—0.02	11.542	883.9	883.8	0.01
24.935	998.4	998.4	0	3.074	877.6	877.5	0.01
19.800	996.0	996.2	—0.02	2.975	877.2	877.4	—0.02
15.630	994.5	994.4	0.01				
13.695	993.4	993.5	—0.01		**250.00°C**		
9.225	991.7	991.6	0.01	30.438	838.0	838.4	—0.04
7.063	990.5	990.6	—0.01	25.416	833.3	833.7	—0.05
5.055	989.9	989.7	0.02	19.232	827.6	827.7	—0.02
1.470	988.0	988.1	—0.01	14.010	822.2	822.4	—0.02
1.426	988.3	988.1	0.02	11.449	819.3	819.6	—0.04
				9.075	816.8	817.0	—0.03
	100.00°C			5.253	812.4	812.7	—0.04
28.164	983.4	983.4	0	5.134	812.5	812.5	0
19.530	979.6	979.6	0				
16.000	978.4	978.0	0.04		**300.00°C**		
11.160	976.0	975.7	0.03	30.655	764.5	765.0	—0.07
9.478	975.1	974.9	0.02	24.687	755.3	755.8	—0.06
6.113	973.7	973.4	0.03	18.292	744.9	744.9	0
1.924	971.6	971.4	0.02	13.930	736.8	736.6	0.03
1.453	971.5	971.1	0.04	11.491	731.9	731.6	0.04
				8.882	726.4	725.9	0.07
	125.00°C			8.811	726.1	725.7	0.06
28.224	964.9	964.8	0.01				
20.810	961.4	961.2	0.02		**325.00°C**		
13.804	958.0	957.8	0.02	29.512	716.0	—	—
7.467	954.9	954.6	0.03	27.360	711.2	—	—
1.785	952.0	951.6	0.04	24.333	704.8	—	—
				18.590	690.3	—	—
	150.00°C			17.828	687.9	—	—
31.120	945.6	945.4	0.02	13.898	676.1	—	—
29.791	945.2	944.7	0.05	13.206	673.5	—	—
21.973	941.1	940.6	0.05				
18.571	939.1	938.8	0.03		**350.00°C**		
14.003	936.9	936.3	0.06	29.736	658.1	—	—
13.090	936.1	935.8	0.03	27.681	651.0	—	—
7.385	933.0	932.6	0.04	25.123	642.6	—	—
6.456	932.7	932.0	0.07	23.068	633.0	—	—
1.850	930.0	929.4	0.06	21.298	625.9	—	—
1.828	929.9	929.4	0.05	19.158	613.3	—	—
				17.343	602.4	—	—
	200.00°C			16.524	595.2	—	—
28.960	895.4	895.9	—0.05				
27.676	895.0	895.0	0				
21.931	891.2	891.2	0				

Table 2 (*continued*)

$P,$ MPa	$\rho_{exp},$ kg/m³	$\rho_{calc},$ kg/m³	$\delta\rho,$ %	$P,$ MPa	$\rho_{exp},$ kg/m³	$\rho_{calc},$ kg/m³	$\delta\rho,$ %
				Concentration 20 g/kg of water			
				6.616	808.0	808.2	—0.03
	75.00°C			5.800	807.0	807.3	—0.04
29.378	993.8	993.8	0	4.865	806.5	806.2	0.04
20.120	989.9	989.9	0	4.543	805.6	805.8	—0.03
13.583	987.1	987.1	0	4.212	806.5	805.5	0.12
6.076	983.9	983.8	0.01				
1.582	981.8	981.8	0		**300.00°C**		
	100.00°C			29.188	755.6	756.3	—0.10
31.198	978.6	978.6	0	25.412	750.2	750.5	—0.04
20.603	973.9	973.9	0	25.319	750.7	750.4	0.05
14.522	971.3	971.1	0.02	20.913	743.1	743.1	0
6.739	967.2	967.5	—0.03	17.979	737.2	737.9	—0.09
1.621	965.2	965.0	0.02	15.110	732.8	732.5	0.05
				12.966	728.8	728.2	0.09
	125.00°C			10.797	723.7	723.6	0.01
30.228	959.7	959.7	0	10.288	723.1	722.5	0.08
22.302	956.0	955.9	0.01	9.512	720.8	720.8	0
13.798	951.9	951.7	0.02	9.269	721.3	720.3	0.12
6.914	948.5	948.2	0.03	9.044	719.2	719.8	—0.08
1.519	945.7	945.4	0.03				
	150.00°C				**325.00°C**		
26.628	937.3	937.1	0.02	30.456	710.8	—	—
17.764	932.6	932.4	0.02	30.277	711.0	—	—
9.189	927.9	927.7	0.03	23.619	695.9	—	—
1.901	923.8	923.5	0.03	23.283	695.5	—	—
				21.505	691.6	—	—
	200.00°C			18.633	683.8	—	—
28.854	890.0	890.0	0	18.476	682.9	—	—
28.160	889.2	889.5	—0.04	18.075	681.5	—	—
22.307	885.1	885.6	—0.06	13.080	666.4	—	—
19.411	883.7	883.6	0.01	12.840	664.9	—	—
19.348	883.1	883.6	—0.05	12.680	664.1	—	—
12.524	878.3	878.7	—0.05				
11.472	878.1	878.0	0.01		**350.00°C**		
11.249	877.5	877.8	—0.03	30.621	653.2	—	—
4.035	872.6	872.4	0.02	29.704	650.7	—	—
3.271	871.5	871.8	—0.04	25.418	635.1	—	—
2.591	871.1	871.3	—0.02	25.040	633.4	—	—
				24.513	631.8	—	—
	250.00°C			20.750	613.8	—	—
29.344	830.8	831.4	—0.07	20.672	612.9	—	—
28.254	830.5	830.4	0.02	19.763	608.4	—	—
22.703	824.7	825.2	—0.05	17.348	591.6	—	—
20.956	823.2	823.4	—0.03	17.161	590.6	—	—
17.850	820.0	820.4	—0.04				
10.696	812.5	812.8	—0.04				
9.385	811.0	811.4	—0.04				

Comparison of calculated (using the above equation) densities of investigated solutions with experimental results for temperatures below 300°C is given in Table 1. It is seen that difference in calculated and measured values up to 250°C does not exceed the density measurement error and in most cases is even smaller.

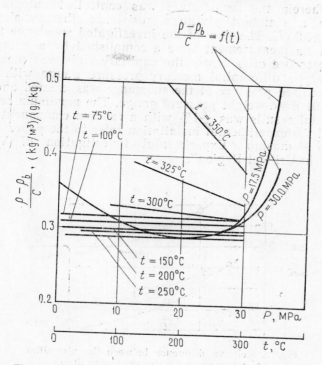

$$\frac{\rho - \rho_b}{C} = f(t)$$

Fig. 1. The values of $A = (\rho - \rho_w)/c$ in equation (1) *vs.* pressure and temperature

At temperatures above 280°C the values of A of equation (1) depend not only on temperature, but on pressure as well. Nevertheless, equation (1) with an error not exceeding 0.1% can be used for calculation of the density of solution at temperatures up to 300°C and pressures up to 30 MPa.

2. VISCOSITY EXPERIMENTS

On the VTI-developed experimental installation for investigating viscosities of liquids and gases, measurements were made of dynamic viscosity coefficients of aqueous boron solutions. The experimental data have been obtained on the viscosity of solution at concentrations 0.5, 1.5, 2.5 and 4% (mass), at temperatures up to 350°C and pressures up to 30 MPa.

Previously, this installation was used to investigate viscosity of ordinary and heavy water in a wide range of parameters of state [6, 7].

The experimental installation employs the capillary method. The 0.3-mm dia and about 500-mm long platinum capillary was used for the experiments. The test section with a capillary was placed in a liquid thermostat wherein the temperature was controlled automatically and measured by a standard platinum resistance thermometer with an accuracy of ±0.03°C. The flow of the investigated substance through the capillary and its measurement were accomplished by a pump-flowmeter device. The pressure drop across the capillary ends was measured by a compensation-type differential mercury pressure gauge with a movable elbow. The required pressure of the substance was created and measured by a Class 0.01 dead-weight pressure gauge. The maximum relative error for experimental results was ±1% with a random component of ±0.2%. For details of the experimental installation refer to the paper of S. L. Rivkin and S. N. Romashin "Experimental Investigation of Viscosity of Heavy Water in Critical Region".

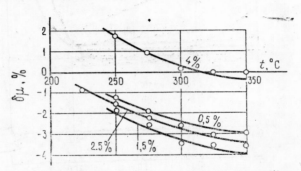

Fig. 2. Relative difference between the viscosities of solutions and water *vs.* temperature

The comparison of viscosities of solutions and water was made as a percent difference between viscosities of the solution and water, i. e., $\delta\mu = (\mu - \mu_{H_2O})/\mu_{H_2O} \cdot 100\%$. The values for the viscosity of water are taken from the tables of [8]. The calculations have shown that within the accuracy of the experiment the above-stated relative difference does not depend on pressure in the considered range of parameters of state and concentrations. The averaged values of $\delta\mu$ vs. temperature and concentration are given in Figs. 2 and 3. One can see that the viscosity of the aqueous boron solutions has a complex dependence on temperature and concentration. In greater part of the investigated range of concentrations the viscosity of the solutions is lower as compared to that of water. It is also seen that the dependence of $\delta\mu$ on temperature is close to a linear one at temperatures up to 300°C (Fig. 2), while $\delta\mu$ vs. concentration is almost parabolic (Fig. 3).

It has been found in the course of the experiments that the viscosity of aqueous boron solutions irreversibly reduces with "overheating". For instance, the dynamic viscosity coefficient of 1.5-% solution at 50°C, 20 MPa is 573.9×10^{-6} Pa·s, while after the solution was heated up to 350°C it reduced to 552.1×10^{-6} Pa·s. That is, the viscosity after the solution was heated up by 300°C was found to be reduced by 4%. Since

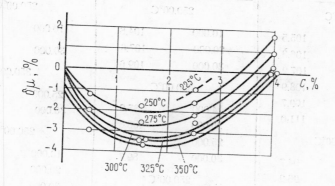

Fig. 3. Relative difference between the viscosities of solutions and water *vs.* concentration

the experiments were conducted on isotherms starting from lower temperatures it was possible not only to investigate the viscosity of the solutions in the usual manner but to simultaneously study the observed effect of "overheating". To quantify this new effect a series of runs were conducted with the same portion of the solution as follows: after measuring the viscosity at a certain "reference" temperature it was measured at a higher temperature and, finally, at the same "reference" temperature. As "reference" temperatures, the temperatures of 50, 250, and 275°C were used.

The experimental data for the dynamic viscosity coefficient of aqueous boron solutions are tabulated in Table 3 in the sequence corresponding to that of the experiments. The values of the effect of the irreversible reduction of viscosity with "overheating" can be easily determined by the results of measurements made in the above-mentioned sequence and are given in Table 4. The columns of Table 4 contain: "reference" temperature (t), concentration of the solution, pressure, coefficient of dynamic viscosity before "overheating" (μ), temperature of "overheating" (t_{oh}), coefficient of dynamic viscosity after "overheating" (μ_{oh}), the value of the effect of viscosity reduction $\Delta\mu = (\mu_{oh} - \mu)/\mu\ 100\%$, the value of "overheating" $\Delta t = t_{oh} - t$, and average value of $\Delta\mu(\Delta t)$.

It is seen from the table that within the range of temperatures from 250 to 350°C the value of $\Delta\mu$ depends only on "overheating" and at 100°C "overheating" lies within —1.4 up to —2%.

Table 3. Experimental data on dynamic viscosity of aqueous boron solutions (accounting the effect of "overheating")

P, MPa	$\mu \cdot 10^6$, Pa·s
Concentration 0.5%	
250.00°C	
7.000	105.5
10.000	106.3
15.000	107.2
20.000	108.9
25.000	109.7
30.000	111.0
275.00°C	
10.000	94.7
15.000	96.2
350.00°C	
25.000	70.5
250.00°C	
20.000	106.7
350.00°C	
25.000	70.4
Concentration 1.5%	
250.00°C	
7.000	104.5
10.000	105.4
15.000	106.6
20.000	108.2
25.000	109.8
30.000	110.8
275.00°C	
10.000	94.4
20.000	97.2
25.000	98.1
30.000	99.5

P, MPa	$\mu \cdot 10^6$, Pa·s
Concentration 1.5%	
250.00°C	
10.000	104.6
20.000	107.6
30.000	109.6
300.00°C	
14.000	85.2
20.000	86.8
25.000	88.4
30.000	89.7
250.00°C	
10.000	104.4
20.000	107.1
30.000	109.3
300.00°C	
14.000	84.9
20.000	87.1
25.000	88.3
250.00°C	
10.000	104.5
20.000	106.9
30.000	109.4
275.00°C	
10.000	93.6
20.000	96.5
30.000	98.9
325.00°C	
20.000	77.8
25.000	79.5
30.000	81.4

P, MPa	$\mu \cdot 10^6$, Pa·s
Concentration 1.5%	
250.00°C	
10.000	104.1
30.000	109.3
350.00°C	
25.000	70.0
30.000	72.6
250.00°C	
10.000	104.0
20.000	106.7
30.000	109.3
50.00°C	
20.000	552.1
30.000	551.8
Concentration 1.5% fresh solution	
50.00°C	
20.000	573.9
30.000	575.0
Concentration 2.5%	
225.00°C	
5.000	117.4
10.000	119.2
15.000	120.6
25.000	122.3
30.000	124.3
250.00°C	
7.000	104.7
10.000	105.8
20.000	108.5
30.000	111.1

Table 3 (*continued*)

P, MPa	$\mu \cdot 10^6$, Pa·s	P, MPa	$\mu \cdot 10^6$, Pa·s
Concentration 2.5%		Concentration 4%	
275.00°C		250.00°C	
10.000	94.6	7.000	107.8
20.000	97.5	20.000	111.2
25.000	98.7	25.000	112.4
30.000	100.1	30.000	113.8
300.00°C		300.00°C	
20.000	87.7	14.000	87.9
25.000	89.4	25.000	91.9
30.000	90.8	30.000	93.2
325.00°C		250.00°C	
25.000	80.0	7.000	107.4
30.000	81.8	20.000	110.9
Concentration 4%		325.00°C	
250.00°C		18.000	79.4
7.000	108.7	20.000	80.4
10.000	109.7	25.000	82.7
15.000	110.5	250.00°C	
20.000	112.1	20.000	110.3
25.000	113.0	350.00°C	
275.00°C		25.000	72.7
10.000	97.4	30.000	75.2
15.000	99.0	250.00°C	
20.000	100.9	20.000	110.0
25.000	101.9		
30.000	103.3		

The maximum value of $\Delta\mu$ of —3.9% is observed with "over-heating" $\Delta t = 300$°C.

The error of $\Delta\mu$ value is about $\pm 0.4\%$ (double random error for the experimental data).

It is worth noting that in plotting the curves of Figs. 2 and 3 use was made of the viscosities obtained at initial heating, i. e., not accounting the effect of irreversible reduction of viscosity with "over-heating".

Table 4. The values of irreversible reduction of viscosity of aqueous boron solutions with "overheating"

t, °C	Conc., %	P, MPa	$\mu \times 10^6$, Pa·s	t_{oh}, °C	$\mu_{oh} \times 10^6$, Pa·s	$\Delta\mu$, %	Δt, °C	$\Delta\mu$, %
50	1.5	20	573.9	350	552.1	−3.8	300	−3.9
		30	575.0		551.8	−4.0		
	0.5	20	108.9	350	106.7	−2.0	100	−2.0
		10	105.4		104.6	−0.8		
		20	108.2	275	107.6	−0.6	25	−0.8
		30	110.8		109.6	−1.1		
		10	105.4		104.4	−0.9		
	1.5	20	108.2	300	107.1	−1.0	50	−1.1
		30	110.8		109.3	−1.4		
		10	105.4	325	104.1	−1.2	75	−1.3
		30	110.8		109.3	−1.4		
250		10	105.4		104.0	−1.3		
		20	108.2	350	106.7	−1.4	100	−1.4
		30	110.8		109.3	−1.4		
		7	108.7		107.8	−0.8		
		20	112.1	275	111.2	−0.8	25	−0.7
		25	113.0		112.4	−0.5		
		7	108.7	300	107.4	−1.2	50	−1.2
	4	20	112.1		110.9	−1.1		
		20	112.1	325	110.3	−1.6	75	−1.6
		20	112.1	350	110.0	−1.9	100	−1.9
275	1.5	10	94.4		93.6	−0.8		
		20	97.2	300	96.6	−0.6	25	−0.7
		30	99.5		98.9	−0.6		

The irreversible reduction of the viscosity of the aqueous boron solution requires further investigation.

REFERENCES

1. Rivkin S. L., Akhundov T. S. (1962): *Teploenergetika* 1. (1963): *Teploenergetika* 9.
2. Rivkin S. L., Troyanovskaya G. V. (1964): *Teploenergetika* 10.
3. Rivkin S. L., Akhundov T. S., Kremenevskaya E. A., and Assadulaeva N. N. (1966): *Teploenergetika* 4.
4. Rivkin S. L., Akhundov T. S. (1962): *Teploenergetika* 5.
5. Rivkin S. L., Akhundov T. S. (1963): *Nuclear Energy* 14 (6).
6. Rivkin S. L., Levin A. Ya., and Izrailevsky L. B. Viscosity of Water and Vapour, Standards Publishing House, Moscow, 1979.
7. Rivkin S. L., Levin A. Ya., Izrailevsky L. B., Kharitonov K. G., Romashin S. N. *Proc. of the 9th Int. Conf. on the Properties of Steam*, Munich, 1979. Perg. Press, 1980, 375-381.
8. Rivkin S. L., Alexandrov A. A., Teplophysical Properties of Water and Steam Energy, Publishing House, 1980.

Experimental Study of Sound Velocity
in the Aqueous Solutions of Sodium Hydroxide Over
a Wide Range of Temperatures
and Concentrations

A. A. ALEKSANDROV, V. S. OKHOTIN,
A. I. KOCHETKOV, and G. G. KUZNETSOV

Moscow Power Engineering Institute, Moscow, USSR

The investigation of sound velocity was carried out in aqueous solutions of sodium hydroxide at the atmospheric pressure over the range of mass concentrations 0-80% and the temperature range 293-403 K.

Fig. 1 shows the schematic diagram of the installation used for measuring sound velocity. The measurements were based on the echo-pulse method having high precision and reproducibility. Previously, this method had been used by the authors for measuring sound velocity in ordinary and heavy water [1, 2].

Industrial high-precision devices were used for these measurements. A square-wave generator excited a piezoelectric quartz transducer (with a resonance frequency of 2.5 MHz). Electric pulses, induced in the receiving piezotransducer by acoustic pulses which repeatedly passed through the sample, were amplified and fed to the oscillograph. The amplification factor of the amplifier run into 120 dB. The frequency match of the pulses passed through the sample was obtained by adjusting the trigger pulse repetition frequency of the oscillograph. Because of stringent demands imposed upon the trigger pulse generator, the latter was chosen from the generators of the same type according to

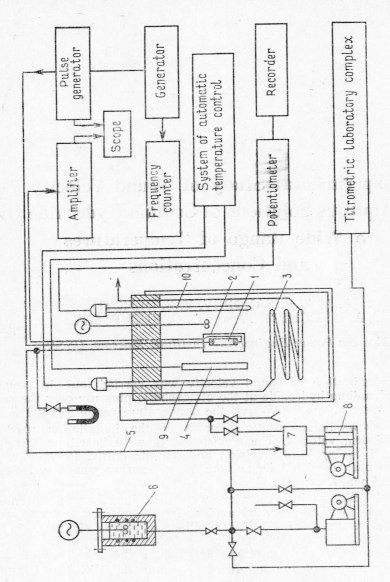

Fig. 1. Block diagram for measuring sound velocity

the least value of the frequency instability. The magnitude of pulse repetition frequency during the match was used for the calculation of sound velocity.

The working section with acoustic cell *1* set in quartz vessel *2* was placed in a liquid thermostat. The vessel was filled with an investigated sample through pipe-line *5* connected with solution preparation system *6*.

For creating and maintaining a required temperature of the experiment the thermostat was provided with heater *4* and for creating low temperatures it had coil *3* through which, depending on the temperature required, either an alcoholic solution from thermostat *7* cooled by refrigerator *8* or tap water circulates.

To maintain a desired temperature, an automatic temperature control unit was used. Platinum resistance thermometer *9* served as a temperature-sensitive element.

The temperature of the liquid in the thermostat was maintained accurate to $\pm0.02°C$.

The temperature control and measurement were made by platinum resistance thermometer *10* of the first class.

The acoustic measurements were carried out in the equilibrium state which was determined by the constancy of the pulse repetition frequency at the coincidence or, which is the same, by sound velocity (within the limits of measurement accuracy).

Before and after carrying out the measurements, samples of the sodium hydroxide solution were taken from the working section and quantitative analysis with the help of a titrometric laboratory complex was carried out. The complex makes potentiometric titration with the automatic recording of the titration curves on the plotter. The uncertainty of the concentration determination is 0.03%. The comparison of the solution concentrations before and after the experiment showed that within the limits of measurement accuracy the concentration value didn't change.

Sound velocity was calculated according to the results of the pulse match frequency measurement. As the measurements of pulse match frequency were carried out by means of the measurements of the ratio of the measured frequency to the base one, the calculated expression for sound velocity has the form

$$W = \frac{2l_t \cdot f_b}{\bar{n}} - \Delta W_{dif} \tag{1}$$

where l_t — the length of the spacer at a temperature of the experiment; f_b — the base frequency (in our case 10 MHz); $\bar{n}$ — the mean frequency ratio of 10 measurements; ΔW_{dif} — the diffraction correction determined by the calculation according to [3].

The experimental data of sound velocity are presented in Table 1 and Fig. 2.

Table 1. Experimental

T, K	W, m/s	T, K	W, m/s	T, K	W, m/s
Ordinary water		C = 2.03%		C = 6.04%	
				293.15	1628.66
		293.15	1531.95	303.15	1645.29
293.15	1482.14	303.15	1555.09	313.15	1656.59
298.15	1496.38	313.15	1572.08	323.15	1663.42
303.15	1508.42	323.15	1583.70	333.15	1666.55
308.15	1519.29	333.15	1590.47	338.15	1666.75
313.15	1528.39	338.15	1591.90	343.15	1665.98
318.15	1536.09	343.15	1592.94	348.15	1664.60
323.15	1542.12	348.15	1592.83	353.15	1662.20
328.15	1546.66	353.15	1591.28	363.15	1655.45
333.15	1550.57	363.15	1587.55	373.15	1646.80
338.15	1552.91	373.15	1580.51		
343.15	1554.75			C = 7.02%	
348.15	1555.41	C = 3.03%		293.15	1651.53
353.15	1555.17			303.15	1666.46
358.15	1554.05	293.15	1554.64	313.15	1676.27
363.15	1552.32	303.15	1576.16	323.15	1682.01
368.15	1549.98	313.15	1591.83	328.15	1683.63
373.15	1546.72	323.15	1602.39	333.15	1684.42
		333.15	1608.33	338.15	1684.18
		343.15	1610.40	343.15	1683.11
		348.15	1610.12	353.15	1678.46
C = 0.54%		353.15	1608.57	363.15	1671.68
		363.15	1604.05	373.15	1662.08
293.15	1495.00	373.15	1596.39		
303.15	1520.69			C = 8.05%	
313.15	1539.47	C = 3.99%		293.15	1675.65
323.15	1552.63			303.15	1688.79
333.15	1560.67	293.15	1580.27	313.15	1697.23
338.15	1563.06	303.15	1600.43	323.15	1701.99
343.15	1564.34	313.15	1614.56	328.15	1703.32
348.15	1564.79	323.15	1624.05	333.15	1703.82
353.15	1564.25	333.15	1629.11	338.15	1703.28
363.15	1560.69	338.15	1630.04	343.15	1701.53
373.15	1554.18	343.15	1629.99	353.15	1696.39
		348.15	1629.33	363.15	1688.99
		353.15	1627.77	373.15	1680.08
		363.15	1621.94		
C = 1.02%		373.15	1615.11	C = 9.02%	
293.15	1506.73			293.15	1701.37
303.15	1531.72	C = 5.18%		303.15	1713.06
313.15	1550.62			313.15	1721.01
323.15	1563.21	293.15	1608.66	318.15	1723.53
333.15	1570.73	303.15	1626.78	323.15	1725.06
338.15	1573.18	313.15	1638.99	328.15	1725.74
343.15	1573.94	323.15	1647.15	333.15	1725.42
348.15	1574.11	333.15	1650.95	338.15	1724.32
353.15	1573.46	338.15	1651.38	343.15	1722.44
363.15	1569.73	343.15	1650.99	353.15	1716.68
373.15	1563.24	348.15	1649.43	363.15	1708.15
		353.15	1647.52	373.15	1697.86
		363.15	1641.11		
		373.15	1633.08		

data of sound velocity

T, K	W, m/s	T, K	W, m/s	T, K	W, m/s
C=9.63%		C=18.16%		C=41.08%	
293.15	1715.99	293.15	1903.50		
298.15	1721.64	298.15	1902.81	293.15	2312.36
303.15	1726.46	303.15	1901.45	313.15	2267.39
308.15	1730.29	313.15	1897.88	323.15	2244.18
313.15	1733.42	323.15	1892.41	333.15	2222.13
323.15	1736.63	333.15	1884.56	343.15	2199.77
328.15	1737.06	343.15	1875.28	353.15	2175.38
333.15	1736.56	353.15	1864.28	363.15	2149.43
338.15	1735.38	363.15	1851.43	373.15	2127.53
343.15	1733.15	373.15	1838.49	393.15	2080.98
353.15	1726.82	378.15	1831.10		
363.15	1718.15			C=45.70%	
373.15	1707.01	C=20.19%		293.15	2351.92
		298.15	1951.88	303.15	2328.46
C=11.86%		303.15	1948.94	313.15	2306.08
		313.15	1942.75	323.15	2282.50
293.15	1768.45	323.15	1934.10	333.15	2259.12
298.15	1772.32	333.15	1925.01	343.15	2236.61
303.15	1775.61	343.15	1913.99	353.15	2213.45
308.15	1777.80	353.15	1902.52	363.15	2189.89
313.15	1778.97	363.15	1892.21	373.15	2167.52
318.15	1779.46	368.15	1885.87	393.15	2121.55
323.15	1779.61	373.15	1879.19		
328.15	1778.94			C=51.08%	
333.15	1777.72	C=25.01%		293.15	2404.19
343.15	1773.15	293.15	2051.12	303.15	2377.99
353.15	1765.42	303.15	2039.65	323.15	2322.09
373.15	1744.35	313.15	2027.95	333.15	2296.66
		323.15	2015.47	343.15	2276.78
C=14.05%		333.15	2001.98	353.15	2253.48
		343.15	1988.43	363.15	2232.42
293.15	1815.06	373.15	1939.46	373.15	2209.67
303.15	1818.99			383.15	2185.54
308.15	1819.73	C=28.48%			
313.15	1819.89	293.15	2118.15	C=65.35%	
323.15	1818.90	303.15	2105.53		
333.15	1815.17	313.15	2088.92	328.15	2376.95
343.15	1808.73	323.15	2074.20	333.15	2362.61
353.15	1800.27	333.15	2057.84	343.15	2338.31
363.15	1790.20	343.15	2041.64	353.15	2313.73
373.15	1778.86	353.15	2021.84	363.15	2289.52
		363.15	2006.76	373.15	2266.45
C=16.22%		373.15	1987.34	393.15	2216.15
		383.15	1968.25	403.15	2191.95
293.15	1863.95				
298.15	1864.20	C=34.58%		C=72.18%	
303.15	1863.94	293.15	2236.05		
308.15	1863.28	303.15	2215.42	338.15	2361.68
313.15	1862.36	313.15	2194.31	343.15	2349.68
323.15	1858.92	323.15	2173.12	353.15	2328.09
333.15	1853.90	333.15	2152.66	363.15	2307.24
343.15	1846.48	343.15	2131.16	373.15	2295.84
353.15	1835.97	353.15	2109.77	383.15	2274.62
363.15	1824.62	363.15	2088.66	393.15	2252.73
373.15	1811.55	373.15	2065.73	403.15	2231.78
		383.15	2045.44		

Fig. 2 shows that with increasing NaOH concentration, the maximum velocity shifts towards low temperatures. Furthermore, with the concentration increase, the maximum smoothes out and disappears. When $C > 25\%$, the solution behaves as ordinary liquid, i. e. the temperature coefficient of sound velocity in all the temperature range is only negative.

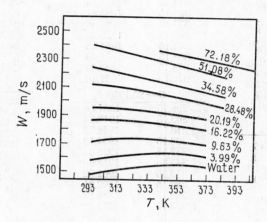

Fig. 2. Sound velocity in the aqueous solution of sodium hydroxide as a function of temperature at different concentrations (Table 1)

The installation was also used for measuring sound velocity in ordinary water in the same temperature range. The discrepancy between the values of sound velocity in water and the data [4] did not exceed 0.02%.

The discrepancy between the values of sound velocity in the 1.05% solution of sodium hydroxide at the atmospheric pressure and the data [5] does not exceed 0.33% and is of the same type as the discrepancy of data for distilled water obtained therein.

The root-mean-square uncertainty of the sound velocity data given in this paper is estimated at 0.04% (taking into consideration the uncertainties of temperature and concentration measurements at $T = 293$ K, $C = 1.05\%$).

The experimental values of sound velocity were approximated as the difference between the velocity in a sodium hydroxide solution and the velocity in distilled water. The approximated equation has the form

$$W(C, T) - W_{H_2O}(T) = C_* \sum_{i=0}^{4} \sum_{j=0}^{4} a_{ij} \left(\frac{T}{T_0} - 1 \right)^i \left(\frac{C_*}{C_* + 1} \right)^j \qquad (2)$$

where W_{H_2O} — the sound velocity in distilled water [6], m/sec; $C_* = C/100$ — the reduced concentration of the solution; $T = t + 273.15$ —

304

the temperature, K; $T_0 = 347.3$ K; the values of coefficients a_{ij} are presented in Table 2.

Table 2. Coefficients of the equation (2)

i	$j=0$	$j=1$	$j=2$	$j=3$	$j=4$
0	1917.088	—2286.52	16441.20	—61572.90	57600.46
1	1548.412	7747.749	—26162.57	26269.70	—39655.56
2	20986.77	—281338.5	1828759	—5132077	5010496
3	—71805.77	1347533	—9877129	29306950	—29805250
4	40485.91	—	—	—	—

REFERENCES

1. Aleksandrov, A. A., Larkin D. K., and Kochetkov, A. I. (1977): *Proceedings of 7th Sympos. Thermophysical Prop.*, Washington.
2. Aleksandrov, A. A. and Kochetkov, A. I. (1979): *Teploenergetika*, **9**, 65-66.
3. McSkimin, H. J., (1961): *Journal of the Acoustical Society of America*, 33, 539.
4. Del Grosso, V. A. and Mader, C.W. (1972): *Journal of the Acoustical Society of America.* 52(2), 1442-1446.
5. Mikhailov, I. G. and Fen Zhau. (1960): *Vestnik LGU*, **16**, 22-41.
6. Aleksandrov, A. A., Okhotin, V. S., Ershova, Z. A., and Matveev, A. B. (1981): *Energetika*, **7**, 120-125.

Measurements of the Thermal Diffusivity of Pure Fluids and the Diffusion Coefficient of Binary Mixtures by Dynamic Light Scattering

P. JANY and J. STRAUB

Lehrstuhl A für Thermodynamik, Technische Universität München, Arcisstr. 21, D-8000 München 2, Federal Republic of Germany

ABSTRACT

Transport properties such as thermal diffusivity and diffusion coefficient are usually measured by methods producing stationary or nonstationary gradients in the sample. In the last decade the method of dynamic light scattering was developed which does not require macroscopic gradients in temperature and concentration any longer. Microscopic fluctuations of these quantities cause fluctuations of the refractive index and accordingly scattering of the incident light in all directions. From the time autocorrelation function of the intensity of the scattered light one can obtain information about the temporal behaviour of the fluctuations and therefrom about the transport properties. They are calculated only by means of the decay of the autocorrelation function, the refractive index of the sample and the scattering angle without the necessity for a calibration. A single measurement takes only a few minutes.

The classical range of application is the vicinity of the gas-liquid critical point or the consolute critical point because of the high scattered intensity.

In this paper it is proved that by means of an appropriate choice of the experimental apparatus the method is also applicable in an extended

range of state. This is shown by our measurements of the thermal diffusivity of sulfur hexafluoride and the diffusion coefficients in aqueous solutions.

KEYWORDS

Dynamic light scattering; critical point; thermal diffusivity; aqueous solution; diffusion coefficient.

INTRODUCTION

Since the basic theory of light scattering was developed more than half a century ago by Debye, Einstein, Mie, Rayleigh and Smoluchowski the immense utility for the investigation of molecular structure is well known. An experimental application became only possible with the advent of laser as an intensive and coherent light source twenty years ago. In this contribution we consider only the central band at the incident light frequency of the total power spectrum called "Rayleigh-scattering". This is caused from spontaneous local fluctuations of the thermodynamic properties of state in the fluid. The effect is scattered light in all directions which contains informations about the molecular transport processes.

One of the purposes of this paper is to explain the determination of the transport coefficients by the analysis of scattered light. Furthermore we will show advantages and disadvantages as well as the classical range of validity of the method. Our own experimental apparatus will point out that there are some possibilities to extend this range. Finally some results of our measurements are presented.

THEORY

The permanently existing random motion of all molecules causes local fluctuations of the thermodynamic properties of state even in the macroscopic state of equilibrium. Because of their influence on the refractive index the fluctuations produce an optical inhomogeneity of the fluid. Therefore secondary waves of light are not extinguished by interference neither in other directions than that of incident light. An exact description of the theory of light scattering is presented by Berne and Pecora 1976, Cummins and Pike 1973 or by Chu 1974. Here we will just outline a short summary.

In a light scattering experiment corresponding to Fig. 1, $\vec{k}_i$, $\vec{k}_s$, θ and $\vec{q} = \vec{k}_i - \vec{k}_s$ represent the wave vectors of incident and scattered light, the scattering angle and the scattering vector, respectively. Assuming that the light frequency ω is constant the calculation of the scattering vector leads to

$$|\vec{q}| = \frac{4\pi n}{\lambda_l} \sin\left(\frac{\theta}{2}\right) \tag{1}$$

where n is the refractive index of the fluid and $\lambda_{L_{\bullet}}$ is the wavelength of the light. Thus the scattered electric field at a large distance R from the scattering volume is

$$\vec{E}_s(\vec{R}, t) = \frac{\vec{E}_0}{4\pi\varepsilon_0} \exp(i\vec{k}_s\vec{R}) \int \exp(i(\vec{q}\vec{r} - \omega t)) \Delta\varepsilon(\vec{r}, t) d^3\Gamma \qquad (2)$$

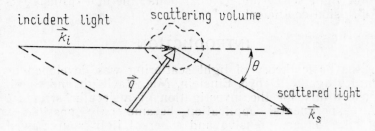

Fig. 1. Scattering geometry

In this expression $\vec{E}_0$ is the amplitude vector of the incident electric field, t is the time and ε is the dielectric constant, fluctuating about its equilibrium value; $\varepsilon(\vec{r}, t) = \varepsilon_0 + \Delta\varepsilon(\vec{r}, t)$. In light scattering experiments only the spatial Fourier component of the total spectrum due to the scattering angle θ is detected. Using the Fourier transformation

$$\Delta\tilde{\varepsilon}(\vec{q}, t) = \int \exp(i\vec{q}\vec{r}) \Delta\varepsilon(\vec{r}, t) d^3\Gamma \qquad (3)$$

and the scattered intensity $I_s(\vec{R}, t) \sim |\vec{E}_s(\vec{R}, t)|^2$, we obtain

$$I_s \sim |\Delta\tilde{\varepsilon}(\vec{q}, t)|^2$$

The temporal behaviour of the intensity of scattered light is determined by the temporal behaviour of the fluctuations in ε.

The reasons for those fluctuations are fluctuations in the properties of state. Their influence on the scattered intensity is explained in literature by means of the descriptive equations. Qualitative predictions of $\Delta\varepsilon$ are possible considering the effect of fluctuations in density ρ and in concentration x:

$$\Delta\varepsilon = (\partial\varepsilon/\partial\rho)_x \Delta\rho + (\partial\varepsilon/\partial x)_\rho \Delta x \qquad (4)$$

The fluctuations in density can be split up in statistically independent fluctuations of the variables entropy and pressure. Because of their different effects they can be determined separately. Fluctuations in pressure cause a frequency shift in the scattered light as thermally stimulated waves of sound (Brillouin-scattering) while the fluctuations in entropy decay as fluctuations in temperature without a shift in frequency. According to Onsager's regression hypothesis (Onsager 1931), the microscopic fluctuations statically regress back to equilibrium due to

the same equations that describe macroscopic relaxation processes. Therefore the laws of Fourier and Fick are applicable

$$\frac{\partial}{\partial t}\Delta T(\vec{r}, t) = a\nabla^2(\Delta T(\vec{r}, t)) \tag{5a}$$

$$\frac{\partial}{\partial t}\Delta x(\vec{r}, t) = D\nabla^2(\Delta x(\vec{r}, t)) \tag{5b}$$

where ∇ is the Nabla-operator, a is the thermal diffusivity and D is the diffusion coefficient. Introducing the spatial Fourier transformation the solution of the differential equations leads to

$$\Delta\tilde{T}(\vec{q}, t) = \Delta\tilde{T}(\vec{q}, 0)\exp(-aq^2t) , \tag{6a}$$

$$\Delta\tilde{x}(\vec{q}, t) = \Delta\tilde{x}(q, 0)\exp(-Dq^2t) \tag{6b}$$

Since the fluctuations represent a stochastic process it is not possible to observe them isolated. Hence the time autocorrelation function is calculated from the scattered intensity

$$c(t^*) = \lim_{t'\to\infty}\frac{1}{t'}\int_0^{t'} I_s(t)I_s(t+t^*)\,dt . \tag{7}$$

The autocorrelation function contains information about the influence of a value I_s at any moment on the same value t^* afterwards. The behaviour of this function depends on the physical process; in this case it decays like an exponential function with a relaxation time of $2aq^2$ or $2Dq^2$. Thus both transport coefficients, thermal diffusivity a and diffusion coefficient D are determined by the time autocorrelation function of the scattered intensity.

EXPERIMENTAL APPLICATION

Compared to usual methods, the outstanding advantage of this experimental technique is the absence of any macroscopic gradients in the test substance. Therefore, as a consequence of maintaining the thermodynamic equilibrium measurements are practicable also in the vicinity of critical points or in the two-phase region. However, two effects can limit the application: decreasing optical transparency of the fluid and insufficient production of scattered intensity. Increasing scattered intensity is attainable with an augmentation of the intensity of the incident light or with increasing fluctuations in dielectric constant ε. According to Eq. (4) there are some possibilities (see also Reile et al. 1984):

— Large fluctuations in the density $\Delta\rho$ observed near gas-liquid critical points;

— Large fluctuations in the concentration Δx which occur near consolute critical points;

— Increasing factors $(\partial\varepsilon/\partial x)_\rho$ due to a large difference between the refractive indices of the two components of a mixture.

By our own measurements it is shown that the method is applicable also in an extended range of state around the critical point. Moreover

are presented results from our measurements of diffusion coefficients of binary mixtures with similar refractive indices of both components far from consolute critical points.

EXPERIMENTAL APPARATUS

The experimental apparatus to measure transport coefficients is schematically illustrated in Fig. 2 (see also Reile 1981). An Argon-Ion laser is used as an intensive and coherent light source. In the beam expander with a spatial filter (1) the laser beam is widened and a

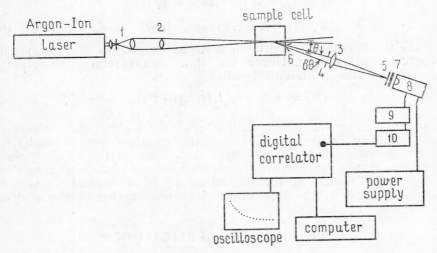

Fig. 2. Schematic diagram of experimental apparatus

homogeneous distribution of the intensity over the profile is obtained. The lens (2) focuses the laserbeam on the center of the sample cell. The components (3) to (8) of the light scattering optics are attached to a revolving table, the focal point being part of its axis. In that way the scattering angle θ is adjustable optionally and determined within an angle range of six seconds. The lens (3) focusses the scattered light from the focal point on the pinhole (5) which determines the effective scattering volume, while (4) defines the angular uncertainty $\delta\theta$ of the collecting system. The pinhole (6) is ought to screen the stray light from the windows of the sample cell because it affects measurements especially if scattering angles are very small. Only light with laser frequency can pass the interference filter (7) and is focussed from the Fourier trans-form lens to the photomultiplier cathode. In the phototube (8) the incident electric field is transformed into output current to be amplified (9). The discriminator (10) produces a sequence of single pulses which are passed into the digital correlator to calculate the time autocorrelation function according to Eq. (7) for a display on the oscilloscope. Online

310

data are stored on disk for computer evaluation of the transport coefficient afterwards.

An extended range of application of the experimental apparatus described above is provided by high laser power, a high sensitive and low noisy detector, small scattering angles and therefore a sufficient length of the sample cell and additionally by using pinhole (6).

The refractive index of the fluid is needed for the calculation of the scattering vector and the density. Measurements are done by a simple refractometer shown in Fig. 3. Furthermore, pressure and temperature of the fluid are measured. Sample cells of different size are available.

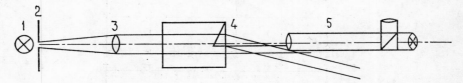

Fig. 3. Schematic diagram of refractometer: *1* — Hg-spectral lamp; *2* — slit; *3* — lens; *4* — prisma; *5* — autocollimation telescope

EXPERIMENTAL RESULTS

(1) Mass diffusivity. In the literature one can find some measurements of diffusion coefficients of binary mixtures near their consolute critical point by means of dynamic light scattering. Therefore we tried to avoid such ranges of state and investigated potassium chloride (KCl) in aqueous solution. The experiments proved to be very difficult because of the aggressiveness of the liquid. Repeatedly generated particles caused inconvenient stray light. For that reason cleaning of the liquid was often required. Fig. 4 shows the measured diffusion coefficients versus temperature. Good agreement with data obtained from literature (Landolt-Börnstein 1969) is visible.

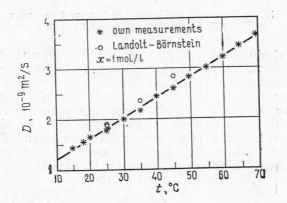

Fig. 4. Diffusion coefficient of H_2O/KCl *versus* temperature

In addition, ethanol (C_2H_5OH) in aqueous solution was investigated. In this mixture the variation of the refractive index with concentration is especially small. Fig. 5 shows not only the refractive index but also measured diffusion coefficients versus concentration. One can see the agreement of our results with literature data (Landolt-Börnstein 1969) and also the limit of measurement series near the maximum amount of refractive index due to $(\partial n/\partial x) \to 0$.

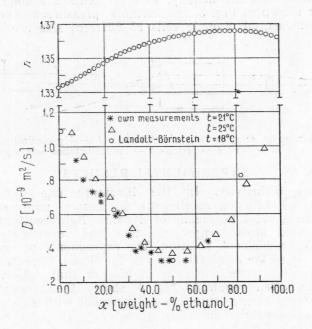

Fig. 5. Diffusion coefficient and refractive index of H_2O/C_2H_5OH *versus* concentration

(2) Thermal diffusivity. Fig. 6 shows the thermal diffusivity a of water versus temperature t and pressure p (Schiebener 1984). It is evident that variations of a for the liquid are very small. The gaseous phase should have a certain relation to the perfect gas behaviour. Therefore in our opinion the most interesting range of the whole area, with greatest variations in a is the two-phase region and the vicinity of the critical point. Therefore we investigated an extended range of state around the critical point. Measurements were carried out along both phases of the coexistence curve, along the critical isochore and along some isotherms with $T \lessgtr T_c$. For example, we present our recent measurements with sulfur hexafluoride (SF_6). A range of $|T-T_c|$ up to 35 K and ρ/ρ_c from 0.15 up to 2 was investigated. Before, temperature differences of 60 K were obtained in other fluids.

A logarithmic description is suitable due to the large variations of thermal diffusivity near the critical point. Fig. 7 shows our measurements of a versus the reduced density ρ/ρ_c except for the critical isochore. The singularity of the critical point can be recognized very well whereby the behaviour of isotherms depends strongly on temperature difference $T-T_c$.

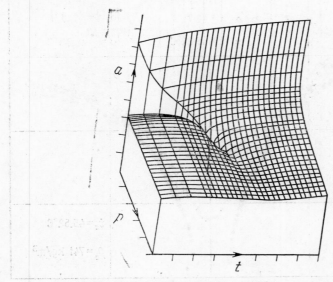

Fig. 6. Thermal diffusivity of H_2O *versus* temperature and pressure

Usually the properties of fluids near critical points are presented double logarithmic versus the reduced temperature difference $\tau = |T-T_c|/T_c$. Fig. 8 shows our results of critical isochore and both phases of coexistence curve. The comparison with data of Braun et al. 1970 and Ford, Benedek 1965 seems to be satisfactory.

According to power laws we tried to fit our data with

$$a = a_0\tau^\varphi. \tag{8}$$

The amount of exponent φ depends slightly on the fitted range. The use of all data leads to $\varphi = 0.83 \pm 0.01$ for the critical isochore and 0.88 ± 0.01 for the gaseous phase. For the liquid phase φ is 0.89 ± 0.01, provided that the exponent is widened to $\varphi + \varphi^*\tau$. The values are higher than data reported in the literature, e. g. Sengers and Keyes 1970. We remark, however, that their exponents are obtained from an evaluation of $\lambda_c/\rho c_p$ data where λ_c is the thermal conductivity, describing the critical anomaly without the background conductivity λ_b. Our exponent should be smaller after multiplication of our thermal diffusivities with $\lambda_c/(\lambda_c+\lambda_b)$. Unfortunately, such data for SF_6 are not available.

In favour of a short contribution we will renounce a presentation of other measurements such as refractive index, pressure and density.

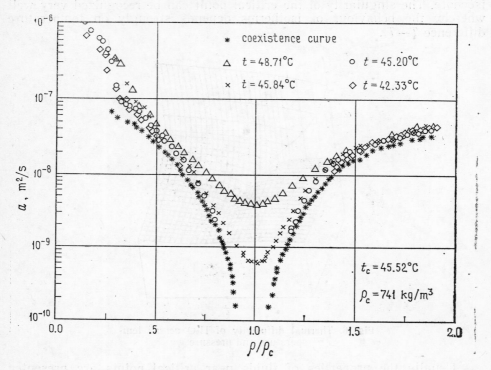

Fig. 7. Thermal diffusivity of SF$_6$ *versus* reduced density

CONCLUSIONS

The dynamic light scattering is a suitable method to measure transport properties such as thermal diffusivity or diffusion coefficient. Although the classical range of application is the vicinity of critical points because of high scattered intensity, measurements are feasible in an extended range of state by means of an appropriate choice of the experimental apparatus. In doing so the greatest advantages of the method are its quickness, the fact of thermodynamic state of equilibrium of the sample and a relatively simple evaluation of data without the necessity for a calibration.

Measurements with pure water as the sample fluid were not carried out owing to the insufficient sample cells with regard to high pressures and temperatures. Certainly measurements with water should be possible since the availability of the method on polar substances has been previously proved by Grabner et al. 1978.

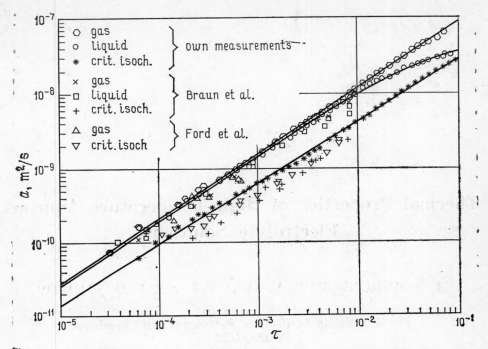

Fig. 8. Thermal diffusivity of SF₆ versus reduced temperature difference

REFERENCES

1. Berne, B. J.; Pecora, R. (1976): Dynamic Light Scattering, John Wiley & Sons, New York-London-Sydney-Toronto.
2. Braun, P.; Hammer, D.; Tscharnuter, W.; Weinzierl, P. (1970): *Physics Letters,* 32a(6) 390-391.
3. Chu, B. (1974): Laser Light Scattering, Academic Press, New York-San Francisco-London.
4. Cummins, H. Z.; Pike, E. R. (1973): Photon Correlation and Light Beating Spectroscopy, Plenum Press, New York-London.
5. Ford, N. C.; Benedek, G. B. (1965): *Physical Review Letters,* 15(16) 649-653.
6. Grabner, W.; Vesely, F.; Benesch, G. (1978): *Physical Review A,* 18(5) 2307-2314.
7. Landolt—Börnstein (1969): Zahlenwerte und Funktionen, Springer-Verlag, Berlin-Heidelberg-New York.
8. Onsager, L. (1931): *Physical Review* 37, 405-426 and 38, 2265-2279.
9. Reile, E. (1981): Messung der Temperaturleitfähigkeit reiner Fluide und binärer Gemische mit Hilfe der dynamischen Lichtstreuung in der weiteren Umgebung des kritischen Punktes, Thesis Technical University Munich.
10. Reile, E.; Jany, P.; Straub, J. (1984): *Wärme- und Stoffübertragung* 18, 99-108.
11. Schiebener, P. (1984): Private communication, Technical University Munich.
12. Sengers, J. V.; Keyes, P. H. (1970): *Physical Review Letters,* 26(2) 70-73.

Thermal Properties of High Temperature Aqueous Electrolyte Solutions

T. S. AKHUNDOV, M. V. IMANOVA, and A. D. TAHIROV

The Azerbaijan Institute of Petroleum and Chemistry,
Baku, USSR

The paper deals with investigating density and pressure in saturated vapours of aqueous solutions of KCl, KBr, and KI by means of a constant-volume cylindrical-type piesometer (28 cm³). The bottom point of the piesometer is connected with a capillary tube of 0.5 mm in diameter. The other end of the capillary leads to the room temperature zone and is connected to a mercury seal — a device fixing the volume of a liquid being studied. This device, in its turn, is connected through a separatory tank to a Class 0.05 dead-weight pressure-gauge testers MP-600 and MP-60. To fill the piesometer with the involved liquid and to withdraw it when transmitting it from one equilibrium state into another, one more capillary is attached to the piesometer. This capillary is led to the room temperature zone as well and is connected to the valve. All the units, the studied liquid and mercury being in contact with, are made of IXI8H10T steel. The piesometer thermostating is carried out in a liquid thermostat equipped with an axial-flow pump, a control heater and a temperature regulator permitting constant temperature regulation with an error of ±0.01°. The temperature is measured with a standard platinum resistance thermometer. The resistance of the thermometer is determined by the compensation method using the potentiometer P363/I and a standard resistance coil. Maximum error of temperature measurements

Table 1

Potassium chloride — 20 g/l (P, MPa; T, K; ρ, kg/m³)

P	ρ	P	ρ	P	ρ	P	ρ
$T=298.15$		$T=348.15$		$T=373.15$		$T=398.15$	
39.920	1027.5	39.633	1005.2	39.902	990.0	39.577	970.4
35.605	1025.4	35.399	1003.4	37.234	988.6	35.134	968.5
28.532	1022.1	25.549	998.4	28.102	984.3	25.332	963.8
24.159	1020.0	20.800	996.0	23.359	981.9	21.550	961.6
19.923	1017.7	15.651	993.7	18.924	979.4	16.703	959.1
15.287	1015.9	10.152	991.2	14.307	977.3	13.148	957.4
11.001	1014.1	6.664	989.7	9.359	975.0	9.107	955.5
4.634	1011.5	3.320	988.0	6.008	973.3	5.634	953.8
0.651	1009.8	0.347	986.5	0.904	970.8	0.588	951.1
				0.1005	970.5	0.2302	950.8
$T=423.15$		$T=448.15$		$T=473.15$		$T=498.15$	
39.907	950.7	39.849	928.5	39.758	905.0	39.901	878.6
36.739	949.0	37.056	927.2	36.250	902.9	37.227	876.5
27.279	944.0	29.199	922.4	27.225	896.8	29.227	870.0
24.685	942.4	24.498	919.7	22.648	893.6	24.750	866.4
20.452	939.9	20.151	917.2	18.502	890.5	20.101	862.4
14.901	936.8	15.628	914.1	14.450	887.7	16.134	859.0
11.760	935.2	11.449	911.7	10.323	884.4	11.799	855.2
7.137	932.6	6.752	908.4	6.378	881.2	7.724	851.5
0.687	928.9	1.851	905.0	2.500	878.1	3.501	847.9
0.4721	928.8	0.8852	904.2	1.542	877.4	2.530	847.0
$T=523.15$		$T=548.15$		$T=573.15$		$T=598.15$	
39.852	848.8	39.700	816.3	39.802	780.0	39.801	739.8
36.701	846.1	35.901	812.5	37.177	777.5	37.500	736.0
27.724	837.9	28.018	803.8	30.002	768.0	31.850	726.8
24.152	834.5	24.919	800.0	26.788	763.2	28.234	720.1
19.605	830.1	21.572	795.8	22.349	756.1	25.122	714.2
15.621	826.2	17.855	791.1	19.350	751.2	21.701	706.5
11.634	822.0	14.302	787.0	16.002	745.7	19.000	700.1
8.077	818.1	10.923	782.9	13.004	740.6	15.602	691.8
4.998	814.6	7.399	778.2	9.303	733.6	12.811	683.5
3.945	813.5	5.900	776.5	8.522	732.1	11.959	681.0

Table 1 (*continued*)

Potassium chloride — 50 g/l (P, MPa; T, K; ρ, kg/m³)

P	ρ	P	ρ	P	ρ	P	ρ
$T=298.15$		$T=348.15$		$T=373.15$		$T=398.15$	
39.867	1045.0	39.753	1022.9	39.499	1007.6	39.998	990.1
36.321	1043.2	35.752	1020.7	35.703	1005.8	37.067	988.6
27.002	1039.0	26.176	1016.1	27.499	1001.7	28.900	984.0
23.176	1037.5	20.749	1013.2	23.107	999.1	24.403	981.3
18.724	1035.4	15.403	1011.3	17.999	996.5	19.653	978.9
14.250	1033.7	11.987	1009.9	13.816	994.4	14.249	976.2
9.932	1031.9	8.404	1008.2	9.319	992.5	9.724	973.9
4.803	1029.5	4.850	1006.8	4.776	989.9	5.178	971.8
0.452	1027.6	0.353	1004.9	0.853	988.0	0.723	969.6
				0.0992	987.6	0.2272	969.2
$T=423.15$		$T=448.15$		$T=473.15$		$T=498.15$	
39.803	970.0	39.654	948.0	39.857	925.5	39.900	899.1
35.779	967.8	35.399	945.1	36.566	923.3	36.402	896.6
26.202	962.2	27.219	940.0	27.750	917.0	26.889	888.7
21.701	959.4	22.774	937.5	23.349	913.8	21.843	884.6
15.899	956.2	18.669	935.0	19.103	910.8	17.178	880.8
13.472	954.8	13.887	932.3	14.427	907.5	11.903	876.3
8.837	952.4	9.861	930.0	10.001	904.4	8.987	874.0
4.753	950.0	6.322	927.7	5.837	901.3	6.424	872.1
0.750	948.1	2.403	925.6	2.500	899.2	3.351	869.5
0.4659	947.7	0.8735	924.6	1.522	898.5	2.496	868.9
$T=523.15$		$T=548.15$		$T=573.15$		$T=598.15$	
39.632	869.8	39.775	837.6	39.797	802.6	39.843	765.2
35.242	865.7	36.976	834.9	36.400	798.9	37.504	762.0
27.073	858.3	29.288	827.2	30.102	791.4	31.604	753.3
22.887	854.4	26.249	824.0	26.681	787.5	28.857	749.1
18.901	850.7	22.173	818.9	23.263	782.6	26.123	745.0
15.299	847.2	19.298	816.0	19.700	777.7	23.702	740.8
11.452	843.9	15.250	811.7	16.147	772.4	19.656	733.4
8.244	840.7	11.248	806.8	12.834	767.0	16.343	726.4
5.221	838.0	7.202	802.5	9.419	760.7	13.101	719.0
3.893	836.6	5.823	801.0	8.410	759.0	11.802	715.9

Table 2

Potassium bromide — 20 g/l (P, MPa; T, K; ρ, kg/m³)

P	ρ	P	ρ	P	ρ	P	ρ
$T=298.15$		$T=348.15$		$T=373.15$		$T=398.15$	
39.904	1028.5	39.467	1005.3	39.653	989.9	39.857	971.5
35.251	1026.4	36.056	1003.9	36.447	988.4	35.303	969.3
25.259	1022.3	26.457	999.9	28.208	984.9	26.499	965.0
23.004	1021.4	21.645	998.0	23.350	982.5	22.377	963.1
19.675	1019.8	16.382	995.7	16.223	979.1	18.179	961.0
14.688	1017.9	12.131	993.6	11.634	976.9	13.903	958.8
10.259	1016.0	7.924	992.0	7.763	975.0	9.724	956.5
4.667	1013.6	3.683	990.0	5.227	973.9	5.760	954.6
0.474	1011.5	0.557	988.5	0.903	971.5	0.688	951.9
				0.1008	971.1	0.2309	951.5
$T=423.15$		$T=448.15$		$T=473.15$		$T=498.15$	
39.867	950.7	39.727	928.7	39.904	904.9	39.881	877.5
36.122	949.0	34.898	926.1	37.133	903.1	36.278	875.1
27.204	944.6	26.557	921.3	28.064	897.2	28.450	869.3
23.569	942.9	21.701	918.4	23.432	894.4	21.222	863.9
18.831	940.2	17.003	915.5	19.600	891.8	17.350	860.8
13.149	937.5	12.744	913.6	15.319	889.0	13.689	857.7
9.005	935.4	8.262	910.4	11.701	886.6	10.551	855.0
5.423	933.5	5.170	908.5	7.789	883.9	6.587	851.3
0.924	930.8	1.368	906.1	2.334	879.2	3.035	847.9
0.4735	930.5	0.8877	905.7	1.547	878.6	2.537	847.5
$T=523.15$		$T=548.15$		$T=573.15$		$T=598.15$	
39.804	848.0	39.650	814.9	39.650	779.4	39.921	737.6
36.555	845.2	36.418	811.8	34.901	773.1	35.702	730.4
27.849	837.6	27.903	802.5	28.924	764.9	30.134	720.1
23.785	834.0	24.250	798.6	25.674	760.0	26.648	713.7
20.001	830.5	20.684	794.3	22.721	755.6	23.899	707.5
16.508	827.0	16.773	789.4	19.133	749.4	20.684	700.1
13.174	823.6	13.350	785.5	16.050	744.1	17.820	692.8
9.442	819.8	9.419	780.4	12.769	738.0	15.179	685.4
4.683	814.1	6.402	776.2	9.221	730.3	12.450	676.9
3.957	813.7	5.917	775.6	8.546	728.9	11.993	675.6

Table 2 (*continued*)

P	ρ	P	ρ	P	ρ	P	ρ

Potassium bromide — 50 g/l (P, MPa; T, K; ρ, kg/m³)

P	ρ	P	ρ	P	ρ	P	ρ
$T=298.15$		$T=348.15$		$T=373.15$		$T=398.15$	
39.921	1049.4	39.604	1026.1	39.803	1009.2	39.833	991.8
36.343	1048.2	34.658	1024.2	35.798	1007.9	35.179	989.7
26.919	1044.6	25.852	1020.5	27.372	1004.3	27.383	986.2
22.350	1042.5	23.234	1019.1	23.181	1002.5	23.724	984.4
17.248	1040.3	17.503	1016.9	18.750	1000.6	18.553	982.0
12.807	1038.2	11.854	1014.2	15.127	999.1	12.150	978.9
8.419	1036.4	7.679	1012.5	10.698	996.9	6.802	976.1
4.604	1034.8	4.197	1010.8	6.366	995.0	3.489	974.6
0.427	1032.6	0.559	1008.9	0.725	991.9	0.598	973.1
				0.0999	991.7	0.2288	972.9
$T=423.15$		$T=448.15$		$T=473.15$		$T=498.15$	
39.851	971.2	39.755	949.4	39.757	925.0	39.849	899.0
36.534	969.6	36.649	948.0	36.343	923.1	35.777	896.3
29.299	966.2	29.569	943.8	28.853	918.5	26.884	889.8
25.289	964.3	25.186	941.4	24.222	915.4	22.343	886.8
20.581	962.0	21.037	939.0	20.197	912.9	19.254	884.3
15.427	959.5	15.319	935.8	15.796	910.0	14.873	880.8
10.832	957.1	10.311	932.9	12.001	907.5	10.624	877.2
6.361	954.6	4.688	929.3	7.422	904.4	6.688	873.9
1.349	951.8	1.576	927.0	2.178	900.5	3.159	870.3
0.4692	951.5	0.8797	926.6	1.533	900.0	2.514	869.6
$T=523.15$		$T=548.15$		$T=573.15$		$T=598.15$	
39.847	869.0	39.799	837.6	39.750	802.1	39.723	762.5
37.159	866.9	36.303	833.9	35.631	796.5	36.501	756.9
29.058	860.6	28.823	825.8	28.787	787.5	31.189	747.6
24.623	856.5	25.585	822.1	26.176	783.7	28.076	742.1
20.902	853.6	21.832	818.0	23.069	779.6	24.114	735.0
17.499	850.1	17.900	813.6	17.776	771.5	21.271	729.4
14.504	847.0	13.417	808.5	15.730	768.2	18.187	722.9
10.122	842.5	9.724	803.6	12.124	761.4	14.759	715.0
5.685	837.2	6.399	799.0	9.113	756.1	12.255	708.9
3.921	834.9	5.864	798.4	8.469	754.8	11.885	708.0

Table 3

Potassium iodide — 20 g/l (P, MPa; T, K; ρ, kg/m³)

P	ρ	P	ρ	P	ρ	P	ρ
$T=298.15$		$T=348.15$		$T=373.15$		$T=398.15$	
39.619	1027.6	39.683	1005.8	39.619	990.6	39.601	971.5
35.277	1026.1	35.701	1004.1	35.671	988.4	33.652	968.7
27.254	1022.4	27.059	1000.3	27.776	984.4	24.304	963.5
23.205	1020.5	19.921	997.0	22.759	981.9	20.361	961.7
18.634	1018.2	14.652	994.9	18.754	980.0	16.178	959.6
14.107	1016.2	9.979	992.5	14.428	977.9	12.250	958.1
11.223	1015.0	6.850	991.2	9.169	975.4	7.424	955.4
4.350	1012.5	2.984	989.1	4.717	973.5	3.196	953.0
0.369	1010.6	0.382	987.6	0.673	971.7	0.611	951.5
				0.1010	971.5	0.2314	951.3
$T=423.15$		$T=448.15$		$T=473.15$		$T=498.15$	
39.611	951.1	39.186	927.5	39.676	903.8	39.427	876.9
34.602	948.2	34.104	924.5	36.258	901.5	35.184	873.8
27.150	943.8	24.888	919.0	27.680	895.5	26.602	866.5
20.711	940.1	21.166	916.6	22.927	892.2	21.833	862.8
15.707	937.6	18.122	915.0	17.389	888.1	17.848	859.7
12.418	935.8	13.908	912.9	13.150	884.9	13.417	856.2
8.379	933.7	8.680	910.1	9.142	882.0	9.569	853.3
4.726	931.8	5.331	908.4	5.353	879.2	6.311	850.5
0.943	929.6	1.299	906.0	1.940	876.6	2.808	846.9
0.4745	929.4	0.8897	905.8	1.550	876.4	2.543	846.7
$T=523.15$		$T=548.15$		$T=573.15$		$T=598.15$	
39.701	848.7	39.702	814.6	39.650	777.1	39.847	735.1
37.334	846.6	36.350	811.5	36.403	773.6	37.704	732.0
29.240	838.8	28.824	803.0	30.254	764.9	31.619	720.8
25.750	835.5	25.684	798.8	26.250	758.6	27.421	711.9
22.092	831.4	22.133	794.4	22.633	752.9	24.376	704.8
19.234	828.6	18.200	789.5	18.954	746.5	21.322	696.9
16.046	825.4	14.159	784.2	15.402	739.8	18.576	689.4
8.450	817.5	10.211	778.5	12.066	733.7	15.680	680.7
4.383	811.9	6.350	772.4	8.875	727.1	12.304	669.4
3.965	811.2	5.931	771.5	8.566	726.5	12.020	668.5

Table 3 (*continued*)

P	ρ	P	ρ	P	ρ	P	ρ
$T = 298.15$		$T = 348.15$		$T = 373.15$		$T = 398.15$	
39.702	1050.4	39.698	1027.0	39.168	1010.3	39.651	992.1
36.230	1048.8	36.350	1025.7	33.879	1008.2	35.632	990.1
28.333	1045.0	26.236	1021.0	22.818	1003.1	26.033	985.2
24.584	1043.2	20.198	1018.5	18.354	1000.9	21.080	982.5
21.200	1041.8	17.004	1016.7	14.833	999.3	15.574	980.0
15.253	1038.7	13.611	1015.1	10.687	997.2	12.751	978.6
9.189	1036.0	8.408	1012.6	6.750	995.0	8.952	976.4
4.007	1033.5	3.009	1009.9	3.624	993.6	3.221	973.5
0.352	1031.8	0.350	1008.5	0.587	991.9	0.840	972.0
				0.1005	991.6	0.2301	971.5
$T = 423.15$		$T = 448.15$		$T = 473.15$		$T = 498.15$	
39.675	971.4	39.621	949.0	39.604	923.9	39.570	896.3
35.123	968.8	34.300	945.7	35.353	921.0	36.837	894.2
25.334	963.2	23.951	939.0	25.557	914.4	29.078	888.3
20.708	960.6	19.187	936.2	21.458	911.7	23.711	884.0
15.727	958.6	15.006	933.6	17.605	909.1	20.504	881.2
9.442	954.4	10.889	931.1	13.177	905.8	15.950	878.1
7.134	953.8	7.924	929.2	9.316	903.2	11.703	874.5
2.903	950.7	5.108	927.4	5.650	900.6	6.869	870.6
0.962	949.5	1.369	924.5	1.897	898.0	3.146	867.7
0.4719	949.1	0.8848	924.3	1.542	897.6	2.529	867.2
$T = 523.15$		$T = 548.15$		$T = 573.15$		$T = 598.15$	
39.647	868.1	39.658	835.0	39.681	798.7	39.750	756.2
36.353	865.0	36.117	831.8	36.707	794.4	37.172	752.2
28.224	857.7	28.628	823.5	30.273	785.6	30.873	740.5
23.653	853.4	24.563	818.6	26.801	780.2	27.666	733.9
19.612	849.1	20.451	813.4	22.637	774.1	23.723	724.7
15.163	844.9	16.398	808.0	18.590	767.3	21.009	718.2
11.051	840.7	12.751	802.6	14.846	761.0	18.841	712.0
7.240	836.0	9.749	798.0	11.337	754.5	16.250	704.6
4.354	832.2	6.251	792.1	8.842	749.2	12.387	692.7
3.944	831.4	5.898	791.5	8.518	748.7	11.954	691.1

Potassium iodide — 50 g/l (P, MPa; T, K; ρ, kg/m³)

is no more than ±0.02°. The error of pressure measurements, taking into account the corrections for the oil, mercury and studied liquid level differences in capillary tubes, is no more than ±0.05%. The piesometer volume with regard to all the necessary corrections is determined with an error not more than ±0.02%. The total mean-square error in density measurements is ±0.065%. Experimental procedure is described in detail in [1]. The results of measuring saturated water vapour density and and pressure at a temperature up to 598.15 K correlate with the data presented in [2] within the limits of estimated errors. The density data for solutions containing only 20 and 50 g of salt per litre are given in Tables 1, 2 and 3. The pressure measurements of saturated vapours were carried out on experimental isotherms entering the two-phase zone of state variables from the side of the liquid. The density values of the saturated liquid at temperature 373.15—598.15 K were obtained by a graphical extrapolation of experimental isotherms according to the corresponding pressure of saturation (Tables 1, 2, 3; the final line). Experimental data on vapour pressure of P_s of KCl, KBr, and KI up to 270, 500, 600 g/l respectively and on density ρ_s up to 50 g/l are given by equations:

$$P_s = P_w 10^{ac} \tag{1}$$
$$\rho_s = \rho_w + (a_0 + a_1 T + a_2 T^2) C \tag{2}$$

where P_w and ρ_w are the pressure of saturated vapour, MPa, and water density, kg/m³ [2]; C is the concentration of the solution, g/l, according to mass; a, a_0, a_1, a_2, n, w are the coefficients the values of which for certain electrolytes are given in Table 4; T is the temperature, K.

Equations (1) and (2) describe the experimental data with an error not more than 0.2%. Isotherm of 598.15 K for KCl and KBr is an exception. Here the error of equation (2) reaches 1.5 and 0.8% respectively.

Table 4.

Denomination	$a \times 10^4$	n	a_0	$a_1 \times 10^2$	$a_2 \times 10^5$
Potassium chloride	−1.52983	1.05	1.316	−0.3874	0.528
Potassium bromide	−0.87779	1.09	1.411	−0.3708	0.461
Potassium iodide	−0.44753	1.13	1.463	−0.3652	0.400

REFERENCES

1. Rivkin S. L. and Akhundov T. S. (1962): *Thermoenergetics*, No. 1, pp. 57-65.
2. Vukalovich M. P., Rivkin S. L., and Aleksandrov A. A. (1969): 408 p. Tables of thermophysical properties of water and water vapour.

11*

Isochoric Heat Capacity of Aqueous Sodium Chloride and Sodium Hydroxide Solutions

Kh. I. AMIRKHANOV, M. M. BOCHKOV,
V. I. DVORYANCHIKOV, B. A. MURSALOV, and G. V. STEPANOV

*Institute of Physics of the Daghestan Branch of the USSR
Academy of Sciences, Makhachkala, USSR*

The isochoric heat capacity of liquids and gases is directly related to internal energy and therefore opens wide possibilities for studying them both theoretically (hydration, association, etc.) and practically (calculation of necessary thermodynamic characteristics).

The present work discusses a study of isochoric heat capacity and density of aqueous solution of sodium chloride and sodium hydroxide conducted at: concentration 1, 5, 10 and 20 weight percent; temperature 373-773 K; and specific volume $V = 0.901 \times 10^{-3} — 2.393 \times 10^{-3}$ m^3/kg.

The measurements were made in the high-temperature adiabatic calorimeter by the method of continuous heating along isochores. The calorimeter developed by one of the authors utilizes a high-sensitive semiconducting thermocouple (Cu_2O) placed in the clearance between the calorimeter vessel and the thick-wall membrane (for details see [1, 2]). The thermocouple monitors the adiabatic conditions and at the same time transmits pressure to the more thick external membrane.

During measurements the adiabatic conditions and thermostating of calorimeter system were automatically maintained with the aid of temperature controller BPT-3. The temperature was measured with compact platinum resistance thermometer HTC-10 (calibration in IPTS-68 with an error of 0.01 K). Solutions of necessary concentrations were prepared from material of mark ЧДА and using bidistillate.

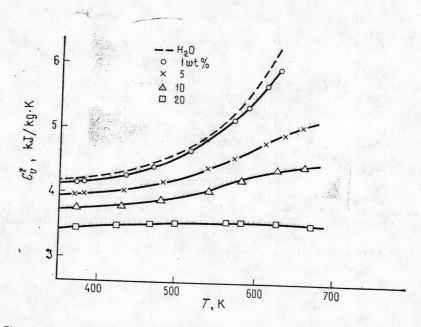

Fig. 1. Temperature dependence of isochoric heat capacity of NaCl-H₂O system in a two-phase state

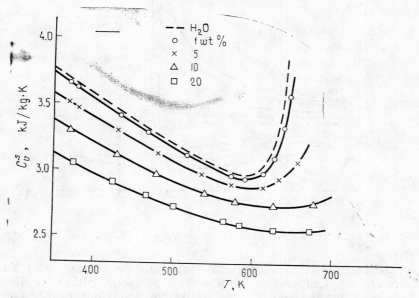

Fig. 2. Temperature dependence of isochoric heat capacity of NaCl-H₂O system in a single-phase state

The vicinity of the boundary curves was investigated from the liquid phase side when the system changed from a two-phase (heterogeneous) C_v^t to a single-phase (homogeneous) C_v^s state. The total error in determining C_v amounted to 0.3-2.5% in different ranges of the constitution diagram.

Experimental results of aqueous solutions of sodium chloride and water are given in Figs. 1 and 2, as dependences of C_v^t and C_v^s on temperature along the boundary curves are also shown.

An analysis of the behaviour of heat capacity shows that C_v^t increases with temperature. However, the nature of variation of heat capacity is ambiguous: in more concentrated solutions this increase is essentially small. In the single-phase range a heat capacity minimum is observed on the phase equilibrium curve. This, typical for water property, is more distinct in solutions with 1 and 5 mass % concentration and disappears in solutions with large concentration. Obviously, this is associated with the conservation of residual structures in the solution, which conform to pure water, and is due to the destructive effect of the dissolved ions.

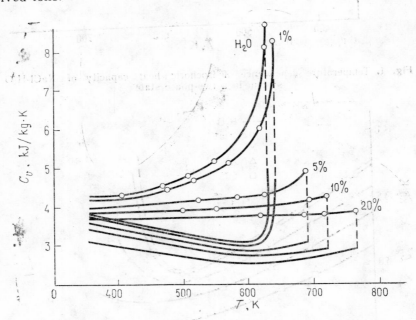

Fig. 3. Dependence of C_v on temperature of NaOH-H$_2$O system on boundary curves

Fig. 3 illustrates dependences of heat capacity of water and NaOH-H$_2$O solutions on temperature on the liquid-vapour phase equilibrium line. As seen from the figure, the qualitative behaviour of C_v^t and C_v^s of the NaOH—H$_2$O system is analogous to that of sodium chloride

solutions. The dotted line shows the difference between the heat capacity in the homogeneous and heterogeneous states at high temperatures, which significantly decreases with increase in concentration of NaOH.

The experimental technique enables the temperature dependence of density on the phase equilibrium curve to be determined correct to 0.06-0.2% after fixing the jump of heat capacity. The results obtained for density do not contradict the reported PVT data [3].

The enthalpy and entropy of solutions were calculated using the experimental and literature data on saturated vapour pressure [3, 4]. The calculations were made by the equation:

$$H = H_0 + \int_{T_1}^{T_2} \left(\frac{C_s}{T} - V \frac{dP}{dT} \right) dT;$$

$$S = S_0 + \int_{T_1}^{T_2} \frac{C_s}{T} dT$$

As point of control was taken the temperature 273 K to which corresponds the zero entropy of water [1]. The data on isochoric heat capacity and density were extrapolated to 273 K and a zero value was assigned to thermodynamic functions: entropy and enthalpy at this point.

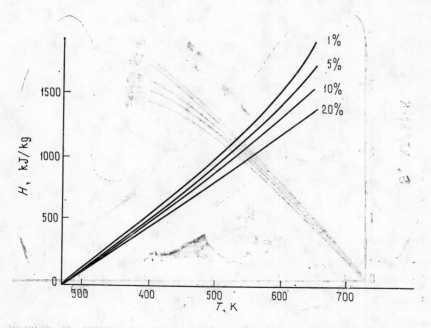

Fig. 4. Enthalpy of aqueous NaCl solutions at different concentrations

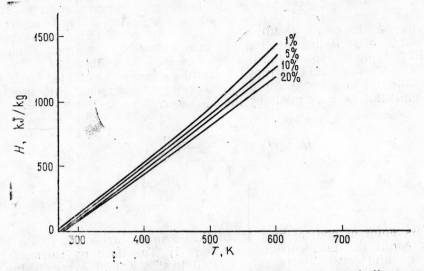

Fig. 5. Enthalpy of aqueous NaOH solutions at different concentrations

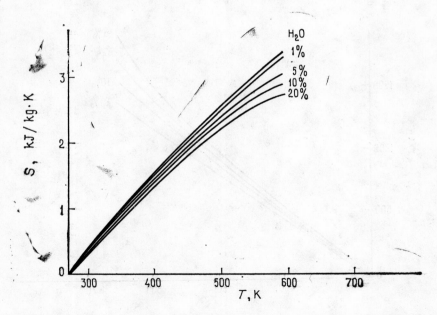

Fig. 6. Entropy of NaOH-H₂O system versus temperature at different concentrations

The dependences between the computed values of enthalpy and temperature at different concentrations of the systems are shown in Figs. 4 and 5. As seen, enthalpy grows with temperature, but with the increase in the electrolyte concentration this growth slows down.

Figure 6 shows how enthalpy along the boundary curves of sodium hydroxide varies with temperature; this has been compared with water. As is seen, the additions of alkalies do not qualitatively change the behaviour of enthalpy, but mainly affect its quantitative value.

Thus, the results obtained enable us to conclude that in concentrated solutions the variation in the above given thermodynamic functions is appreciably less than in dilute solutions.

REFERENCES

1. Kh. I. Amirkhanov, G. W. Stepanov, and B. G. Alibekov. Isochoric heat capacity of water and vapour. Publishing House of the Daghestan branch of the USSR Academy of Sciences, Makhachkala, 1969.
2. B. A. Mursalov, M. M. Bochkov, and G. V. Stepanov. The thermal physical properties of liquid and gases. Publishing House of the Daghestan branch of the USSR Academy of Sciences, Makhachkala, 1979, 70-74.
3. I. Kh. Khaibullin and N. M. Borisov. The experimental study of thermal properties of water and vapour solution of sodium chloride and calcium. *Thermal physics of high temperatures,* 1966, v. 4, N 4, 518-523.
4. W. W. Wospennikov. The study of thermodynamical properties of sodium hydroxide water system. Autoreferat of dissertation, 1980.

Behavior of Density and Viscosity
of Water + Alcohol Mixtures

HIRONOBU KUBOTA, YOSHIYUKI TANAKA,
and TADASHI MAKITA

*Department of Chemical Engineering, Kobe University,
Kobe 657, Japan*

HIROSHI KASHIWAGI

*The Graduate School of Science and Technology,
Kobe University, Kobe 657, Japan*

ABSTRACT

The density and the viscosity of binary mixtures of methanol, ethanol, 1-propanol, and 2-methyl-2-propanol with water, and methanol with 2-methyl-2-propanol were investigated as functions of temperature, pressure, and composition. The density was measured by a high-pressure burette method in the temperature range 283.15 to 348.15 K and at pressures up to 200 MPa. The viscosity measurements were performed by a falling-cylinder method in the same temperature range but at pressures up to 120 MPa. The uncertainties in the density and the viscosity are estimated to be less than 0.09 and 2%, respectively. The isotherms of both properties of the mixtures at each composition can be represented as a function of pressure by Tait equation type expressions. The composition dependence at constant temperature and pressure exhibits a maximum in viscosity and a minimum in thermodynamic properties such as a partial molar volume. Some theoretical interpretations on water+alcohol mixtures are discussed.

1. INTRODUCTION

Water is one of the most strongly and most regularly hydrogen-bonded liquids [1], and has various unusual properties in comparison with "normal" liquids. Aqueous mixtures of nonelectrolytes exhibit some anomalies in the composition dependence of thermophysical properties. In particular water+alcohol mixtures are some of the most interesting systems for investigating complicated molecular interactions between unlike components because alcohols have hydrophilic and hydrophobic properties. High pressure studies on the mixtures provide valuable information on the static and dynamic molecular motion. The purpose of this work is to investigate the behavior of the physical properties of binary aqueous mixtures containing alcohols under pressure. The density and the viscosity of four mixture systems have been measured in the temperature range 283.15 to 348.15 K at pressures up to 200 MPa for density and 120 MPa for viscosity. The systems measured are completely miscible mixtures: water+methanol (MeOH) [2], +ethanol (EtOH) [3], +1-propanol (n-PrOH), and +2-methyl-2-propanol (t-BuOH). The density of MeOH+t-BuOH mixtures was also measured at 323.15 K and up to 200 MPa. The experimental results for both properties and derived thermodynamic properties are discussed as functions of pressure, temperature and composition.

2. EXPERIMENTAL

The apparatus and experimental procedures were essentially the same as used in earlier works [2-4]. Only the principles of the method are described here.

Density Measurements. Density was determined by measuring the change in length of a mercury column in a high-pressure burette, connected to the sample liquids of a known weight in a pressure vessel. The position of a magnetic float on the surface of the mercury column was determined outside the burette by a linear differential transformer. The pressure vessel consisted of coaxial double cylinders: the inner cylinder was a thin-walled sample container of volume 83 cm³. The uncertainty of the measured density is estimated to be less than 0.09%.

Viscosity Measurements. Viscosity was measured by a falling-cylinder viscometer. A blue glass plummet with hemispherical ends fell through a smoothly bored Pyrex glass tube placed coaxially in a pressure vessel with a pair of optical windows. The falling time of the plummet was determined by a time-interval counter using a He-Ne gas laser beam which passed through the windows to a phototransistor. The measurements were performed on a relative basis by using the reference viscosity values for water [5, 6] and the "viscosity standard liquid" JS 5 provided from the National Research Laboratory of Metrology, Japan. The uncertainty of the measured viscosity is estimated to be less than 2%.

Materials. The pure alcohols were reagent-grade materials of the highest purity supplied from commercial sources, stated to be better

than 99.5%. The alcohols and water were purified more than twice by fractional distillation and were degassed by crystallization in a vacuum. The mixtures were prepared by weighing. The uncertainty of the mole fraction is less than 0.0001.

3. RESULTS

We briefly summarize the behavior of the density and the viscosity of pure components. Figs. 1 and 2 illustrate the pressure dependence of

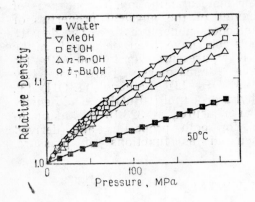

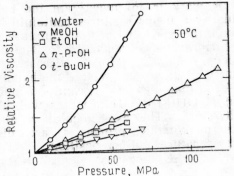

Fig. 1. Pressure dependence of the relative density of water and alcohols

Fig. 2. Pressure dependence of the relative viscosity of water and alcohols

relative quantities, the ratio of the density and the viscosity at high pressures to those at atmospheric pressure at 323.15 K. Both properties increase with increasing pressure except the viscosity of water at temperatures below 298.15 K where the viscosity decreases at the initial compression. We compared the experimental data with literature values for water, methanol, ethanol, and their mixtures [2, 3]. The present results are in good agreement with reliable data in the literature.

Composition Dependence. The molar volume and the viscosity of each mixture at 298.15 K and at atmospheric pressure are shown in Figs. 3 and 4, respectively. Some literature values [7-12] are also plotted in the figures. The agreement with the present results is satisfactory within the combined uncertainty. Since water+alcohol mixtures are highly associated owing to the hydrogen bonds, their composition dependence of the viscosity deviates greatly from ideality.

Figs. 5 and 6 show the isobaric variation with composition of the molar volume and the viscosity of water+t-butanol mixtures at 323.15 K, respectively. Behaviors similar to the figures are found in the other mixture systems. The isobars of the molar volume against mole fraction are slightly concave, although this tendency is not distinct because of the small scale of the figure. The excess molar volume

V^E, which is a measure of non-ideality of real mixtures, was evaluated. V^E is always negative over the entire experimental range and becomes less negative at the higher pressures. The isobars of V^E become more asymmetric in their composition dependence as the hydrophobic group in the alcohol molecule becomes larger.

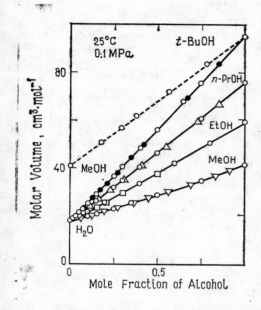

Fig. 3. Composition dependence of the molar volume of water + alcohol mixtures at 298.15 K and 0.1 MPa: ○ This work ▽; △ Mikhail [8, 9]; ● Nakanishi [10]; □ Yusa [7]

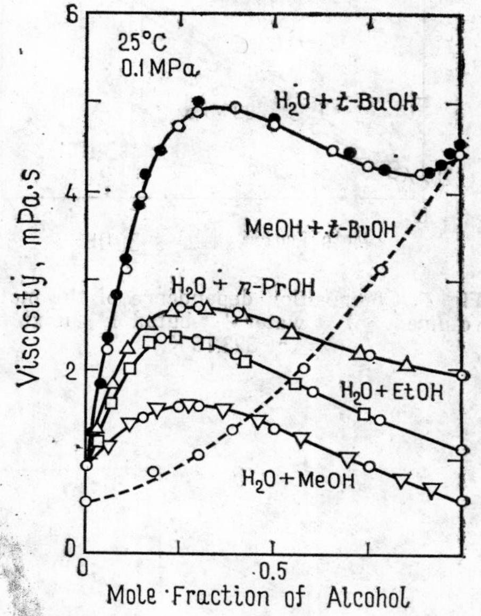

Fig. 4. Composition dependence of the viscosity of water + alcohol mixtures at 298.15 K and 0.1 MPa: ○ This work ▽; △ Mikhail [8, 9]; ● Broadwater [12]; □ Kikuchi [11]

Each isobar of the viscosity at low temperatures and at low pressures has a maximum near a mole fraction of 0.3 to 0.4. As shown in Fig. 4, only the isobars of t-butanol solution have two extrema: a maximum near a mole fraction of 0.3 and a shallow minimum near 0.9. This minimum diminishes gradually as temperature increases. Since the freezing point of t-butanol is very low in the alcohols studied, low molecular mobility of t-butanol-rich mixtures would affect the viscosity.

Pressure Dependence. Both the density and the viscosity increase with increasing pressure and decrease with increasing temperature. The isotherms of the density against pressure appear orderly as a function of composition, whereas those of the viscosity intersect with one another as shown in Fig. 7 for t-butanol+water mixtures at 323.15 K. The strong non-ideality in the composition dependence of the viscosity at low pressures results in such a complicated shape in the figure.

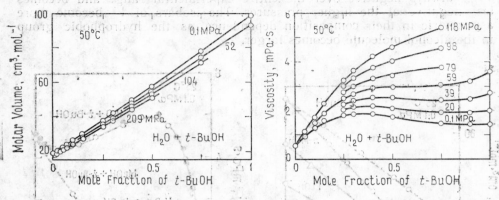

Fig. 5. Composition dependence of the molar volume of water + t-butanol mixtures at 323.15 K

Fig. 6. Composition dependence of the viscosity of water + t-butanol mixtures at 323.15 K

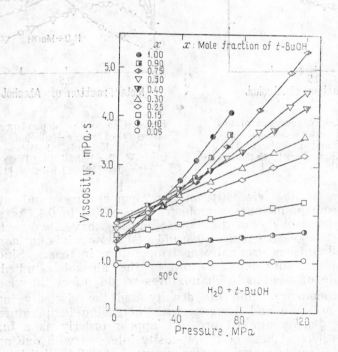

Fig. 7. Pressure dependence of the viscosity of water + t-butanol mixtures at 323.15 K

The Tait equation was used to represent the density ρ along isotherms under pressure:

$$(\rho-\rho_0)/\rho = C \ln[(B+p)/(B+p_0)] \tag{1}$$

where ρ_0 is the density at a reference pressure p_0, often taken at atmospheric pressure, and B and C are empirical constants dependent on the mixture and temperature. This equation could reproduce the present results within an average deviation of 0.10% and a maximum deviation of 0.20% over the entire experimental range.

An expression similar to the Tait equation was tested to correlate the viscosity η with pressure:

$$(\eta-\eta_0)/\eta = A \ln[(D+p)/(D+p_0)] \tag{2}$$

where η_0 is the viscosity at p_0. This isothermal equation was found to give a good representation of the present results with an average deviation less than 0.9% and a maximum deviation less than 2% except for water at low temperatures.

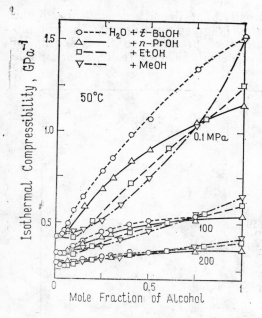

Fig. 8. Composition dependence of the isothermal compressibility of water + + alcohol mixtures at 323.15 K

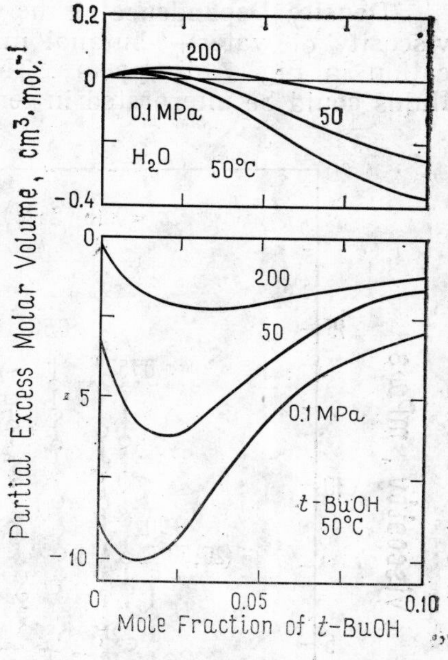

Fig. 9. Composition dependence of the partial excess molar volume of water + t-butanol mixtures at 323.15 K

Isothermal Compressibility and Partial Molar Volume. The isothermal compressibility $\beta_T = \rho^{-1}(\partial\rho/\partial p)_{T,x}$ was calculated by differentiating the Tait equation with pressure. The dependence of the isothermal compressibility on composition at 323.15 K is shown in Fig. 8. The isothermal compressibility decreases with increasing pressure and increases

with increasing temperature for each of the mixtures except for pure water below 319 K where $(\partial \beta_T / \partial T)_{p,x}$ is negative as is well-known. A definite minimum is observed near a mole fraction of 0.06 to 0.15 at low temperatures and low pressures, but these extrema diminish gradually as the temperature and pressure become higher.

Partial molar quantities of a system are adequate intensive properties which provide information on the contribution of each component to the extensive properties of the system. The partial molar volume $\bar{V}$ was evaluated for the present results. The partial excess molar volume of water+t-butanol $\bar{V} - V^\circ$ is illustrated in Fig. 9, where V° is the molar volume of the pure component. Each isobar of alcohol has a minimum at a low alcohol concentration, while that of water has a small maximum at the same concentration. The mole fraction of the extrema is nearly equal to that for the isothermal compressibility. The minimum of alcohol becomes less negative with increasing pressure. The composition of the extrema shifts slightly to higher alcohol-fraction at high pressures.

Density Dependence of the Viscosity. The density dependence of the viscosity of water+t-butanol mixtures is given in Fig. 10. The complication in Fig. 7 disappears. The effect of pressure on the viscosity of fluids could be interpreted in terms of density. Therefore, it is reasonable

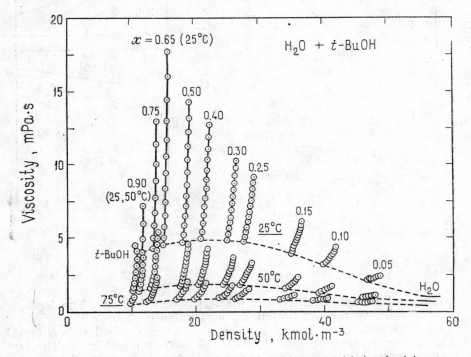

Fig. 10. Density dependence of the viscosity of water + t-butanol mixtures

that the change in viscosity with pressure is expressed as a function of density instead of pressure from a theoretical and an empirical point of view. Using the present results, we have examined the following expressions based on the hard-sphere model which relates the viscosity to the density of liquids:

(1) Rough hard-sphere model [13]

$$1/\eta = 0.02358\,(\pi mkT)^{-1/2}\sigma^2[3.7043 - 6.3355n\sigma^3 + 2.6718\,(n\sigma^3)^2]\tag{3}$$

where η is in mPas, m is the molecular mass in g, n is the number density in cm^{-3}, σ is the hard-sphere diameter in cm, and β is an adjustable parameter. The parameters σ and β were simultaneously determined by statistical regression of the data along isotherms. The agreement of the model with the experimental data of higher-alcohol-rich mixtures is not satisfactory. This suggests a limitation of the RHS model for liquids composed of aspherical molecules [14].

(2) Free volume expression by Dymond [15]

$$\ln(\eta V^{2/3}/\sqrt{MT}) = E + F[V_i/(V - V_i)]\tag{4}$$

where T is absolute temperature. We determined the coefficients E, F, and V_i as constants which depend on temperature and composition for each mixture. This equation reproduces the present results within average deviations of 0.04 to 1.4%. The variation of V_i with composition was not the systematical one expected from the concept of free volume. It is difficult to obtain physically-meaningful coefficients for associated liquids. Therefore, V_i should be regarded as one of the constants of Eq. (4) obtained by statistical regression.

4. DISCUSSION

No one has yet proposed a quantitative theory of aqueous solutions of nonelectrolytes [1]. Therefore, we can discuss only a qualitative interpretation on the variation of thermophysical properties with temperature, pressure, and composition.

(1) Alcohol molecules in aqueous solutions give strong influence on the water structure. When a small amount of alcohol is added to pure water or a water-rich mixture, the alcohol molecules substitute for water ones. The hydrophobic group of the alcohol can be accommodated interstitially in the bulky hydrogen-bond network. The presence of hydrophobic molecules enhances the degree of hydrogen bonding of pure water around it. This structure promotion results in a minimum in composition dependence of the pVT-relations. Further addition of alcohol, in contrast, leads to the breaking of the hydrogen-bond network and the larger compressibility of alcohol becomes dominant. Restricted molecular motional freedom due to the hydrophobic hydration and the shape of alcohol affect the variation of the viscosity with composition. It is found that t-buthyl group has an adequate size to develop the hydrogen bond.

(2) At temperatures below 313 K and at pressures below 200 MPa, both temperature and pressure have a parallel effect on water structure:

the initial compression leads to a significant distortion and/or disruption of the hydrogen-bond network. On the other hand, at higher temperatures and pressures, their effects are just the opposite due to increased packing of the molecules like normal molecular liquids [16].

(3) Alcohols also behave as normal liquids, although their shape and size significantly affect the thermophysical properties of the system.

The above relations affect the molecular interaction of aqueous solution in a complicated manner, which results in the anomalous behavior of thermophysical properties specific to aqueous solutions of nonelectrolytes.

The authors would like to express their appreciation to Messrs. M. Morikawa and H. Fujiwara for their careful efforts in the measurements.

REFERENCES

1. Rowlinson, J. S., Swinton, F. L. (1977): "Liquids and Liquid Mixtures", Butterworth, London.
2. Kubota, H., Tsuda, S., Murata, M., Yamamoto, T., Tanaka, Y., Makita, T. (1979): *Rev. Phys. Chem. Jpn.,* **49**, 60.
3. Tanaka, Y., Yamamoto, T., Satomi, Y., Kubota, H., Makita, T. (1977): *Rev. Phys. Chem. Jpn.,* **47**, 12.
4. Kubota, H., Tanaka, Y., Makita, T. (1983): *J. Soc. Mat. Sci. Jpn.,* **33**, 107 (in Japanese).
5. IAPS (1974): "Dynamic viscosity of water substance", 8th ICPS.
6. Watson, J. T. R., Basu, R. S., Sengers, J. V. (1980): *J. Phys. Chem. Ref. Data,* **9**, 1225.
7. Yusa, M., Mathur, G. P., Stager, R. A. (1977): *J. Chem. Eng. Data,* **22**, 32.
8. Mikhail, S. Z., Kimel, W. R. (1961): *J. Chem. Eng. Data,* **6**, 533.
9. Mikhail, S. Z., Kimel, W. R. (1963): *J. Chem. Eng. Data,* **8**, 323.
10. Nakanishi, K., Kato, N., Maruyama, M. (1967): *J. Phys. Chem.,* **71**, 814.
11. Kikuchi, M., Oikawa, E. (1967): *Nippon Kagaku Zasshi,* **88**, 1259 (in Japanese).
12. Broadwater, T. L. (1970): *J. Phys. Chem.,* **74**, 3802.
13. Chandler, D. (1975): *J. Chem. Phys.,* **62**, 1358.
14. Kashiwagi, H., Makita, T. (1982): *Int. J. Thermophys.,* **3**, 289.
15. Dymond, J. H., Brawn, T. A. (1977). Proc. 7th Symp. Thermophys. prop., pp. 660, ASME, New York.
16. DeFries, T., Jonas, J. (1977): *J. Chem. Phys.,* **66**, 896.

Phase Equilibrium for the Ternary System Methanol-Water-Lithiumbromide

K.-F. KNOCHE and W. RAATSCHEN

*Rheinisch-Westfälische Technische Hochschule Aachen,
Lehrstuhl für Technische Thermodynamik, Schinkelstr. 8,
D-5100 Aachen, Federal Republic of Germany*

SUMMARY

For the ternary working fluid methanol(1)-water(2)-lithiumbromi-de(3) phase equilibrium has been calculated by a thermodynamic consistent method using Gibbs free energy. The correlation procedure is based on the coefficients of the three possible binary mixtures and a term taking the real behaviour of the ternary system into consideration. Vapor phase concentration has been evaluated as a function of liquid phase composition and temperature.

INTRODUCTION

During the last 10 years research work and development of heat pump systems have increased. Especially for the working pair H_2O-LiBr and CH_3OH-LiBr the advantages and disadvantages have been discussed by several authors [1], [2], [3], [4]. The ternary mixture of methanol, water and lithiumbromide shows a better absorption compared with binary systems and therefore promises a better chance for application in industrial absorption heat pumps [5].

Based on a proposal of Hala [6], the activity coefficients for the three binary mixtures have been evaluated. An additional term was added, taking the interaction forces of the ternary system into account.

The concentration ranges up to salt concentrations of about 60 wt% and temperatures up to 130°C, where vapor pressures exist below one bar.

To describe the activity coefficients of a mixture containing an electrolytic component and a non-electrolytic solvent, Hala proposed a generalized excess function ΔG^E composed of two parts;

$$\Delta G^E = nRT(Q_a + Q_b) \tag{1}$$

Q_a considers the long range forces of the electrolytical component and Q_b describes the phase equilibrium of non-electrolytical systems. If the Margules equation is used

$$Q_b = x_i x_j (A_{ji} x_i + A_{ij} x_j) \tag{2}$$

METHANOL(1)-LITHIUMBROMIDE(3)

Lithiumbromide is non-volatile. Therefore the binary mixture is characterized by an activity coefficient γ_1 of methanol

$$p_1 = \gamma_1 p_{01} x_1 \tag{3}$$

where p_1 denotes the partial pressure, p_{01} the vapor pressure of pure methanol and x_1 the analytical mole fraction of methanol in the liquid phase. Though Hala proposed to use ionic concentrations, defined by Eq. (4), we found negligible differences using stoichiometric concentrations. Therefore we renounced to work with stoichiometric or analytic mole fractions and activity coefficients

$$x_{1\pm} = \frac{x_1}{x_1 + 2x_3} \tag{4}$$

Vapor pressures can be correlated with sufficient accuracy by the Antoine equation

$$\ln \frac{p_{0i}}{p^+} = a + bT^{-1} + cT^{-2} + dT^{-3} \tag{5}$$

Table 1 represents Antoine vapor pressure constants for methanol(1) and water(2).

Table 1. Antoine vapor pressure constants of pure methanol and water [3]

	Methanol (1)	Water (2)
a	18.06388	18.2708
b/k	-2.84592×10^3	-3.42691×10^3
c/k^2	-3.74341×10^5	-3.28520×10^5
d/k^3	2.18867×10^7	9.92618×10^6
$p^+ = 1$ mbar		

For Q_a and Q_b Hala proposed the following equations

$$Q_a = c_{13}x_3^{1.5} + x_3 A_{31} \tag{6}$$

$$Q_b = x_1 x_3 (A_{31}x_1 + A_{13}x_3) \tag{7}$$

Now the Gibbs free excess energy change relates to

$$\Delta G^E = nRT(c_{13}x_{13}^{1.5} + x_3 A_{31} + x_1 x_3 (A_{31}x_1 + A_{13}x_3)) \tag{8}$$

For evaluating the three parameters A_{13}, A_{31} and c_{13}, we use the relationship

$$\Delta G^E = RT(n_1 \ln \gamma_1 + n_3 \ln \gamma_3) \tag{9}$$

With

$$\ln \gamma_1 = \frac{1}{RT} \frac{\partial}{\partial n_1} (\Delta G^E) \tag{10}$$

we get

$$\ln \gamma_1 = -0.5 c_{13}x_3^{1.5} + x^2{}_3 (A_{13} + 2x_1 (A_{31} - A_{13})) \tag{11}$$

For the correlation of vapor pressures 155 data points from Renz [3] were used. These data are of excellent quality and conform with our own measurements. Also data of pure methanol [3] were added. The liquid concentration ranges from pure methanol ($x_1 = 1$) up to the solubility limit at $x_1 \approx 0.64$ between 40 and 100°C. All the measured vapor pressures are below 1 bar. The three coefficients A_{13}, A_{31} and c_{13} are considered temperature dependent.

$$A_{13} = A_{131} + A_{132} \ln T$$

$$A_{31} = A_{311} + A_{312}T$$

$$c_{13} = c_{131} + c_{132}T$$

By minimizing the relative deviation of the partial pressure using a least square fitting, best approximation was obtained with

$A_{131} = 1287.92$	$A_{312} = 0.780671$
$A_{132} = -212.667$	$c_{131} = -500.658$
$A_{311} = -309.777$	$c_{132} = 1.29754$

The mean deviation

$$(p_{1, exp} - p_{1, calc})_{mean} = 4.38 \text{ mbar}$$

$$\left(\frac{p_{1,exn} - p_{1,calc}}{p_{1,exp}}\right)_{mean} = 1.71\%$$

is satisfactory, taking into account, that only 6 parameters are used for a pressure change from 0 up to 1000 mbar. With eight coefficients it is possible to reduce the mean relative deviation to 1.58%.

WATER(2)-LITHIUMBROMIDE(3)

The same correlation procedure has been used for $H_2O(2)$-$LiBr(3)$. 69 vapor pressure data of Renz [3] and of pure water [7] have been used

for the approximation. The activity coefficient of water is expressed by equation (12)

$$\ln \gamma_2 = -0.5c_{23}x_3{}^{1.5} + x^2{}_3(A_{23} + 2x_2(A_{32} - A_{23}))\tag{12}$$

With the same temperature dependence as it was used for the binary system of $CH_3OH(1)$-$LiBr(3)$, the following coefficients and deviations were obtained:

$A_{23} = A_{231} + A_{232}T$		$A_{32} = A_{321} + A_{322}\ln T$	
$A_{231} =$	22.4821	$A_{322} =$	72.2586
$A_{232} =$	51.7117×10^{-4}	$c_{231} =$	-97.0271
$A_{321} =$	-457.862	$c_{232} =$	0.185859

$$(\Delta p_2)_{mean} = 3.26 \text{ mbar} \sim \text{rel.} = 1.71\%$$

METHANOL(1)-WATER(2)

For the mixture of the two non-electrolytic solvents the three suffix Margules equation is used

$$\ln \gamma_1 = x^2{}_2 A_{12} + 2[(A_{21} - A_{12})x_1]\tag{13}$$
$$\ln \gamma_2 = x^2{}_1 A_{21} + 2[(A_{12} - A_{21})x_2]\tag{14}$$

A_{12} and A_{21} are functions of temperature. According to our measurements, the temperature dependence of A_{12} and A_{21} could be correlated quite well by equation (15).

$$A_{ij} = A_{ij1} + A_{ij2}{}^*T + A_{ij3}{}^*T^2\tag{15}$$

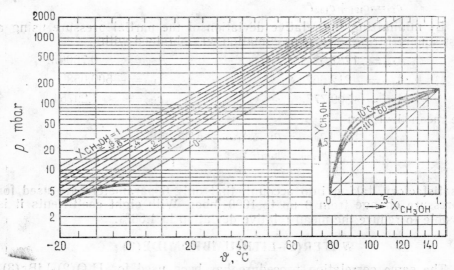

Fig. 1. Vapor pressure and vapor composition of CH_3OH-H_2O

Many literature data are available for the methanol(1)-water(2) system. For our selected data (Ref. 8) we obtained the following coefficients:

$$A_{121} = -0.10203 \times 10^2 \qquad\qquad A_{211} = -0.18931$$
$$A_{122} = 0.60965 \times 10^{-1} \qquad\qquad A_{212} = 0.48705 \times 10^{-2}$$
$$A_{123} = -0.84623 \times 10^{-4} \qquad\qquad A_{213} = -0.84042 \times 10^{-5}$$

with the deviations:

$$(\Delta p_{rel})_{mean} = 1.14\% ; \qquad (\Delta y_{1,\ rel})_{mean} = 1.51\%$$

Fig. 1 shows the $\lg p$, $1/T$-diagram with the corresponding x-y-behaviour.

THE TERNARY SYSTEM
METHANOL(1)-WATER(2)-LITHIUMBROMIDE(3)

Combining Q_a and Q_b of all three binary mixtures, we obtain the following approximation for the Gibbs excess enthalpy:

$$\frac{\Delta G^E}{nRT} = cx_3^{1.5} + (A_{31} + A_{32}) x_3 + x_1 x_3 (A_{31} x_1 + A_{13} x_3)$$
$$+ x_1 x_2 (A_{21} x_1 + A_{12} x_2) + x_2 x_3 (A_{32} x_2 + A_{23} x_3) \qquad (16)$$

The constant c has been correlated in such a way, that the vapor pressure deviations in both mixtures of $CH_3OH(1)$-$LiBr(3)$ and $H_2O(2)$-$LiBr(3)$ are minimized.

Using a quadratic function of temperature for the parameter A_{13} and A_{31}, the following coefficients and errors were obtained:

$$A_{131} = 0.19884 \times 10^3 \qquad\qquad A_{311} = -0.28168 \times 10^3$$
$$A_{132} = -0.56460 \qquad\qquad\qquad A_{312} = 0.82144$$
$$A_{133} = 0.29851 \times 10^{-3} \qquad\qquad A_{313} = -0.30954 \times 10^{-3}$$
$$c_1 = c_{121} = c_{131} = -360.0$$
$$c_2 = c_{122} = c_{132} = 0.92 \qquad\qquad c = c_1 + c_2 T$$

The mean relative error of the partial pressure is 2.74% for p_1. The coefficients for $H_2O(2)$-$LiBr(3)$ are

$$A_{231} = 0.87593 \times 10^3 \qquad\qquad A_{321} = -0.45871 \times 10^2$$
$$A_{232} = -0.41567 \times 10^1 \qquad\qquad A_{322} = -0.42424$$
$$A_{233} = 0.49938 \times 10^{-2} \qquad\qquad A_{323} = 0.12950 \times 10^{-2}$$

with a mean relative deviation for p_2 of 2.71%.

An empirical correction term with only one adjustable parameter is used to consider interactions between the three components. The final equation for ΔG^E is

$$\frac{\Delta G^E}{nRT} = cx_3^{1.5} + (A_{31} + A_{32}) x_3 + x_1 x_3 (A_{31} x_1 + A_{13} x_3)$$
$$+ x_1 x_2 (A_{21} x_1 + A_{12} x_2) + x_2 x_3 (A_{32} x_2 + A_{23} x_3)$$
$$+ x_1 x_2 x_3 (0.5\ \Sigma A_{ij} - c_{123}) \qquad (17)$$

343

After differentiation γ_1, γ_2 and γ_3 relate to

$$\ln \gamma_1 = -0.5\,cx_3^{1.5} + 2\,A_{21}x_1x_2 + 2\,A_{31}x_1x_3\,(1-x_1)$$
$$+ x_2x_3\,(0.5\,\Sigma A_{ij} - c_{123}) - 2\,A_{12}x_1x_2^2 - 2\,A_{13}x_1x_3^2$$
$$- 2\,A_{21}x_1^2x_2 - 2\,A_{32}x_2^2x_3 + A_{13}x_3^2 + A_{12}x_2^2$$
$$- 2\,x_1x_2x_3\,(0.5\,\Sigma A_{ij} - c_{123}) - 2\,A_{23}x_2x_3^2 \qquad (18)$$

$$\ln \gamma_2 = -0.5\,cx_3^{1.5} + 2\,A_{12}x_1x_2 + x_1x_3\,(0.5\,\Sigma A_{ij} - c_{123})$$
$$+ 2\,A_{32}x_2x_3 - 2\,A_{12}x_1x_2^2 - 2\,A_{13}x_1x_3^2 - 2\,A_{23}x_2x_3^2$$
$$- 2A_{31}x_1^2x_3 - 2A_{21}x_1^2x_2 - 2A_{32}x_2^2x_3 + A_{23}x_3^2$$
$$+ A_{21}x_1^2 - 2\,x_1x_2x_3\,(0.5\,\Sigma A_{ij} - c_{123}) \qquad (19)$$

$$\ln \gamma_3 = 1.5\,cx_3^{0.5} + A_{31} + A_{32} + A_{31}x_1^2 - 2\,A_{31}x_1^2x_3 - 0.5\,cx_3^{1.5}$$
$$+ 2\,A_{13}x_1x_3 - 2\,A_{13}x_1x_3^2 - 2\,A_{21}x_1^2x_2 - 2\,A_{12}x_1x_2^2$$
$$+ A_{32}x_2^2 - 2\,A_{32}x_2^2x_3 + 2\,A_{23}x_2x_3 - 2\,A_{23}x_2x_3^2$$
$$+ x_1x_2\,(0.5\,\Sigma A_{ij} - c_{123}) - 2\,x_1x_2x_3\,(0.5\,\Sigma A_{ij} - c_{123}) \qquad (20)$$

While the liquid phase of this ternary system consists of all three components the vapor phase only contains methanol and water.

Experimental data of the ternary working fluid have been obtained in a dynamic phase equilibrium apparatus with an integrated "Cottrell-Pump". More than 80 $p.T.x.y$-data have been measured in a concentration range up to salt concentrations of 50 weight %; the temperature range was 25° to 120°C with a maximum pressure of 1.4 bar.

With the experimental data it is only possible to evaluate activity coefficients of $CH_3OH\,(1)$ and $H_2O\,(2)$ according to equation (3).

$$py_i = \gamma_i x_i p_{0i} \qquad (21)$$

p is the total vapor pressure of the mixture and y_i is the mole fraction of the vapor phase. Following equations (18) and (19)

$$\ln \frac{\gamma_1}{\gamma_2} = A_{12}x_2^2 - A_{21}x_1^2 - 2\,x_1x_2\,(A_{12} - A_{21})$$

$$- x_3\{2\,A_{32}x_2 - 2\,A_{31}x_1 + x_3\,(A_{23} - A_{13})$$
$$+ (x_1 - x_2)\,(0.5\,\Sigma A_{ij} - c_{123})\} \qquad (22)$$

For the parameter c_{123} a linear function of temperature has been chosen

$$c_{123} = c'_{123} + c''_{123}T$$
$$c'_{123} = 530.98 \qquad\qquad c''_{123} = -1.3257$$

The mole fraction y_1 of methanol in the vapor phase can be evaluated by equation (23)

$$y_1 = \frac{\dfrac{\gamma_1}{\gamma_2}}{\dfrac{\gamma_1}{\gamma_2} + \dfrac{x_2}{x_1}\dfrac{p_{02}}{p_{01}}} \qquad (23)$$

The mean relative deviation for y_1 of 1.19% is quite satisfactory.

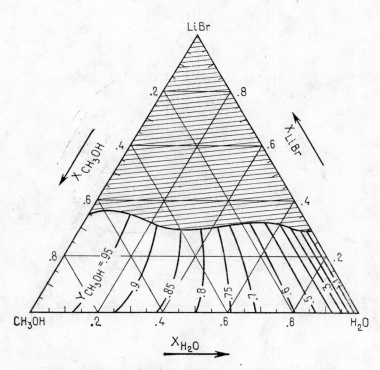

Fig. 2. Vapor composition of CH₃OH-H₂O-LiBr at 60°C

Fig. 2 and Fig. 3 show vapor compositions in the $CH_3OH(1)$-$H_2O(2)$-$LiBr(3)$-system as a function of liquid composition at 60° and 100°C respectively. At 60°C and constant ratio of $CH_3OH(1)$ and $H_2O(2)$ the vapor concentration is nearly independent of the salt concentration. With increasing temperature the affinity of lithiumbromide to water increases, therefore the vapor consists of nearly pure methanol at higher salt concentrations.

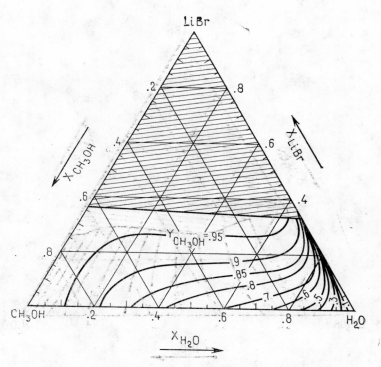

Fig. 3. Vapor composition of CH_3OH-H_2O-$LiBr$ at 100°C

SUMMARY

For the binary mixtures of $CH_3OH(1)$-$LiBr(3)$ and $H_2O(2)$-$LiBr(3)$ vapor pressures have been correlated by an equation for the free excess enthalpy proposed by Hala. With only 6 coefficients good agreement could be obtained with experimental data. The description of the vapor phase of the ternary system $CH_3OH(1)$-$H_2O(2)$-$LiBr(3)$ as a function of liquid composition and temperature was successful.

NOMENCLATURE:

n_i = number of moles of component i

n = sum of moles in the liquid phase

x_i = stoichiometric or analytical mole fraction of component i in the liquid phase

$x_{i\pm}$ = mole fraction defined by equation (3)

y_i = analytical mole fraction of component i in the vapor phase

γ_i = activity coefficient of component i

p_{0i} = vapor pressure of pure component i

p_i = partial pressure of component i

p = total pressure

R = universal gas constant

T = Kelvin temperature

Q_a = excess function (coulombic part)

Q_b = excess function (non-coulombic part)

ΔG^E = Gibbs free excess energy change

A_{ij} = empirical constants

c_{ij} = empirical constants

REFERENCES

1. Mc Neely, L. A.: Thermodynamic Properties of Aqueous Solutions of Lithium Bromide, ASHRAE, PH-79-3, No. 3.
2. Eichholz, H.-D.: Korrelation experimentell bestimmter Zustandsgrößen elektrolytischer Methanollösungen als Arbeitsmittel für Absorptionswärmepumpen mit einer Fundamentalgleichung, Forschungsberichte des Deutschen Kälte- und Klimatechnischen Vereins, DKV-Bericht Nr. 7. Stuttgart, 1982.
3. Renz, M.: Bestimmung thermodynamischer Eigenschaften wässriger und methylalkoholischer Salzlösungen, Forschungsberichte des Deutschen Kälte- und Klimatechnischen Vereins, DKV-Bericht Nr. 5, Stuttgart, 1980.
4. Löwer, H.: Thermodynamische und physikalische Eigenschaften der wässrigen Lithiumbromid-Lösung, Dissertation, Karlruhe, 1960.
5. Knoche, K. F., Raatschen, W.: Thermodynamic and Physical Properties of the Ternary System CH_3OH-H_2O-LiBr. Will be published as VDI-Paper, Proceedings of the International VDI-Seminar ORC-HP-Technology, ETH Zürich, 10.-12. Sept., 1984.
6. Hala, E.: Vapour Liquid Equilibria in Systems of Electrolytic Components, I. Chem. E. Symposium Series No. 32 (1969: Instn. chem. Engrs, London).
7. N.N.: VDI-Steam Tables, 6th Revised and Amplified Edition (8), Springer-Verlag Berlin, 1963.
8. Gmelin, J., Onken, U.: Vapour-liquid Equilibrium Data Collection of Aqueous — Organic Systems, Chemistry Data Series, Vol. I, Part 1.
8.1 Uchida, S., Ogawa, S. et al.: *Chem. Eng.* **17**, 191 (1953).
8.2 Othmer, D. F., Benenati, R. F.: *Ind. Eng. Chem.* **37**, 299 (1945).
8.3 Olevsky, V. M., Golubev, I. F.: *Tr. Giap, VYP,* Bd. 6, 45 (1956).
8.4 Butler, A. V., Thompson, D. W., Lennan, W. N.: *J. Chem. Soc.* (London) 1933, 674.
8.5 Ratcliff, G. A., Chao, K. C.: Can. *J. Chem. Eng.* **47**, 148 (1969).

Measurement of Thermal Conductivity of Tetrafluoropropanol-Water Mixtures

J. YATA, T. MINAMIYAMA, and H. KATAOKA

Department of Mechanical Engineering, Kyoto Institute of Technology, Matsugasaki, Sakyo-ku, Kyoto 606, Japan

ABSTRACT

Fluoroalcohol-water mixtures have recently attracted wide attention as the working fluids in the various types of the power plants. However, knowledge of the thermophysical properties of these mixtures is rather limited at the present time.

In this paper, the thermal conductivity of tetrafluoropropanol-water mixtures $(H(CF_2CF_2)CH_2OH + H_2O)$ in the various concentrations including the azeotropic mixture, has been measured, in the temperature range approximately from 25 to 200°C and in the pressure range up to 50 MPa, by using the steady-state coaxial cylinder method. The accuracy of the present results is estimated to be better than 2%. The measured values of the thermal conductivity of the mixtures are correlated by a modified Filippov-type equation, with the values of pure tetrafluoropropanol by this work and those of water by IAPS, in the whole range of measurement.

INTRODUCTION

Recently, binary mixtures of various fluids have attracted wide attention as the working fluids in the power plants and the heat pump

systems. However, knowledge of the thermophysical properties of binary mixtures is rather limited, and both experimental and theoretical investigations in this field are necessary.

The thermal conductivity of fluids is essentially important in the heat transfer calculations in industry. Following the previous paper on the aqueous solutions of trifluoroethanol [1], the present paper provides the experimental results of the thermal conductivity of the aqueous solutions of tetrafluoropropanol and clarifies the temperature-pressure-concentration dependence of the thermal conductivity of the solutions.

EXPERIMENTAL

Experimental method and apparatus are very similar to those used in the previous papers [2, 3, 4], and the details are described there. The cross-section of the coaxial cylinder cell is shown in Fig 1. The nominal width of the test fluid layer is 0.5 mm. Heat is generated in the inner heater which is made of the sheathed nichrom wire inside the inner cylinder. The temperature drop across the fluid layer between the inner and the outer cylinders is measured by three pairs of the sheathed chromel-alumel thermocouples inserted in the cylinders. Two pairs of the thermocouples are used to measure the conduction heat loss in the axial direction. The cell is inserted into the high pressure vessel which is made of Inconel X750, and the vessel is placed in the thermostated bath filled with oil (Hidol 1000). The temperature of the bath is controlled at prescribed temperatures within ±0.01 K.

The purity of 2,2,3,3-tetrafluoropropanol-1 is 99.46% and the other contents are C_3F_7H (0.22%) and $H(CF_2CF_2)_2CH_2OH$ (0.29%).

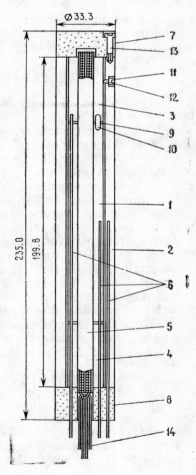

Fig. 1. Co-axial thermal conductivity cell: *1* — inner cylinder; *2* — outer cylinder; *3* — upper guard cylinder; *4* — lower guard cylinder; *5* — inner heater; *6* — thermocouples; *7* — upper alumina insulator; *8* — lower alumina insulator; *9* — mica spacer; *10* — alumina piece; *11* — brass screw; *12* — alumina pin; *13* — brass screw; *14* — compensative heater

The thermal conductivity λ (Wm^{-1}K^{-1}) measured by a coaxial cylinder method is given by the equation,

$$\lambda = \frac{Q}{\Delta T \cdot B} \tag{1}$$

where Q (W) is the amount of heat transferred through the fluid layer by conduction and ΔT (K) is the temperature measured between the fluid layer. B (m) is the geometric constant defined by

$$B = \frac{2\pi l}{\ln(r_2/r_1)} \tag{2}$$

where l is the length of the inner cylinder and r_1 and r_2 are radii of the inner and outer cylinders, respectively. The uncertainty in Q, ΔT and B is estimated to be less than 0.2, 0.6 and 1.1%, respectively. Thus, the total uncertainty in λ is less than 2%. As the uncertainty in the geometric constant is not small, the thermal conductivity of toluene, which is one of the most promising liquids for thermal conductivity standard is measured as described in the following.

EXPERIMENTAL RESULTS

Measurement of aqueous solutions usually requires a long time, because of three parameters, namely temperature, pressure and concentration. Thus, it is necessary to assure the constancy of the measuring apparatus during the whole series of the measurement. Therefore, the thermal conductivity of toluene has been measured approximately before the first series and after the last series of the aqueous solutions. The constancy of the apparatus is assured by the good reproducibility of the values for toluene. At the same time, the measured values agree with the recommended values for toluene by Nagasaka and Nagashima [5] within ±1%.

Measurements for aqueous solutions of tetrafluoropropanol (as referred to TFP) are carried out for TFP concentrations 15.0% (2.35 mol%), 30.0% (5.52 mol%), 55.0% (14.3 mol%), 72.5% (azeotropic mixture, 26.4 mol%), 86.0% (45.6 mol%) and 100%, in the temperature range approximately from 25 to 200°C and at pressures of 0.1 (and at saturation pressure close to 0.1 MPa), 2.0, 10.0, 20.0, 30.0, 40.0 and 50.0 MPa. The experimental results of the thermal conductivity as the function of temperature for five aqueous solutions and pure TFP are shown in Figs. 2-7. The results are also tabulated in Table 1.

EQUATION FOR AQUEOUS SOLUTIONS

Up to now, the thermal conductivity of aqueous solutions has been measured by a number of workers. Representative works were reported by Riedel [7] and Filippov [8], and a comprehensive study on temperature-pressure-concentration dependence for aqueous solutions of ethanol was reported by Popov and Malov [9].

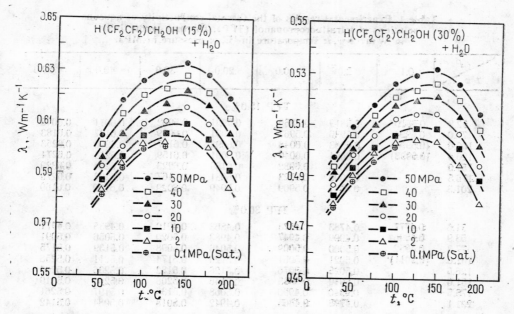

Fig. 2. Experimental results of thermal conductivity of aqueous solution of TFP **(TFP 15%)**

Fig. 3. Experimental results of thermal conductivity of aqueous solution of TFP (TFP 30%)

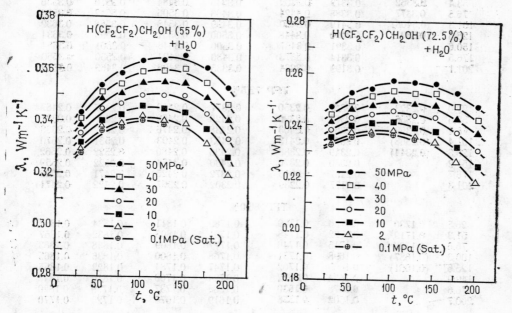

Fig. 4. Experimental results of thermal conductivity of aqueous solution of TFP (TFP 55%)

Fig. 5. Experimental results of thermal conductivity of aqueous solution of TFP (TFP 72.5%)

Table 1. Experimental results of the thermal conductivity of aqueous solutions of tetrafluoropropanol (TFP). Thermal conductivity in W m⁻¹ K⁻¹, t: temperature in °C, p: pressure in MPa.

P / t	0.1	2.0	10.0	20.0	30.0	40.0	50.0
TFP 15.0%							
52.9	0.5810	0.5819	0.5869	0.5931	0.5962	0.6011	0.6043
77.0	0.5924	0.5936	0.5967	0.5991	0.6039	0.6127	0.6183
100.6	(0.5994)	0.6003	0.6048	0.6092	0.6137	0.6200	0.6253
126.4	(0.5993)	0.6004	0.6040	0.6091	0.6159	0.6218	0.6274
151.0		0.6033	0.6082	0.6159	0.6223	0.6284	0.6329
176.5		0.6000	0.6023	0.6081	0.6167	0.6203	0.6276
201.3		0.5834	0.5909	0.5949	0.6029	0.6100	0.6169
TFP 30.0%							
31.6	0.4777	0.4783	0.4823	0.4863	0.4910	0.4935	0.4971
53.9	0.4896	0.4909	0.4947	0.4983	0.5020	0.5056	0.5091
77.3	0.4952	0.5000	0.5009	0.5045	0.5098	0.5139	0.5175
101.6	(0.5011)	0.5021	0.5061	0.5120	0.5174	0.5211	0.5258
126.6		0.5025	0.5076	0.5128	0.5182	0.5236	0.5285
151.6		0.5040	0.5085	0.5146	0.5208	0.5266	0.5319
176.5		0.4940	0.4990	0.5068	0.5135	0.5197	0.5261
201.1		0.4796	0.4865	0.4942	0.5018	0.5084	0.5142
TFP 55.0%							
29.1	0.3278	0.3286	0.3315	0.3350	0.3382	0.3408	0.3432
53.4	0.3362	0.3379	0.3402	0.3345	0.3484	0.3510	0.3536
76.6	0.3377	0.3385	0.3424	0.3466	0.3506	0.3542	0.3583
101.3	(0.3408)	0.3419	0.3460	0.3499	0.3548	0.3588	0.3632
126.3	(0.3391)	0.3400	0.3448	0.3500	0.3551	0.3593	0.3634
150.6		0.3391	0.3437	0.3500	0.3548	0.3601	0.3654
176.5		0.3314	0.3373	0.3430	0.3494	0.3550	0.3660
201.1		0.3199	0.3264	0.3335	0.3406	0.3469	0.3529
TFP 72.5%							
29.0	0.2323	0.2327	0.2350	0.2377	0.2410	0.2437	0.2467
52.6	0.2356	0.2365	0.2396	0.2426	0.2459	0.2490	0.2521
76.4	0.2359	0.2369	0.2403	0.2438	0.2476	0.2512	0.2538
101.4	(0.2368)	0.2376	0.2413	0.2453	0.2493	0.2531	0.2570
126.3	(0.2344)	0.2349	0.2388	0.2436	0.2480	0.2522	0.2562
151.2		0.2330	0.2376	0.2430	0.2474	0.2522	0.2559
176.7		0.2257	0.2315	0.2372	0.2425	0.2471	0.2522
201.3		0.2177	0.2239	0.2302	0.2367	0.2422	0.2471
TFP 86.0%							
26.5	0.1730	0.1736	0.1759	0.1789	0.1811	0.1834	0.1852
51.3	0.1719	0.1722	0.1748	0.1781	0.1805	0.1832	0.1857
75.6	0.1708	0.1715	0.1746	0.1777	0.1809	0.1838	0.1863
100.3	(0.1687)	0.1698	0.1731	0.1768	0.1800	0.1836	0.1865
125.6	(0.1661)	0.1665	0.1702	0.1747	0.1784	0.1822	0.1855
149.9		0.1634	0.1672	0.1724	0.1768	0.1810	0.1848
176.6		0.1578	0.1630	0.1688	0.1737	0.1779	0.1820
200.7		0.1502	0.1558	0.1619	0.1675	0.1725	0.1770

Table 1 (*continued*)

P / t	0.1	2.0	10.0	20.0	30.0	40.0	50.0
				TFP 100.0%			
26.7	0.1267	0.1271	0.1293	0.1320	0.1340	0.1365	0.1381
50.7	0.1242	0.1251	0.1275	0.1302	0.1327	0.1352	0.1374
75.3	0.1217	0.1223	0.1253	0.1282	0.1311	0.1339	0.1366
100.5	(0.1179)	0.1187	0.1220	0.1256	0.1287	0.1319	0.1346
125.4	(0.1129)	0.1140	0.1181	0.1222	0.1263	0.1297	0.1331
150.5		0.1100	0.1143	0.1189	0.1232	0.1271	0.1308
176.1		0.1034	0.1086	0.1141	0.1189	0.1232	0.1271
200.7		0.0968	0.1027	0.1091	0.1141	0.1187	0.1231

Values in parentheses are at saturation pressure close to 0.1 MPa.

At present, theoretical study on the thermal conductivity of binary liquids is not satisfactory, and empirical equations have been proposed. The thermal conductivity of binary liquids including aqueous solutions is generally well correlated by Filippov equation as follows

$$\lambda = \lambda_1 c_1 + \lambda_2 c_2 - a|\lambda_1 - \lambda_2|c_1 c_2 \tag{3}$$

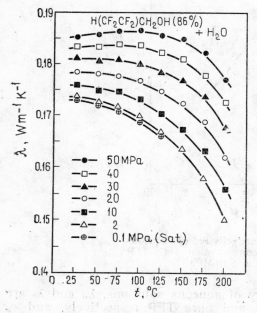

Fig. 6. Experimental results of thermal conductivity of aqueous solution of TFP (TFP 86%)

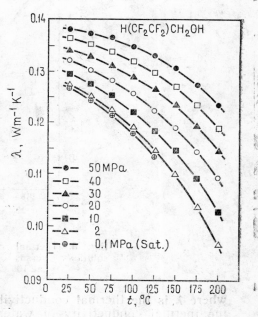

Fig. 7. Experimental results of thermal conductivity of pure tetrafluoropropanol (TFP 100%)

where λ is the thermal conductivity of binary mixture, λ_1 and λ_2 are the thermal conductivity of the pure liquids, and c_1 and c_2 are the mass fractions of the pure liquids. In the small temperature range, α is assumed to be a constant whose value depends on the binary system being investigated. In the wider temperature range, however, it is assumed to be temperature dependent.

In Fig. 8, the experimental results of the thermal conductivity of the aqueous solutions of TFP versus concentration of TFP in percent at 100°C and 10 MPa, as an example, are shown. As is clear from Fig. 8, Eq. (3) is not suitable for the present case, that is, for the aqueous solutions of organic liquid with a large molecular weight as tetrafluoropropanol. Thus, a modified equation in the following is applied for the present solutions

$$\lambda = \lambda_w c_w + \lambda_t c_t - \alpha (\lambda_w - \lambda_t) c_w c_t (c_t - \beta) \tag{4}$$

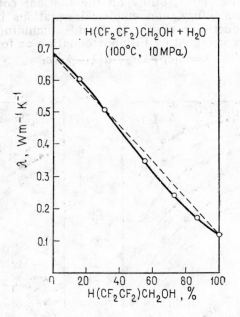

Fig. 8. Thermal conductivity of aqueous solutions versus concentration of TFP at 100°C and 10 MPa

where λ is the thermal conductivity of aqueous solutions, λ_w and λ_t are the thermal conductivity of water and pure TFP respectively, and c_w and c_t are the mass fractions of water and TFP respectively. α is a function of temperature, and β is a constant.

The thermal conductivity of water and pure TFP is correlated with the recommended values of IAPS 1977 [6] and the measured values by this work as follows

$$\lambda_w = \sum_{0}^{6} a_{0,j} \cdot t^j + p \sum^{2} a_{1,j} \cdot t^j + p^2 \sum^{2} a_{2,j} \cdot t^j \qquad (5)$$

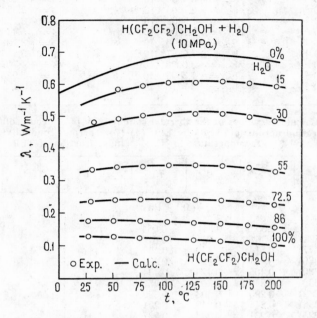

Fig. 9. Comparison of experimental results of thermal conductivity with calculated values at 10 MPa

$$\lambda_t = \sum_{j=0}^{5} b_{0,j} \cdot t^j + p \sum_{j=0}^{2} b_{1,j} \cdot t^j + p^2 \sum_{j=0}^{2} b_{2,j} \cdot t^j \qquad (6)$$

where t is temperature in °C and p is pressure in MPa. The functional form of α is assumed as follows

$$\alpha = d_0 + d_1 \cdot t + d_2 \cdot t^2 \qquad (7)$$

and the coefficients are determined by using the experimental results. The numerical values of the coefficients that appear in Eqs. (4)-(7) are tabulated in Table 2.

By using Eqs. (4)-(7), the experimental results of the aqueous solutions can be correlated with the average deviation of 0.6% and the maximum deviation of 2.5%. Thus, the equations above can be applicable for the present aqueous solutions with any concentration with the uncertainty of the order of 2%. In Fig. 9, the experimental results at 10 MPa, as an example, are compared with the calculated values by the

equations above, and the calculated values at selected temperatures and pressures for TFP concentration from 0 to 100% are tabulated in Table 3.

Table 2. Numerical values of the coefficients in Eqs. (4)-(7).

$a_{0,0} = 5.6200 \times 10^{-1}$	$a_{2,0} = -6.4886 \times 10^{-7}$	$b_{1,1} = -2.2115 \times 10^{-6}$
$a_{0,1} = 2.1663 \times 10^{-3}$	$a_{2,1} = 2.7005 \times 10^{-8}$	$b_{1,2} = 2.5215 \times 10^{-8}$
$a_{0,2} = -1.4250 \times 10^{-5}$	$a_{2,2} = -1.9462 \times 10^{-10}$	$b_{2,0} = -3.1106 \times 10^{-6}$
$a_{0,3} = 6.0667 \times 10^{-8}$	$b_{0,0} = 1.2900 \times 10^{-1}$	$b_{2,1} = 5.4499 \times 10^{-8}$
$a_{0,4} = -2.2000 \times 10^{-10}$	$b_{0,1} = -1.2000 \times 10^{-4}$	$b_{2,2} = -3.7078 \times 10^{-10}$
$a_{0,5} = 3.2000 \times 10^{-13}$	$b_{0,2} = 7.6667 \times 10^{-7}$	$d_0 = 5.6364 \times 10^{-1}$
$a_{1,0} = 5.6466 \times 10^{-4}$	$b_{0,3} = -8.0000 \times 10^{-9}$	$d_1 = 1.7075 \times 10^{-3}$
$a_{1,1} = -2.9787 \times 10^{-6}$	$b_{0,4} = 1.3333 \times 10^{-11}$	$d_2 = -6.9557 \times 10^{-6}$
$a_{1,2} = 2.4726 \times 10^{-8}$	$b_{1,0} = 3.7710 \times 10^{-4}$	$\beta = 0.250$

Table 3. Calculated values of the thermal conductivity of aqueous solutions of tetrafluoropropanol (TFP) by Eqs. (4)-(7). Thermal conductivity in W m^{-1} K^{-1}, t: temperature in °C, c_t: mass fraction of TFP.

c_t \ t	0	25	50	75	100	125	150	175	200
				0.1 MPa					
0.0	(0.562)	0.608	0.641	0.664					
0.1	(0.522)	0.564	0.594	0.614					
0.2	(0.477)	0.514	0.540	0.558					
0.3	(0.430)	0.461	0.483	0.497					
0.4	(0.380)	0.405	0.423	0.434					
0.5	(0.330)	0.349	0.362	0.371					
0.6	(0.282)	0.295	0.303	0.309					
0.7	(0.236)	0.244	0.248	0.251					
0.725	(0.225)	0.231	0.235	0.237					
0.8	(0.194)	0.197	0.199	0.199					
0.9	(0.158)	0.158	0.157	0.155					
1.0	(0.129)	0.126	01.24	0.121					

CONCLUSION

The thermal conductivity of the aqueous solutions of tetrafluoropropanol is measured in the temperature range up to 200°C and in the pressure range up to 50 MPa, with the accuracy better than 2%. The experimental results are correlated by a modified Filippov equation, as a function of temperature, pressure and concentration, with the average deviation of 0.6%.

ACKNOWLEDGEMENT

The authors are very grateful to Messrs. T. Kitano, M. Fukuda, K. Ohsako, and M. Konjya for their assistance in the experiment, and to Daikin Industry Co. for supplying the high-purity tetrafluoropropanol.

This work was partly supported through the Grant-in-Aid by the Ministry of Education, Science and Culture of Japan (No. 58460109).

REFERENCES

1. Yata, J., Minamiyama, T., and Kataoka, H. (1983): *20th Symposium of Heat Transfer Society of Japan,* 457-459, (in Japanese).
2. Yata, J., Minamiyama, T., Tashiro, M., and Muragishi, H. (1979): *Bulletin of the Japan Society of Mechanical Engineers,* **22**, 1220-1226.
3. Yata, J., Minamiyama, T., and Kajimoto, K. (1979): *Bulletin of the Japan Society of Mechanical Engineers,* **22**, 1227-1233.
4. Yata, J., Minamiyama, T., Kim, T., Yokogawa, N., and Murai, H. (1979): *Proceedings of the 9th International Conference on the Properties of Steam,* 431-438.
5. Nagasaka, Y. and Nagashima, A. (1980): *Proceedings of the First Japan Symposium on Thermophysical Properties,* 67-70, (in Japanese).
6. The International Association for the Properties of Steam, (1977): Release on Thermal Conductivity of Water Substance.
7. Riedel, L. (1951): *Chemie Ingenieur Technik,* **23**, 465-469.
8. Filippov, L. P. (1968): *International Journal of Heat and Mass Transfer,* **11**, 331-345.
9. Popov, V. N. and Malov, B. A. (1969): *Teploenergetika,* **16**(6), 87-89.

OUTSTANDING INVESTIGATORS

James **WATT**
1736-1819

Was the first to establish the relationship between pressure
and temperature of saturated steam

Oscar KNOBLAUCH
1862-1946

Participated in the International Steam Table Conferences
in 1929 and 1930

Richard MOLLIER
1863-1935

Published in 1904 the enthalpy-entropy-diagram for graphical
representation of thermodynamic changes of state,
called the "Mollier-Diagram" since that time,
and the first steam table in 1906.
Participated in the International Steam Table Conferences
in 1929 and 1930

Hugh Longbourne CALLENDAR

1863-1930

Participated in the International Steam Table Conference in 1929

Werner KOCH

1885-1950

Participated in the International Steam Table Conferences
in 1929, 1930 and 1934

Frederick George K E Y E S
1885-1976

Participated in the International Steam Table Conferences
in 1930, 1934, 1954, and 1963

Sir Alfred Charles Glyn EGERTON
1886-1959

Participated in the International Steam Table Conferences
in 1930, 1934, and 1956

Ernst SCHMIDT
1892-1975

Participated in the International Steam Table Conferences since 1934

375

Ladislav MIŠKOVSKÝ
1893-1953

Participated in the International Steam Table Conferences
in 1929 and 1930

377

Dudley Maurice N E W I T T
1894-1980

Participated in the International Conferences on the Properties
of Steam in 1954, 1956, and 1963

Sugao **SUGAWARA**

1896-1983

Participated in the International Conferences on the Properties of Steam in 1963 and 1968

Mikhail Petrovich V U K A L O V I C H
1898-1969

Participated in the International Conferences on the Properties
of Steam in 1956, 1963, and 1968

Joseph Henry KEENAN
1900-1977

Participated in all the International Conferences on the Properties
of Steam from 1929 to 1974

W. W. CAMPBELL

Participated in the International Conferences on the Properties
of Steam in 1956, 1963, and 1968

AUTHOR INDEX

Achtermann H. J. 29
Akhundov T. S. 316
Aleksandrov A. A. 299
Alvarez J. 240
Amirkhanov Kh. I. 324
Androsov V. I. 165

Bochkov M. M. 324
Bonhsack G. 121

Chang R. F. 277
Churagulov B. R. 258
Crovetto R. 240

Demianets Yu. N. 41
Dvorjanchikov V. I. 324

Everhart C. M. 277

Fernandez-Prini R. 240

Ganiyev Y. A. 210
Garrabos Y. 77
Gaskova O. L. 249
Gorbaty Yu. E. 41
Grigoryev B. A. 210

Imanova M. V. 316

Jany P. 306

Kashiwagi H. 17, 330
Kataoka H. 348
Kelly J. A. 266
Kharitonova N. L. 228
Knoche K. F. 339
Kochetkov A. I. 299
Kokhanova E. G. 67
Kolonin G. R. 249
Korzhavina N. A. 190
Kremenevskaya E. A. 289
Kritsky V. G. 185
Kubota H. 330
Kuznetsov G. G. 299

Le Neindre B. 77
Levelt-Sengers J. M. H. 277
Lobkova N. V. 196
Lubimov Yu. A. 48
Lvov S. N. 185
Lyashchenko A. K. 258

Makita T. 17, 330
Marshall W. L. 145
Martynova O. I. O. I. 87, **185**
Matsuo S. 17
Minamiyama T. 348
Morrison G. 277
Mursalov B. A. 324

Nabokov O. A. 48
Nagasaka Y. 203
Nagashima A. 203
Novikov B. E. 190

Okhotin V. S. 299

Panakhov I. A. 196
Pepinov R. I. 196
Petrova T. I. 228, 234
Pitzer K. S. 91

Raatschen W. 339
Rastorguev Y. L. 210
Rivkin S. L. 289
Rögener H. 29
Romashin S. N. 289

Safronov G. A. 210
Samoilov Yu. F. 228, **234**
Shironosova G. P. 249
Silkov A. A. 157
Staples B. R. 170
Stepanov G. V. 324
Straub J. 306
Styrikovich M. A. 67
Suzuki J. 203
Suzuki K. 56

Tahirov A. D. 316
Tanaka Y. 330
Traktuev O. M. 289
Tufeu R. 77

Udodov Y. N. 157

Valyashko V. M. 134
Vargaftik N. B. 63
Voljak L. D. 63
Volkov B. N. 63
Vospennikov V. V. 165

Wakeham W. A. 219

Yata J. 348
Yukhnevich G. V. 67
Yusufova V. J. 196

FOR NOTES